HISTORY OF THE ORIGIN OF ALL THINGS

GIVEN BY

THE LORD OUR GOD

Through Levi M. Arnold

1 8 5 2

REVISED BY HIM

Through Anna A. MacDonald

1 9 3 6

Edited by Robert T. Newcomb

VOLUME ONE

Engaging God

In, by and through the *History of the Origin of All Things*

All we need to do to engage God productively in our lives is to purposely invite God and/or Christ into our lives.

Knowing that the living, loving God expresses to us or touches us Spirit to spirit and empowers us by the touch of His Being through the Christ Spirit is the basis of Christian faith.

Seeing Jesus as the Christ Spirit, born among the people of earth to demonstrate the Christ attributes available to *whosoever*, shapes our lives in the Holy Spirit in a transcendent way. Thus, we may look beyond our bodily reality and earthly circumstance to our spiritual being, and believe that we are really in spiritual being right now. Acting out of our spirit center, with Spiritual motivation to express by creative, life-enhancing, productive effort, is all that it takes to establish ourselves in God-life. "God is Spirit, and we must worship Him in spirit and in truth." We are spirit; and, worship is *actively engaging God* at the spiritual level of our being.

As Christians, we are taught that one-seventh of our time (one day a week) and life energy and attention are to be devoted to God. That in resting from natural activities in reverential stillness and attention keeps us in attunement. Yea — keeps the lines of communication open that we may receive our direction and feed our developing souls on and in the substance of spiritual life that streams forth out of the Creator. God is the ultimate magnetic attraction, and our journey toward Him is eternal.

In our present time, 1994, we have externalized our being more than is possible, and still live in the physical, mental, emotional and spiritual balance that is imperative in life. We are a spirit, born of God; we have a soul that is perfectible unto eternity; we live in a body which was created for us as the perfect means of articulating in the earth realm of substance where we presently make *our* place, as *part* of our infinite and eternal life. Can we not, as educated beings, see that letting God reign supreme in our lives, letting truth and love from the living, eternally active Father of Life, flow through our spirits, souls and bodies is the perfect way to realize health, harmony, happiness and joyful productivity?

Life is essentially creative; and here, we create both in our being and beyond it. The most important product is what we create in the spirit/soul being of ourselves and others. The process is perfect to the accomplishment of God's purpose of developing beings to share in the *eternal expression of the infinite potential of life* as it streams out of the God-center of the universe, into created things unto infinity and eternity.

Our *free-will* <u>is</u> the gift that lets us act in the dynamic Life of God and Creation. We are created to be ever established in the source, but can *will* it otherwise. Essential in the use of will, acting with and in volition, is the decision that our souls, by and in our spirit, stay engaged in the God Spirit by Christ, His Spiritual Son, and the Holy Spirit, carrier of all infinite knowledge.

This book, the *History of the Origin of All Things*, stands as a perfect guide. It reflects on all the essentials of life, elucidating life, scriptures and purpose — extending major keys to every searcher. It describes the continued progression beyond earth experience and the importance of making the most of this life in the binding and loosing of the elements of soul that will carry over. The truths are easy to believe and inculcate, for, by and through engagements for the purpose of *essential and successful use of one's life and free-will.*

I searched for such all my life. Now, my purpose is to be involved in the distribution and the practice of its truths.

In Jesus Name. AMEN.

Donald O. Haughey

PREFACE TO THE SECOND EDITION

In presenting this volume of Truth the second time to mankind, I do it hoping and trusting they may find it of priceless value, as it has been to me.

To the hungry and thirsty soul starving for knowledge, asking for light, it will be as a diamond set in pearls of inestimable value.

This book came to me providently. I read, and thanked God for its sublime truths. It has been to me like a cloud, to guide my wandering feet by day, and a pillar of light in the dark hourse of sadness and the night of adversity.

Of late very many calls have been made for this book, and none to be found. I therefore got the right to republish it at my own expense, and with a sincere prayer I send it forth on its mission of Truth to every dark and benighted soul of earth who needs its light.

This book was written by a man of common education, simple, honest, a Quaker in principle. Having fulfilled his mission, twelve years ago he passed on to his home among the angels.

Yours for the truth,

ANNIE GETCHELL, M. D.

1 BEACON STREET, Boston, Mass.

PREFACE TO BOOK THIRD, 1ST SERIES

This book is the third of the first series of the *History of the Origin of All Things,* and is presented as the revelation of many things on which mankind have long desired information. Its truth is unshakable by reason or argument, by other manifestations, or by the production of any other theory. Be wise and understand, for the time of the end is near in which all shall see God as He is, and know Him to be Wise, Happy, Good, and Benevolent, ever acting, eternally existing, and universally present.

Poughkeepsie, Sept. 30, 1852.

This book will be in two parts and three books will be found in one, forming the second series and published, like the preceding series, without desire for profit or any remuneration whatever to the medium for his labor, time, or attention. Freely he receives, freely he gives. He does not rely even upon the proceeds of one book for the publication of the next but he avails himself of the facilities I afford him of advancing the necessary funds for their publication without any risk to the publisher, who has no other interest or control over them than to sell such as are placed for that purpose in his hands.

PREFACE TO THE REVISED EDITION

The time has come when man needs to have this Book, and its followers, brought to his attention afresh. The old form and type make it necessary to republish it in a new dress, so to speak. But truths never change, neither do their Creator nor their application to the life of man. It is man who does the changing, and not The Father, His laws nor His plan; which includes His plan for man's progress as well as for man's redemption.

This is the plan for man's progress, and it must include the necessary experiences in what he considers to be evil, which are necessary for his perceiving and recognizing evil, and how to avoid the things which pull him away from The Father's plan for his final salvation. This means very little to man now because so many years have gone by since it was enunciated and so many men have interpreted it since it was formulated, each one giving it more of his interpretation until it is hard to recognize many of the principles laid down by The Father and reiterated and taught by His Son, Jesus Christ.

The time has come when man must face this truth as he has to face what man calls evil; but The Father calls it lessons which man must learn through experience.

In truth, what man calls "the Devil" is the desire of the soul, for the sake of its own growth, to have these experiences of "evil", for the soul recognizes the fact that it must have them. Man must know evil in order to be able to recognize it, and to *be sure,* he must know it so well that he will not be overcome by it again.

This means experiencing until the soul is so filled with horror which each "evil" thing which pulls man down that he turns from it in disgust. Thus each single individual soul has to go through each phase of evil, until he be proof against any form which it may assume.

This is an explanation of why the idea of a personal "devil" grew up. In truth that "devil" lies in man's own soul, through his free-will. This free-will was given him so he *might learn to choose for himself,* because he knows from his own experience—and not from the experience of any other person—what the evil was, and why he did not deserve it for himself.

Is it not the way a parent trains his child? Does not the parent tell the child what he should do and what he should avoid? What does the child do after he grows old enough to begin to want to do as he wishes to do? Does that child not experiment for himself, and choose which line he will follow?

So it is with mankind from the minute he leaves Paradise, or the Garden of Eden as Paradise is called in the Bible, to begin his practice in using his free-will, which is his knowledge of good and evil. He tries this way and that way, until of himself and through his own efforts, he finds God's way for him to go, the way established for man from the beginning but which he had to *hunt for* and find, through his desire to find it because he had found any other way was wrong, and would lead him away from God's plan instead of toward it.

Now this is "evil" as man sees it, and it is personified in a being man has named "the Devil", while in truth it is only the efforts of man to try many ways before he finds and is willing to follow the ONE WAY—which is the WAY The Father planned for him at first.

This is given as an example of the way God's truths and plan for man, which He has never changed, have been unintentionally changed by man's own comprehension which in turn has been influenced by his desires.

Does this explain why The Father at certain periods, ever since man was given his free-will, has given man a new interpretation, a new leader, a new vision of the truth as man was ready to receive it? Even now there will be few who will be able to accept the truths given through this Book, though it has been given by Him who taught man in person almost 2000 years ago, to His chosen mediums—those whom in this Book we shall call holy mediums to distinguish them from those who do not work for or under Him, though they may be able to interpret the spirit-world to man, from a lower level and from a less high motive. This change in the Book is made thus to distinguish Our Lord's agents from others, and to save the one thus chosen from the implied meaning the word medium now carries to man's mind—again, an example of how man has degraded the true conception by substituting one which suits his own understanding.

So a new printing of this Book will be made, and many there be who will profit by it; profit in joy and gratitude for a truer and a more satisfying interpretation. Still, there will be many who will reject all efforts to give them new light; a new meaning which, by its simplicity, would clear up much for them in their understanding and in their joy in their efforts to find and follow the leadership of Our Lord, who, also, may lead man into The Father's way and help him to keep to that way.

Again, Jesus Christ is personally supervising a slight revision of the Book so it may be more easily read and understood by those who are ready to absorb and accept its statements of truths. He is doing this revision through Anna A. MacDonald, and wishes to thank, through her, all those who have helped by preserving the Book, making it known, and now those who are helping in this work of revision, for He is not ignorant of the effort and time needed in the work, slight as the revision is. But He does wish to say that all who are working upon the Book are of His Own, and have already accepted the truths it portrays.

Given in the shadow of His holy retreat upon earth, near Asheville, N. C., July 25, 1936, by Himself, Jesus Christ.

L. M. ARNOLD: A SKETCH

L. M. Arnold was born in 1813. At the time the books were received he resided in Poughkeepsie, N. Y. Henry Carpenter, whose father was of the same society of Friends in which Arnold worshipped, was interested in the revelations when they were first published, and as a truthseeker became more intimately acquainted with Arnold. Because of that acquaintanceship Arnold was led to employ B. F. Carpenter as a confidential clerk, to learn his business and become interested as a silent partner, and it is to him that we are indebted for information regarding the life and character of the man who received these books.

Arnold's business was the "Poughkeepsie Foundry", an old establishment to which he succeeded. There were about thirty employees producing various castings, chiefly for New York customers. Arnold attended to all correspondence in addition to keeping all accounts. He was a very active, public-spirited citizen, and in 1864 was the chief promoter of the incorporation of the First National Bank.

Descriptions of how he received and recorded the words of his writings are given in very precise detail in the published books, but the reader will have to search through several books to find all the details. In like manner the only way to comprehend fully all that is given on any subject is to search through all the writings. The spiritual jewels are scattered like diamonds in sand.

The preparatory revelations commenced April 5, 1851, in the form of movements of the pencil, answers to questions, and words internally heard. Such writings were continued for a year and a few days, when the first book of The History of the Origin of All Things was commenced, in a small bound blank book, written with lead pencil throughout.

When there were calls for the published book, Carpenter assisted in filling orders, and naturally was inclined to seek for further information regarding statements made in them, especially such as were prophetic, and was surprised to learn that he had not read his works since publication. He had attempted to do so, but each time was informed he better not peruse them.

He was looking forward to the second coming of Jesus of Nazareth, but expected it to be an event that mankind would not recognize at the time it should occur, but later should realize it fully.

He usually attended Sabbath day services at the Hicksight Society of Friends, but he did not converse on religious subjects or take any active part in religious proceedings. As related to his employees he was a very gentle man, frugal and soft-spoken.

He died of typhoid fever, September 24, 1864.

CONTENTS

PART I. The Origin Of All Things

PART II. The Chronological, Geological, Geographical, and Social History of the World, with the Story of Divine Influence.

History Of The Origin Of All Things

PART I.

The Origin

of

All Things

INTRODUCTION

What man is, was, and will be, has always occupied his thought and engaged his deep research. But heretofore he has had few materials for his study, and thus has made small progress in solving the problem upon which his curiosity impelled him to work. Recently the minds of many have experienced the dawn of brighter anticipation, through the communications of spirits. Hitherto, the self-will of the mediums has not been deeply impressed with the opening of the door to let in God's spirit—the Divine Influence, the Holy Spirit—all synonymous terms. This medium has been through a course of discipline which afflicted, grieved, and surprised him, and has experienced the hand of God in many ways to fit him for this work. It is the will of God that he should now proceed more clearly and understandingly, but the way will not be strewn with flowers only. Thorns are the usual accompaniments of roses, and this is a life of probation. And how can men be proved without trial or condemned without sin? He will receive pardon, though, when he is qualified for it. So will all men.

The life of man here is short, hereafter it is eternal. Everlasting to everlasting is his generation. But the last shall be first in glory and first with God. The first shall be last with men. But the glory of the last is eternal, and the glory of the first vanishes as dew before the beams of God's love. All shall be, forever, the children of God and all shall, in the future, see God. But the last shall be first and the first, last. Be then consoled, ye afflicted; be comforted, ye mourners; be joyful, ye who weep; for God rules and reigns beyond the powers of Kings or Popes. Priests, laymen, widowed churches, and joyful fools shall all be cast down, but God shall reign undisturbed in the crash of matter and the fall of universes. His calmness is imperturbable; His will is perfect in wisdom and power; His justice and His mercy go hand in hand; and His love unites all, sustains all, blesses all, and will cause all to glorify, to praise, to honor, and to love Him, who is beyond, above, every finite comprehension, but bows down the majesty of His nature to hear the prayer of an humble and contrite spirit.

CHAPTER I.

In the beginning God created heaven and earth and all things that are therein. He made man to govern the animal creation, to secure the welfare of himself by the exercise of his mental faculties and his spiritual aspirations.

Having placed him in a happy position, God left him to cultivate the earth and to supply his bodily wants by his animal observation. The love of God protected him from fear and from doubt. But the evil of sin entered because man would not be satisfied with the good God had bestowed upon him. So evil came in the shape of desire of change; of love or desire for unknown things. Having so allowed evil to enter into the soul, which till then had been the sanctuary of God, man fell into despondency. Man feared God would repay him evil for evil and place him in a state of unhappiness, because he had not been satisfied and happy when in a superior position. But God never placed man in such an inferior position, but caused His holy love to persuade man to advance, to trust to His (God's) mercy, and to lean on Him for support in every affliction. As man rested on God for support, he was strengthened; as he loved God, he was purified, for the love of God is a consuming fire which eradicates every evil desire, which conquers and turns to dust and ashes every unholy aspiration, every free thought of fallen man. His wishes, when brought into subjection to God's will, will be the emanation of God's spirit.

There is now proceeding from God's spirit, an influence which acts on man, through the spirits of his departed friends; friends who have left the body to exist in spirit-form only. This state-of-the-spirit is a blissful one, compared with that of bodily existence, because man is thereby relieved from the temptations which the bodily nature impels; and having no thought for self only, or no need to have such thought, he delights in doing good to others with himself. Being relieved from bodily temptations, he ceases to sin, and becomes purified by the fire of God's love from the consequences of the sins he has committed in the body. Being purified gradually from these, he ascends to a higher condition in which he possesses a greater measure of God's presence in him. And being so fitted for higher duties he becomes set apart to usefulness of various kinds, such as God in His wisdom imposes as his duties; and these duties are performed as pleasures. They are pleasures and confer upon him the highest happiness of which he is capable.

These duties consist in loving those who are then assigned to his care; in watching over and influencing them under the direction of God, thereby bringing forth in them the fruits of repentance and the desire of good works. This position can not be occupied by one in the body. But those in the body can be placed in an analogous one, that of serving God by promoting the good of fellow men. These are, however, subordinate to and under the direction of the higher spiritual existences, as the love of God operates through a chain of existence; the higher part being more filled with God's glory, with His love and with His power; the lower receiving the influence of the higher, as iron receives the properties of the magnet without becoming the magnet. While so influenced the lower seems sometimes to be equal to the higher; but some change of circumstances, which breaks the connection of the helice, or chain, dispels the illusion. The unaided man then falls back by unholy or inharmonious desires into helplessness; again to be strengthened and revivified by the holy encircling influences of the higher and purer existences.

It is by passiveness, by submission of will, by desire of God's love, that the mind of man while in the body is prepared to receive this evidence of God's love and this manifestation of His power. It is by a faithful observance of the impulses of man's own higher nature that he can attain this state of passiveness, submission, and desire. When imbued with it, he progresses. His position becomes eventually more and more elevated. He becomes stronger and stronger in faith, which impels him more and more to serve God by obeying the highest impulses he feels. He so comes to have a high and holy calling, and being so called, he is a willing servant. He has sacrificed his will as an acceptable offering on the holy altar of God's love. He treads forward in his work, rejoicing as he advances. When the summons to leave the body arrives, he is ready and willing. He is always resigned to the dispensations of God's will, because he feels and knows that God loves him; that God does not afflict him from hate or revenge; but that pity and compassion are the nearest approach to wrath of which God is capable.

God is omnipotent; and He is omniscient. If, then, He knows all, even foreknows all, and His power executes His will even as His will exists, how then could He have wrath, how could He hate, how could He revenge? The moment His will shall have exercised itself, the effect would be accomplished. No struggle would avail, no pity could move, no submission would have time to operate. In an instant of time, and whole creation might be resolved into its original elements, or into nothing—the nothing from which the will of God formed it. He spake, and it was done. He commanded, and it stood fast. The inspired penman has well expressed the instantaneous operation of God's will. He can as easily destroy as create. Man can more easily destroy than create, because both are laborious to him, though one involves more labor than the other. To God, and with God, both are equal.

CHAPTER II

CREATION AND MAN'S IDEA OF IT

We will now return to the first topic; to our real subject: The Origin of All Things. God, having resolved that a creation should be established, willed it. Not instantly in all its parts but by laws of progress. The laws being established, the effects followed inevitably. It is beyond the powers of the finite mind to conceive of the operations of the Infinite. Therefore, the manner of the conception of the idea is to man inexplainable, though the effects can be related.

When matter was chaos, a state to which it had progressed from nothing, it continued to progress into order. Order established the foundations of the astronomical systems, and in order they continued to progress or develop. From one confused chaos, pervading infinite space, there resulted heavenly systems of universes, systems of suns, systems of planets, primary and secondary. These various, these innumerable bodies, pervading infinite space, all experience the Divine care and partake of the Divine love. The love of God is unfolded by degrees to the knowledge of every creature.

These worlds or globes of matter are all habitable, and all inhabited—some by beings higher than man, some by lower orders only, some have things only like men. Does God, then, reserve man, or that family of men placed on this earth, as the only order of beings worthy of His care? Is it for man that all the starry globes, twinkling from afar, exist? Does man require that God should Himself descend from the seat of His power and the home of His love; from pervading the whole universe of universes, illimitable, unimaginable by man, to dwell in one body, to concentrate Himself to one soul and that the soul of a man? What, must we suppose, would be the presumption of a being like man to require this? And yet man believes that only so has God been able to save the beings that have existed on this globe. And how, then, have they been saved on other globes? Not by having other Sons, because then the glory of Him who saved men would be diminished. Not by sending that same Son to other worlds, because that would be condemning that only Son to infinite, if not unending, suffering. For who can conceive of the end of a time that would be necessary to pass through an existence in each of the innumerable worlds that comprise a universe? And that universe is only an atom of the great whole of God's creation.

Oh, man, what pride is thine. What can he not deem worthy of himself, if no sacrifice but the Son of God, an equal with God, is sufficient to satisfy his honor, to exalt his glory, to secure his happiness? Alas, that an idea conceived in ignorance, when the wonders of creation had not been opened to the astonished gaze and admiring mind of men, when the earth was thought to be all the habitable creation, when it was easy to suppose that heaven was above and hell below its great plane, alas, I say, that an idea originating in such dark ignorance of nature and of nature's God should still be maintained by intelligent, educated, and even greatly learned men. This is indeed folly. God loves all His creatures. God has power to save them all. Can He fail to desire to have them saved? Can He fail to have His desire fulfilled? If He can fail, He is not God. God will not be mocked. Such a doctrine is now the doctrine of pride, as it was once the doctrine of ignorance. Let man be humble. Let him be willing to be saved by God's mercy, which has no end. Let him be passive to the holy influences God will throw around him and act in humble reverence of a being as merciful as He is powerful, and as loving as He is great and good.

The end of the matter is—God has placed man here, as He has placed other beings elsewhere, to manifest His glory, to make known His power, to establish His mercy, to exercise His love. He wills them to encounter temptations, to bear afflictions, to sustain labors, to sink under evil, to be raised by His love, to be established by His power, all as the work of His hands; being the part which pertains to the great whole of His universal creation. Man was before this earth was. Man was not created for himself nor for the earth. He was made for God's glory and to enjoy the benevolence of a merciful and all-wise creator. It is God's pleasure that man shall be happy. He wills that man shall be able to enjoy and to appreciate the happiness He wills to confer upon him. It is only by contrast that we can know happiness. It is only by cold we can know heat. It is only by pain we can know pleasure. This life in the body, pleasurable though it is to us on the whole as we experience it, will be the foundation for the superstructure of unutterable bliss that will be enjoyed by every man at some period, when he will look back to this bodily state as one of misery; misery mitigated only by the knowledge that it was dictated by true benevolence, that it was only a preparation for higher states, in which its memory should be present as a contrast. Even as every picture must have shade to contrast its light, so must every mind have unhappiness wherewith to measure its happiness.

This is the true explanation of the usefulness of evil. All are but parts of one stupendous whole, says the poet. All that can ennoble and elevate man is often obscured by the darkness of ignorance and the love of self. But when the body is forsaken, the soul, freed from further temptations, is operated upon by the efforts of others under God's laws, until, little by little, it is freed from the depths of the darkest ignorance. And recognizing its brighter and holier nature, it gladly strives to serve God, which is to serve others. Strictly speaking we can not serve God. He needs no service. An Omnipotent being can not need help. He is happy. We can not make Him more so. We can promote our own good and elevate our happiness, by serving others under His influence; by coming into harmony with His Divine nature. The pure in heart shall see God. As we advance we reach nearer and nearer to His nature, we are more and more purified and refined in our nature and manifestations, until at last we shall be one with God, because we shall have no other will than His, no other desire than to be in harmony with Him. We shall then be Sons of God, reconciled to Him, one with Him—one in power, one in glory, one in honor—because we shall be no other than a part of God Himself; an emanation of Himself united to Himself after a long series of adventures, in which we have experienced and continue to enjoy an individual existence, yet harmonized mysteriously with, and into, the Divine Nature.

CHAPTER III.

THE ORIGIN OF MAN

Having now opened our subject, I will proceed to state the Origin of Man.

Man is a being of various existences, connected with each other by ties of various natures. His origin is God, from whom proceedeth all things. All things are of God, and, in one sense, all things are in God. But, yet, some are more separated from God than others; and though God fills all space and exists at one and the same time in every part of the universe of His creation—pervading every creature, maintaining every life—He still gives His creatures an independence of Him, greater or less, never absolute. It will perhaps be better understood to say, every creature is more or less dependent on Him, though all had originally, when created, points or angles of separation. The course of their existences never is parallel with God, but all are diverging from or approaching Him. Man's course is first divergent from God. Man's spirit, which is the man while the body is merely its clothing, is an emanation from the Deity, a part of the Divine spirit.

It is first placed in a state of quiet happiness, removed from pain, subject to no trials, having no knowledge of affliction or of temptation. Here it is male and female. Not that one being is of both sexes, but that two beings unite to form one harmonious existence in each other. This state exists for a long period; to man's comprehension it would be an eternity. But it is an existence of sameness without emotion. No events mark its progress, or recall and measure its period. The existence is pleasurable. They are as Gods: each as God, so far as being without affliction or unfulfilled desire. But they have not eaten of the tree of knowledge of good and evil. They can not taste that without passing from their harmonious existence. They are pure, and see God. They are innocent, and love Him. They are as children and, being thus passive in the hands of God, they are in heaven. It is the state of Paradise. They are not yet clothed with earthly bodies. They have not even the spiritual body which men possess after the death of the natural body. They exist with constant pleasurable sensations and associations. Each pair is independent of the others. No government is required; there are no crimes to punish, no rights to maintain. God sustains all, is in all, and in Him they move and have their being. But where are they? is asked. They are in God. They are where He is, sensible that they are not God but that they have an individual and parital existence. This is the first state of man; and is also the first state of all beings similar to him that inhabit other globes, as well as of every superior order of beings that exists. These superior orders of beings differ from man by having afterward more wonderful bodies, endowed with higher and more extensive powers. But they assume such bodies as their nature requires—to every seed God giveth its own body. There are various orders of these beings, as all God's creation is composed of varieties; as men differ from each other in race, features, manners and intellect. Yet before God all are equal. They are each and all such as He willed to have them.

In the beginning, God created the worlds that fill or are scattered through infinite space. The lapse of time since the beginning is too remote for man's comprehension. Astronomers tell how long the most distant stars of which they know must have existed, in order that their light should have reached the earth. But how much longer they have existed, or how much further creation extends, none can compute.

After the worlds, or globes of matter, were formed from the chaos of the first creation, man was separated from the perfect oneness he, before that, had had with God. Placed then in Paradise, a long period of tranquil happiness prevailed, undisturbed by aspirations or desires. This may be known as a sabbath of rest; as a time when the morning stars sang together and all the Sons of God shouted for joy. Such was Paradise, and such it is yet to the created, but unborn, spirits of men.

Yet every part of God's creation had impressed upon it, by unerring wisdom, the laws of progress. And though man's path at times diverged from God's own course, it none the less led to higher glory and it ends in the greatest happiness.

The law of progress having aroused in man, while in his quiet state of separate and bipartite existence, a desire for greater knowledge—a desire to experience, to act, instead of merely existing and enjoying—the being passes into a body, prepared under the operation of God's immutable but progressive laws for its reception. It is expelled from Paradise, because Paradise ceases to be all it desires, ceases to render the man happy.

And how did the first spirit or soul of man that left Paradise find a body? In the order of God's creation, a body was prepared. In the fullness of His knowledge, every want was foreseen and the very period in which it should occur known. God formed the body from the dust of the earth. The spirit that had long before emanated from God, entered into the body and conferred upon the insensate mass, life. Man became a living soul. Multitudinous desires sprang into existence; multitudinous difficulties prevented their realization. Disappointment is affliction. Deprivation of present happiness is the inevitable consequence of affliction. Selfish desires are unholy, because they are indulged in at the expense of benevolence. Benevolence consists in doing good to others. Selfishness desires to benefit self only. Its indulgence at the expense of others is often a crime by the laws of man. By God's law it is a sin. It may be pardoned; its legitimate consequences may be overruled. But in general it makes an indelible print on the character, on the spirit or soul, of the actor. Indelible print except by the mercy of God. Yet His mercy is ever ready to exercise itself upon the sinner; and so long as sin exists God's spirit will strive with man.

But God's spirit will not always strive with man. Therefore, the day will come when sin will no longer exist. Then God will have reconciled all to Himself. Sin and death will no longer live, or occur. All will be one with God, through Christ—Christ, or the Messiah, or the Sent of God, for all these terms are synonymous. There is then one God, the Saviour of all men, and Jesus, whom He sent, preached this. Every true prophet has preached it, and will preach it so long as men require preaching. But the time cometh when all shall know God, from the least to the greatest. That is, the time is coming to every individual separately, not to all at once. He will be known of all men. He will write His law in their hearts, and put it in their inmost parts; though not always while they are in the body but certainly at some time. The time will come when men will be willing and desirous to be led and guided by Him; to receive the fulfillment of His gracious promises. For the free will of man is never infringed. He must work out his own salvation. He must be purified like gold in the refiner's fire, like silver in the pot. This is the sure and steadfast promise of God, which is Yea and Amen forever. Blessed be God who confers such happiness on His creatures, who bestows such good gifts on His children—Yea, even the boon of eternal life, eternal happiness, eternal progression in the beams of His all-beneficent power and love. Glory to God in the highest. Peace on earth and good-will to men.

CHAPTER IV.

THE DUTIES MEN OWE TO EACH OTHER

There is now a long course of instruction required upon the duties men owe to each other. But I shall not undertake to preach it, for the time has not come when men are willing to receive it. Selfishness prevails too generally; is too strongly intrenched in the hearts of men. Its outworks are found in the laws and customs of traffic, in the arts and professions, in the habits of workmen. Its strongholds can be reached only by demolishing the defenses it has raised in society, in politics, even in law and in theology. The social system is too much based on self. But yet it contains within itself the laws of progress and will purify itself from error by the aid of good men acting under the inspiration of God, through spirits devoted to this work. Fear not, for I am with you to the end of the world, is still the language of the Sent of God to all who desire to believe in His power and do His work. For already the harvest is ripe. The laborers are now few; but there is the more occasion for their activity. At the eleventh hour the laborers will be many; and each man who works to the end shall receive equally the reward of a faithful laborer. That reward is eternal life. For the last shall be first and the first last. But to him who endureth to the end is laid up a crown of glory; and not to one only, but to all of them that love God, that love His Messiah.

And can any one refrain from loving a Being whose nature is love and who confers only benefits on all? God is love and none can resist the operation of His eternally acting love, that as a consuming fire will overcome and consume every evil thought, every evil desire, every evil imagination, wherever it may exist. Even selfishness, if it could possibly retain its nature while seeing fully the love of God, would, from love of self, desire the love of God, seeing the ineffable happiness and the triumphant glory of the spirits that live and love to serve others. To them, indeed, all else is added. So will it be to all those who seek first the kingdom of God. But the moment that selfishness begins to desire the love of God, it begins to progress to be like God—to be benevolent, as He is benevolent; to be loving, as He is loving; to be kind, as He is kind; to be faithful, as He is faithful; to be willing to help, as He is willing to help; to be in every possible way, more and more divine, more and more in harmony with God, until at last the demon of self abandons the happy man and God is with him, all in all.

But how will the ungodly appear, if the righteous scarcely be saved? Remember the parable of those who worked different hours but all received the same reward. Are any wholly good? No; not one. All have sinned and come short of the Glory of God. Are any wholly bad, or ungodly? No; not one. The divine essence, which is the germ of man, is never extinguished. It is immortal and incorruptible. It may be concealed for a time, yet its aspirations must at last have vent. They will reach the great fountain of itself and of all good. There is a Messiah to every soul, a redeemer to every spirit. Were it not so, I would have been lost to God. All would have wandered without a guide and who could have been saved? God is not partial. His ways are not as man's ways. He is just. All men are equal before Him. If it were not so, His justice would not be manifest, and without justice, He would not be the Deity. For deity is perfection and a being can not be perfect without justice.

God is just. Will He not, then, punish by eternal tortures those rebels against His laws who, living only for self, have delighted in crime and walked in wickedness; who have in truth acted in an ungodly way as if there were no God? Alas, for humanity! It would persuade itself that it is superior to God in mercy and compassion. The most daring rebels are pardoned by human governments and the governors are commended for their humanity. They have acted upon the preaching of the Messiah. They have heaped coals of fire upon the heads of their enemies. They have overcome evil with good. But is God less merciful? Is vengeance more necessary to Him? Is the fear of terrible punishment necessarily ever to be held before the imagination of His enemies to enable Him to overcome their evil with good?

Is man, only, to act upon the heavenly teachings of Jesus the Messiah? Is he, only, to forgive insults and injuries? No, these teachings are heavenly because they inculcate God's order, God's laws, God's rules of justice and mercy. When Jesus taught these doctrines their novelty was startling. He taught as no man had ever taught. Now we commend the teaching, we glorify those whose actions accord with it. But do men believe themselves generally capable of acting in accordance with them? Or do they not rather put them off as beyond their nature; as being too God-like, as pertaining too much to heaven to be practiced on earth? And yet, they have been practiced by man, and men have been commended by their fellow men for it.

How then will you not permit God's justice to be reconciled with mercy? Has God need to protect His station by punishing rebels with eternal torture? Is He so affected by the crimes, or sins of men, that He can never forgive without endangering His reputation for justice? Not so. The actions of men can impair their own present happiness, but God, who sees to the end, does not feel annoyed by the evil, or the sin. Thou fool! cease to do evil, learn to do good. Cease to impute to God action you, yourself, would be ashamed of; punishment that you, yourself, feel incapable of inflicting upon your own erring children. You seem even to desire that God should be unforgiving in His nature if only you, selfish creatures, can be saved by the sacrifice of an innocent victim!

It is no doubt true, very often, that the just suffer for the unjust, but never by God's interference. It is only the work of men. Are these assertions arbitrary and mere assertions? I ask you to go down into the innermost part of your being. There ask the Divinity, that never wholly forsakes man, if they are not true; if they are not in accordance with eternal justice, universal benevolence, all-pervading love and inexhaustible mercy? Look to the example of every good man; look at what is recorded of Jesus, the Son and Sent of God, and see if they are not in accordance with His precepts and His actions. He went about doing good; preaching that men should practice practical virtue. He healed the sick, forgave sins, and prayed for His enemies. He resisted not those who took His life, but forgave them. He calls on you to be perfect, even as your Father in heaven is perfect. He calls on you to be one with Him, even as He was one with the Father. Would you do this? Then act as He did. Sacrifice self. Live for others.

CHAPTER V.

WHAT IS MAN?

What is man, O Lord, that Thou art mindful of him, and the Son of man, that Thou regardest him? exclaimed the Psalmist in olden time. Thou hast created him a little lower than the angels, Thou hast raised him to power and placed him in glory. O God, Thou knowest all things and all things tend to increase Thy glory! Wherewith shall I come before the Lord? Shall I offer my firstborn? or will He be content with thousands of rams and tens of thousands of lambs? Alas! that man should fancy his sacrifices could please God, except as they are the evidence of a willingness to do His will. And what is His will? Be ye perfect, even as your Father in heaven is perfect. Is this a hard saying? Take My yoke upon you, My burden is light. Is this easier? What will you have? Will you have salvation by mercy? or by merit? My friends, be not dismayed. You may be saved by both. Be willing to be saved in God's own way. This is the first requisite. This willingness, I assure you, is far from common. Easy as you think it at first sight, easy as it is in reality, there is a time when it seems hard. It seems hard to yield all our prejudices, to believe that all may be saved as easily as we can be, to believe that none can say, stand back, I am holier than thou. To believe that we may go where others can not, is not believing all men are equal before the eternally just God.

Man is a two-fold being, as he now appears on the earth. He has an animal and a spiritual body; albeit, that was not first which is spiritual, but that which is natural; and afterward that which is spiritual. Now, this text requires explanation for its true interpretation. The first state of man is spiritual, purely so. But, then, in this present state, the animal is first formed and receives its life and intelligence by influx of the spiritual. What, then, shall we say? Was Paul wrong? Was he without knowledge of this fact of pre-existence? No, but he was endeavoring to show the Corinthians that they should live beyond the grave. That those brethren whom they mourned as being dead without having lived to see Christ at His second coming, were not dead really, but were living, sentient, spiritual beings, capable of joy and beyond sorrow, living to praise God, and living to love Christ, whom God had sent to teach them all things the Father had given Him to deliver. Could the Corinthians have borne the full revelation of the past, as well as the glimpse of the future, Paul would have told them of it. But they were, as yet, scarcely delivered from the errors of pagan idolatry and Paul was too well disciplined a soldier of the cross, he was too passive in God's hands to declare what they were unprepared, in that age of the world, to receive with faith. They could hardly believe in the resurrection! How, then, might he declare the more wonderful doctrine of pre-existence? They were edified by what he wrote. They would have been torn by dissension, had he declared too much. Why, then, did he not word this passage differently, so that we, at least, might find an evidence in it that he knew of pre-existence? Thou, who thus parleys with Me, know that Paul was only a man; inspired, if you please, by God, to deliver the truth to mankind. He had his mission. He fulfilled it.

He who declared that one came after him whose shoe latchet he was unworthy to unloose, was greater than Paul, as a prophet. And yet, how little is recorded of John the Baptizer. Paul was not required to unfold doctrines hard to receive. There were already too many speculations upon abstractions. The church was already full of dissensions. He desired to preach Christ crucified and Him only. That is, to declare that Christ, the Messiah so long promised, had come; that He had called men to repent and to seek the living God; that He had been crucified and raised, by the power of God, from the dead so as to appear again to many of His followers; that He still lived, able to aid and comfort His followers, and help them on their way to join Him in the everlasting courts of heaven, where the praises of God continually ascend and where the incense of good works is ever acceptable.

Whence we come and where we go, has been declared by inspired writers long ago. Look for the evidence and you will find it. Yet, it has been overlooked by all the wise and great in the churches thus far. Look at Solomon's declarations. There is one way for all to go, both man and beast. Man dieth like grass, etc. Does not this require elucidation? But I shall not take up every hard saying or puzzling text and explain why it was written. Let it suffice, that when it was written, it was well. It had a purpose and the purpose was accomplished. Let the dead bury their dead. Do you press forward in the way of life eternal; which is, to love God and be His servant. His servants are only required to work for their fellow-men. He needs not their labor for Himself. But He desires to see them happy, and to be happy they must be useful.

Where is the man whose life is pleasurable in idleness? The idle man's body stagnates, his life corrupts. Usefulness is required of every man, in every condition. The more God distributes to one, the more has that one to account for. And will not God be found a hard master in the sense that He will require a strict account? Yea, of every idle word you must give an account. Be wise, then, ye rich in this world's goods. See that nothing is wasted, either of your goods or of your time, but devote them all to the service of your fellow-men. Relieve suffering, find employment, render others useful. Be wise in time, for ye know not when ye shall be called on to account for the deeds done in the body and for the deeds you neglected to do. When you are told, You did it not unto these, therefore, you did it not unto Me, what will be your regrets over your misspent time? In some respects it is far easier to progress here than in spirit form. But do not regard the nice points of doctrine or belief. Take no heed what ye shall put on, but be useful to mankind. View all as your brethren, whom it is a duty to assist. Press onward in this path of progress and you will find peace—peace on earth and good-will to men.

Be not over-careful of the future. Sufficient unto to-morrow will be its own cares and duties. Be faithful now to your present duties. Put not off the time when you will be useful to your fellow-men, to your brethren. Be wise to-day, for you know not what a day may bring forth. Be careful to know your duties, be diligent to perform them. Let each day have all its duties performed within itself, leaving the morrow free of debt. So shall you progress rapidly in good works. You will never need to repeat the saying of the Roman Emperor, I have lost a day, because each day will be saved. God will have been served by you with diligence, love, and reverence. All men have their duties; if riches make them manifold, poverty makes the few arduous.

Progress, then, every one—rich and poor, wise and simple. Be not fearful that you will faint by the way, for you shall have meat you know not of. You shall find it meat and drink to your soul to serve God, and your soul will enlarge and strengthen itself and your body, under this regime. How, then, will you delay to take the cross of Christ, the Messiah, the Son and Sent of God, who called you to this work by precept and example? How will you refuse the help God offers you now through His willing spirits? The help of their counsel, advice, and works—The Father has worked, the Son has worked, the brethren now work. The brethren of Christ, each of whom is Christ because each is one with Him who preached at Jerusalem. All are children of one Heavenly Father, who delights in their joy and beholds with pleasure their efforts to serve Him with a pure heart and willing mind. Would you, too, be a son of God, begotten by His power and saved by His mercy? Then follow Him, who called you affectionately, Come unto Me all ye that labor, and I will give you rest. Come unto Me all ye that thirst, and I will give you drink. Come unto Me all ye that have no homes, and I, who had not where to lay My head, temporally, will give you a place in heavenly mansions. For in heaven there are many mansions. And, however strong may be a man's expectations, he shall not come short of his reward, for the praises of God are bestowed on good servants, and they are made rulers over many.

CHAPTER VI.

THE SOUL OF MAN IN ITS RELATION TO GOD

What is there in the earth, then, that can enchain the soul of man when it properly appreciates its destiny? Is the honor and fame of the greatest, or the glory of the wisest man, commensurate with his aspirations? Does not the soul still ask for more? More! more! more! is ever the prayer of man. Can anything short of perfection satisfy this longing? Can any soul be eternally satisfied with the highest bliss it can now imagine? Not at all; for beyond that happy state lies another glorious mansion—higher, purer, more spiritual and more real. But the greatest pleasure that heaven's residents enjoy is the satisfaction of having performed their duties, of having thereby served mankind and thereby pleased God. Let us think, for a moment, of the state of a man condemned to an eternal sojourn in the body. Even could the body retain its youth, how must he tire of every amusement, how must he long for novelties to explore, for worlds to conquer. Why, then, O man, why dread the hour of separation from the dust of the earth!

Alas! ages of ignorance and dark error have obscured the light within every man's soul; the light of God's love. When man really believes God loves him, he can not but desire to be with God as earnestly as Paul or Jesus ever did. And can you not believe God loves you? Do you fear your sins have turned His love into hate? Or, do you believe there is no God to love or hate? No one to rule the universe that chance has made and which chance preserves? Even then, would not a leap into chance be desirable? Would you not choose to put your hopes in chance that might be for the better; a chance that, at least, would have novelty to recommend it? Annihilation is feared; but what can annihilate a thought, an idea? And can a producer of an immortal be mortal? The idea is not scientific. It is not reasonable. Reason is the gift of God and, well exercised, it will lead you to its Giver. Ponder upon this. Ponder upon all you see of creation's glories; ponder upon all you ask to make you completely happy, and see if any thing but an Infinite can gratify your wants, or display the glories you daily behold.

So common are wonders that you cease to admire. But, now, new wonders are presented to you. It has pleased God, in these latter days, to send us His messengers—angels, spirits, or whatever you please to call us but, certainly, His servants. It has pleased Him to direct us to call you by new methods, by outward signs, by wonders that philosophy can not explain and credulity only can question. Yes, it is only questionable by credulity. Men have, credulously, admitted the assertions of fallible creatures like themselves to be infallible truths but now they refuse to listen because they are told it is dangerous; because it may overthrow their authority, so cherished; their idol, so worshipped. Their pastor, perhaps, threatens, warns, entreats, cries out, wolf! wolf! When, in reality, it is only to a still, small voice that we ask you to listen. We startle you with violent or forcible demonstrations, only to awaken your attention; to break the slumber that enchains the hearts given up to faith in a brother, equally enchained by ages of tradition and invention. Why, then, do we not speak to you in the still voice? Because you are not willing to receive it. Behold, we stand at the door of your heart knocking and asking for admittance. You will not open, or open only a crack, as it were. You stand before us in fear.

But do we preach frightful doctrines? Behold! we preach one God, the Father of All, the Maker of Heaven and Earth and all that is therein, and Jesus Christ, the Messiah, His Son, who was born of a virgin mother, who was crucified, was dead, was buried; who descended into the place of departed spirts; the third day He arose again, He ascended into heaven from whence He shall come to judge the living and the dead. I, or we (for, though many, we are one), believe in the Holy Ghost, the Holy Catholic Church, the Communion of Saints, the Life Everlasting. Do you believe in this? Scarcely. You repeat the form of words, handed down from the beginning, but you pervert the substance, you misconstrue the spirit of them. What do you believe of the Communion of Saints? Will you listen to the Saints when they desire to communicate with you? I only ask you to do that. I only ask you to open the door wide, and wider, to let the King of Glory enter in. What do you believe of the Holy Ghost? The Holy Spirit, as it ought to read, and does read in other languages. Do you believe it can communicate with you? Do you believe it would be desirable that it should do so? Then be willing to open the door of your heart, at which I am standing till My head is wet with dew and My locks with the drops of the night. Open the door, and let the King of Glory enter in. He is ready and waiting for you to be willing, for man shall be left free to choose the good and refuse the evil; or, to choose the evil and leave the good.

What do you believe of the Holy Catholic Church? Do you believe all churches belong to it, or all men, or all your own church members, or only a few selected from your own church? Or, are they not all men who serve God, who are neighbors to those that fall among thieves? O wicked and corrupt generation of vipers! You will neither enter in nor let others pass the open door. You would, if you could, shut the gate of God's mercy and condemn your brethren to eternal tortures. But God is too strong for you. The foundations of His goodness and mercy are laid in His strength, and who shall fight against that? The superstructure is built by Love upon this sure foundation, the rock of faith, and who shall declare Him unrighteous? Who shall deny Him the right to love His children? His children ask Him for bread; will He give them a stone? What will you have—mercy, or justice? Will you choose mercy for yourselves and declare that justice requires others to suffer? What will you require of God—heartless cruelty, or glorious mercy, ineffable love? Alas! that man should be so wise in his own conceit as to make God powerless, desiring to save sinners, yet unable to accomplish it. A God desiring to save mankind and reconcile the world to Himself, yet, after sacrificing His only Son, to be foiled by man's perverseness or the arts of an enemy—an enemy that was either created by Himself, or else has ever existed independent of Him!

CHAPTER VII

THE DESIGN OF MAN'S CREATION

Man was created for the pleasure of the Deity. This truth will be evident to a reflecting mind, because the Creator could have had no other inducement. He had no need of man to insure His empire or happiness, but He could please Himself by occupying His mind with the struggles and pains, the pleasures and raptures, of a free-willed being. He made man and pronounced him good. He formed him for enjoyment and placed him in Paradise. Foreseeing, foreknowing, that the desire of change would come upon the tablet of man's mind, and that with such a desire would come a struggle for its gratification, He provided the means for man to gratify this, his first desire. He knew it would be a blind one; that man would not first ask to know what he would get by the change. But the life of enjoyment, which is aimless, ceases soon to be a life of enjoyment. Then change of some kind is demanded and even a change for the worse would be welcomed. What, then, could God do to make man happier, since he was in Paradise, which is tranquil happiness, but give him a change? He resolved, in infinite wisdom, that it should be to taste of the Tree of Knowledge of Good and Evil. So He placed before him only that way to escape from Paradise; sure that he would, sooner or later, embrace the opportunity afforded him to pass out of the lovely and beautiful garden, where perennial bloom and fruit ever adorned the trees; where the animals were submissive to man, their master; where no violence ever disturbed, no horrors ever appalled the peaceful mind of the primitive man.

This state still exists to many individuals of the human race. They have not yet resolved to taste the fruit of the tree; the tree that will relieve them from confinement, as many have regarded it. They still linger with pleasure by the cool fountains, the beauteous groves, the sparkling lights of the ever fresh and ever new land of Beulah. Happy denizens of a blissful Paradise, alas! All happiness that is not founded on benevolence, is transitory. However long the time man may enjoy Paradise, it can never be an eternity. That world must, at last, be left solitary and uninhabited, unless God shall take pleasure in a new creation. This He may, or may not, do. He knows: but the angels in heaven know not, neither doth any man.

There is now a period of repose prepared by the Almighty Ruler and Creator for Himself. Not that He needs repose but that He pleases to have it. Having labored, or acted, as you may choose to phrase it, He is pleased to rest. His rest will be comparatively short. His laws and foundations might stand eternally if He chose. But He prefers to change, or modify, them. He might continue in a state of rest eternally, but He prefers to make new laws and lay new foundations, to have new creations. Amid all the wrecks of matter, the chaos of destroyed or finished or incomplete works, He is ever able to make new worlds or universes start into being. No variety is too infinite for Him to accomplish. No monstrosity is too great, no sameness, or similarity, too trifling for His powers. He is Almighty. You admit Him to be Almighty, but you do not, and can not, realize the meaning and force of the appellation. But you could better realize it than you do. Try to do this. You can conceive that no other is so powerful, no other one is so good, no other one has so much love, and no other one deserves so much praise as the Deity. But you can not conceive how much of them He has, because He has an infinite quantity, and you are finite.

Then, be not caviling against the order of God's creation and desiring to help Him reform the world! What? you say, not try to reform the world? When so much wickedness exists in it! Not at all, I say. But, remember, I do not tell you to be idle or wasteful of your time or means of doing good. By no means. It is to the full exercise of these that I urge you. Do all the good you can, but don't wait for the time when you can do it on a large scale. Be faithful in small things and you shall have an opportunity to attend to great ones. When you see a brother in distress relieve him. Is he hungry, give him food; is he thirsty, give him drink; is he in prison, visit him; is he sick, perform his duties for him. Inasmuch as ye do these things, even so shall ye have praise from those whose praise is in the love of God. Inasmuch as ye do these things, ye shall have favor with God and be raised to a Sonship; be placed on His right hand, separated unto eternal glory and made a joint heir with Jesus of Nazareth, the Man of Sorrows, the Acquainted with Grief, but the Beloved Son of God, in whom the Father was well pleased, and for whom He bowed down the heavens to declare, in a voice of thunder, Behold My Beloved Son, in whom I am well pleased! O wicked and perverse generation! Ye seek after a sign. But were the wonderful works wrought in your midst that were done by Him in Judea and Galilee, eighteen hundred and more years ago, ye, too, would blasphemously declare, He casts out devils by the prince of devils. Ye, too, would cry out, Away with Him, crucify Him; crucify Him! We will not have Him to rule over us, we are Caesar's subjects.

CHAPTER VIII.

THE FUTURE OF MAN

There is no subject on which more speculation has been exercised, more ingenuity wasted, more labor lost, than the Future of Man. Placed as he was between two states, of which he was entirely ignorant, it was not strange that at first he should have supposed the condition in which he found himself was the only one he ever did, or ever would, enjoy. But it soon pleased God to inform him of the future, so far as to make known to him:—first, its existence; second, its nature; and third, its design. But yet, this was given darkly; and man only saw it as through a glass, darkly. Passing along, the vicissitudes of his journey often caused man to sigh for repose. The glory of the future was reduced to a state of rest, and the activity of God's servants was supposed to consist in praising Him, who alone is beyond praise; in glorifying Him, whose glory can not be exalted. But God was pleased to let man corrupt the light He had thrown into their hearts and understandings, by false lights, by perversions of His revealments and misconstructions of His motives.

Having at sundry times clearly shadowed forth the nature of the future, it now pleases Him to declare it plainly, through an humble seeker of happiness who is willing to be taught and who only desires to receive the truth; who will receive his reward by having his patience exercised, his motives traduced, his efforts to enlighten his fellow-men made abortive. Yet, the truth will prevail. God will save mankind. And the time will come when the knowledge and love of God will cover the earth as the waters cover the sea, His will shall be done on earth as it is in heaven, and His servants shall have their daily bread, that cometh down from heaven, placed before them to be received with joy and thankfulness. This will be a joyful time for all who shall see it; and verily, I say, that there be those standing now in the body who, in the body, shall see it. Blessed be God, that He has chosen the poor of this world and the humble among men for the accomplishment of His ends. He will not appear as men have expected Him. He never has' appeared as they expected Him; and never will He be the mere follower, or fulfiller, of the imaginations of men. Alas! that such blindness persists; that such ignorance defeats and such folly derides the poor in spirit, and the humble followers and seekers of truth.

Then will be an overturning, an overturning, an overturning! The sun will be darkened and the moon will cease to give her light. The stars of heaven will fall to the ground and the heavenly host will disappear, in the day of God's appearing. Then will the mountains be called on to hide men from the face of Almighty God; then will men get down into the depths of ignorance and abase themselves before idols of flesh, who can no more save than idols of wood or stone. Then will the sea give up its dead, the graves be opened and the long since departed from the sons of men will reappear to their astonished vision. But when will these things be? Not in this state of existence! Oh, no. The fowls of the air, the beasts of the field, the fish of the sea, must all be raised when the bodies of men are raised for they are all one flesh. What, then, do I mean to say, and what shall be understood from My dark sayings? I mean, that when the soul shall have left the body to the dust from which it came and to which it must ever, and forever, return, that then men will see the great day of the Lord. Then will appear the sign of the Son of Man coming in the clouds with great glory. Then will the trump sound, and the dead in sin will come forth from their lives of ignorance and darkness.

Then will appear the New Jerusalem, that cometh down from Heaven, arrayed as a bride for her husband. Her gates shall be praise and her streets paved with good works. Her glory shall exceed that of the sun and there shall be no night there. Do you desire to be a resident of this beautiful city? to be the husband of this lovely bride? You may be. You will be. But it is for you to say when. Verily, I say, there be some standing now in the body, who shall not taste of death ere they enjoy its embrace and taste its happiness. Verily, I say, there be some standing in the body now, that shall, when suns and moons and the earth itself shall have passed away, be yet so far from it, that the power of God only shall suffice to bring them to the home He designs for all men. Choose ye now, whether ye will serve God or Baal. Choose ye now, whether ye will be meek and lowly and acquainted with grief; whether ye will walk in His ways and follow His precepts; follow the ways and precepts of Him who suffered at Jerusalem that you might be instructed; who died that you might be saved from the death of sinners, who preached and prayed and worked and suffered that He might teach mankind how to arrive at the New Jerusalem more speedily, and so be saved from punishment for their errors, from suffering for their sins, from death and the grave. This is the reward He hoped for, and you can see how far men have profited by His example. Waste not your time in regret for the past. Behold, the door is open. Press forward now, while it is day, for the night is approaching in which no man can work. All he can do will be to suffer and to let the angels, who have worked, work upon him.

What, then, shall man give to forward his salvation? Give your heart. Let nothing earthly come between you and your God. Commune with Him in spirit. Let your spirit have communion with the saints. Let there be no man to tell you when to pray, or when to work; when to listen, or when to sing. Work when you find work, pray without ceasing, sing when you can, and be content with every dispensation of God's pleasure. Hear Him whenever He speaks to you in your heart, in that still, small voice that is not easily heard. Turmoil and strife will deaden your sensations, and the ears that have been well accustomed to it may easily be dead to the voice of God in the heart. But do not be proud, O ye followers of Jesus! Do not require that God should, Himself, come to you or manifest His own presence plainly to you.

No man can see God and live; and yet men have seen God! How is this? Men see God when they have experienced the elevation of heart that God's spirits can give. They can see Him by the eyes of others and the mirror that reflects His image does it truly. It is, then, a glorious privilege which few, of all that have been born on the earth, have had in the body—of seeing God Himself pictured on the soul of a fellow man who was in the spirit-world. It is such visions which have made the apparent contradiction. It is this saying which has caused so many to stumble that I now explain, when I ask you to be willing to be instructed by God's Son and Sent, by the Holy One of Israel. He was superior to all pride of heart. He was a man who delighted in virtue, who walked in humility, who ever was ready to help the needy or feed the hungry. But God was in Him through the departed spirits of other men, who, before Him, had sacrificed their lives, their fortunes, and their reputations in His service. They had then been found worthy to loose the seals of the Book of Life and wait on God in celestial courts, where every soul is subservient to the higher powers; where principalities, dominions, powers, thrones, and every kind and graduation of them, and of other station, stand before God, willing to serve Him in whatever capacity and in whatever rank He may desire.

Their will must be submissive, or they cannot be His servants. They cannot serve two masters. If God rules, well. If not, they die to His presence and glory; they fall into ignorance, doubt, and perplexity. They can recover only by faith and hope of mercy and help of God's servants, as God wills that His servants shall help one another. By this shall all men know that ye are the servants of God and of His servants. For God's servants have servants under them. But the servants do God's will and work, because His upper servants are harmonious, and united with Him. Let no man, then, despise God's servants. Receive them as angels. They are angels. They are continually praising God, but they praise Him by works. They know when the sinner tires of sin; when he seeks refuge in God. They are ready, then, to meet with him, to sup with him, to be his waiter, to bring him messages from God, and to lead him on his way to the Heavenly City.

But, behold, how men have lately rejected all these servants of God. They have refused to believe them good, even. They have feared they would ask them to cast off the pride of the age, the lust of the flesh, the hope of their fellow-men being doomed to eternal punishment. All these have been urged against them, not openly always, but down in the depths of men's hearts; so deep, sometimes, that the man himself scarcely could perceive them if he looked, which he seldom did. And now what shall we do? We have piped unto you, and ye would not dance; we have played unto you, and ye would not sing; we have urged by affection, we have pleaded by pity, we have wept with those who wept and mourned with mourners. But all, except a few, have rejected us. What will the Master of the feast do, when He finds His tables are not filled? Go out, He will say to us, go out, and compel them to come in! Not by over-bearing their will but by over-mastering their reason. This we must do, and this we shall at once begin to do. Prepare, then, to meet your God.

CHAPTER IX.

THE PREPARATION FOR DEATH

What shall be the preparation for Death is an important question. It is much dwelt upon by many who do not know why they should rejoice at the change, and feared by many who want knowledge of the future. There is among mankind a great want of knowledge of this preparation, and of the want itself. True, men are seen to die and disappear; to vanish, as it were, forever from their accustomed haunts and lay themselves upon the shelf of forgetfulness. Some die willingly, some unwillingly; but nearly all ignorantly or recklessly or defiantly. True, many feel a hope and trust in God's mercy, but it is a hope founded on false conceptions of His justice and mercy. Having had some knowledge experimentally of God's goodness and mercy; having experienced His loving kindness and been pardoned manifold sins; having led many astray in the body and preached falsities to fellow fools, we are now prepared, well-prepared, to declare to men what they want, and what God wants them to have. First, men should have new hearts; second, they should have less pride; third, they should cultivate the knowledge of the wants of their fellow men; fourth, they should have patience to wait for death, and activity to secure the love that God has for them.

To be known as a servant of God, implies that you obey His directions, follow His rules and attend to every intimation He may please to give. This is only what you require of your servants. Why, then, do you complain that God is a hard master, reaping where He has not sown and gathering what He strewed not? Why do you complain that God is not kind, or He would have laid all obstacles away from you, and given you a seat upon His right hand to judge one of the tribes of the earth! My friend, you want a new heart in which the love of God shall be the first principle of action; in which that action shall be exerted in relieving your fellow-men from affliction, from every kind of want, bodily or spiritual. It is these works that Jesus did. It is these that He preached. He healed the sick; He restored the lame, the blind, the halt in every way. He walked in humility, and He died condemned by His generation.

So it may be with you. Do you want to be like Him, to be led to a life of suffering and die a death of agony? No, I hear you say, I can not bear that! How, then, can you follow Him? You call yourself a follower of Christ, do you not? And how do you follow Him? My people have not known Me, they are all gone astray, there is none good, no, not one. Such would be His exclamation now, were you willing to hear Him. Such He will say, whether you hear Him or not. Alas, ye ignorant church members! Ye are no better than church members were two thousand years ago. Your fathers killed the prophets, and you would now offer up the new philosophers on the same altar of ignorance, folly and the love of self, if you had the power. Is it to you, then, that I must address Myself? Yes, it is even you whom I now call upon. I ask you to throw off all dependence on man whose breath is in his nostrils; but not to be afraid of man whose life is a breath everlasting, who, till now, has watched and prayed and desired, with a great desire, the day of the Lord that now approaches. The day so long expected by the Christian church has already dawned and its sun will not go backward.

Come unto Me, then, all ye heavy laden and I will give you rest. You will no longer be required to give tithes of mint and cumin while you devour widows' houses; no more be sent to hell for a want of belief or placed in heaven for a profession of it. No, I only ask works of you, sure that you will not do the works unless you have faith. What, I hear you say, shall it be thought of no consequence what man believe? Shall infidels be saved by morality? Oh, no, my friends; I do not say any such things. I say, again, you can not do the works to which I call you without faith. And then, I say, most firmly and decidedly, that you can not be saved by works. I say nothing can save you but God's mercy. Not that I mean that you would deserve eternal punishment for a few years of venial sins, but that God has chosen to show mercy and He will save you from the consequences of sin. These consequences would extend beyond the grave, and would be felt in other generations were it not for God's infinite mercy, which, being continually exercised, saves you from continuing to commit sin to all eternity.

Is not that a new idea, that you would thus deserve eternal punishment for continual, unending sin? But, blessed be God for His infinite mercy! Though your sins be as scarlet, they shall be as wool. Though your sins be legion, your Father's power and goodness can overthrow them and their consequences. But, does He ask nothing in return from man? Yes, He asks man to be grateful; to love Him; to fear His anger which is His hate, or wrath which is the absence of His love. Nothing more, for God is not subject to passion or fits of rage. No: calmness is never absent from God. Does not the whole creation move in harmony? How, then, should God be angered, as men are, when their work goes wrong? Why should God hate when He can exterminate and destroy? Annihilation is in His power, but it would only be by His absorbing the soul of man; for out of God it proceeded and thus it is, that of all His works only this is immortal. God can not die nor be destroyed. To suppose it, would be folly. Neither can any part of Him die or be destroyed, because, if part could be, the whole might be destroyed.

What, then, do I call you to perform in preparation for death? Not a sudden change of all your habits and emotions and mode of thinking. No, I ask you only to listen to God, who will teach His children of His ways, who will write His law on their hearts, who will put it in their inward parts; who will enter in, and sup with you and you with Him, in a holy and blessed communion, of which no one can partake without the preparation of a willing mind and a changed heart—a heart changed from evil to good; from self to God; from all that relates to earth to such as relates to Heaven. How will you obtain all these blessings, which you will easily admit are most desirable and are also proper preparations for death?

I will tell you what to do. For, since the day in which Pilate asked, What is truth? the question has never been fully answered. Indeed, its answer has hardly been attempted. Shall I venture to declare it now? Will the world, so full of pride, of arrogance, of crime and of self-esteem, receive my answer—given, too, through so humble a medium as I am now using? He is a man whose religious notions have never been approved by any society of men, though he has belonged to one, by a chance; who has departed from the faith he formerly held, and who now is willing to receive truth from heavenly sources without asking the spirit whence it comes or whither it goes; who is willing to believe himself in God's hands without requiring that only God should handle him; who has not felt the love of God in his heart until he was sure that God loved him by reason of His goodness and of His nature; until he had reason and revelation agreeing in his mind to declare, that what is good comes from God, and evil can not proceed from a good and pure source; but that man must be saved by God's mercy, if saved at all, because, of himself, he can do nothing, accomplish no good and withstand no temptation. He is a man who has often and earnestly prayed to know the truth and be brought under its heavenly teachings by Divine aid.

How, then, will I answer you, ye scoffers, when this faithful asker has waited so long unanswered and, while groping, as it were, in the dark, has so often fancied he saw light and has as often found himself deceived and left unsatisfied? But I will address

Myself to him and to you, to high and low, rich and poor, noble and ignoble, priest and flock, believer and unbeliever, until all who are willing to receive the truth in its simplicity, by its internal evidence, shall be convinced; until all who are willing to be more blessed than was the disciple Thomas, shall have received their blessing. Then will God's thunders roar out conviction to stubborn fools and graceless scamps, who choose to serve their lusts; who fear to receive the truth because it would deprive them of their trades; who having long lived in the enjoyment of the world are unwilling to leave it for a dark and doubtful future, unwilling to trust to God's mercy, preferring to trust to a stock of service treasured up for them through the works of fellow-men, who were themselves saved only by God's mercy. Alas; in that day the sun shall be darkened, and the moon shall not give her light, and the stars of heaven shall fall to the ground and disappear forever. And what kind of bodies will these stars and sun and moon prove to be? Naught but the lofty dignitaries of the several hierarchies that usurp power over the bodies and souls of men. Their power was given them for a season but in one hour it shall be destroyed. Then shall the multitude, standing afar off, weep for the destruction that shall come upon all flesh. The ships that go down to Tarshish and those that get the gold of Ophir shall mourn, for no man will buy their merchandise any more. Alas! Alas! they will say, for that great city, which consumed the fruits of the labor of men, which was so splendid and so overgrown, so lofty and so grand in its claims, so powerful and so full of the glory of men.

Now, all these things have been prophesied of before and ye have sought to give the prophecies private interpretations. No scripture is of private interpretation. The letter killeth but the spirit giveth life. But do not suppose that the sense must, necessarily, be obscure because you must take its spirit. On the contrary, reason will greatly aid you in understanding scripture; without reason it can not be understood and without the exercise of reason you can not understand it. True, the wayfaring, or plain, common-sense man, can not err therein, but those who strain for effect and swallow impossibilities, can never be right in their construction.

There are many things I have to say to you, but ye can not bear them now, was said long ago by One who, you admit, knew what was best for those who heard Him. But now, if I say it, you will hardly be satisfied. You will call on Me to stand and deliver. You will try to obtain by force, or by threats, what is not yours; and to make your way to heaven as a thief and a robber. But you can not succeed in any part of this plan. For I shall say only what God authorizes Me to say and that is only what you are prepared to hear with some kind of patience. If I should declare all to you, you would be overwhelmed and would despair of ever being found worthy to receive God's mercy, though it is as infinite as any other part of Him, and as necessary to His perfection as life itself. What, then, is to be done? I shall refer you to the Comforter, the Spirit of God. He shall come to you and teach you all things, even the deep things of God.

And now, it is expedient that I go away to the next part of My subject; or you would never be willing to receive any other Comforter than this outward address of God through your bodily senses. But, when you are willing, the Comforter will take up His abode in your heart and will address Himself to you spiritually. His voice will be heard rebuking you, in the cool of the day. He will be found prompting you to good works, to lofty aspirations, to everything that will elevate and refine, and He will warn you with pleadings of love, ere He leaves you, if you once admit Him to your heart. How long, O wicked and perverse generation, shall I strive with you? I swear, saith God, that I will not always strive with man, whose breath is in his nostrils and who goeth down to the pit. No: I will raise him, even at the expense of his free-will, if necessary. But it shall not be necessary and the day of salvation is at hand! Repent ye, for I will not be put off.

This language, you say, is all assumed; for I have not shown that God has authorized Me to speak for Him. I ought, you say, to do some work, to perform some miracle, to establish some church where the authorized body might speak for God and assume power over men. But I, who am a servant of God, in unity with His other trusted servants, and desirous only to do His will, care not for reproach, or sneers. He saved others, now let Him save Himself! was once said to the greatest miracle-worker the earth ever saw. And how was it received? Father, forgive them: they know not what they do! Alas! that men should in eighteen hundred years have made no more progress than to repeat the sayings of their fathers, who stoned the prophets, and destroyed the sons of the prophets, and cast down the altars they had erected to the living God.

CHAPTER X.

WHERE SHALL THE UNGODLY BE FOUND?

Where shall the ungodly be found, when even the righteous shall scarcely be saved? This is a great question, involving the whole policy of God in His relation to man in the future state. The future state is a state of passiveness to evil and activity of good. In it men get better, but they are not allowed to get worse. They struggle only for good. If evil approaches they receive it not. It is repelled from them as the similar poles of a magnet repel each other. The influence of evil does not act positively. It exists, because the spirit of man has imbued itself in it, and it must be left free itself, by wasting or wearing out, or it must be overcome with good. Overcome evil with good, is an injunction which men must act on in this life and in that which is to come. How this is done, I will briefly sketch in another chapter, for I am resolved now to proceed more methodically, as I find the medium willing, passive and patient.

All truth is deeply hidden by ignorance and by pride. Humility is the condition of truth, and truth is the gift of a benevolent God to its seekers. Truth, then, can be obtained by seeking it, and the only obstacle to its triumphant dominion is the unwillingness of men, in general, to receive it. They are unwilling to confess their errors and to make restitution of the goods they have robbed from others. Who would now follow the example of Levi, who, when his Lord and Master of instruction, sat down at his table in his own house, like unto being in his heart, declared if he had wronged any man he restored unto him fourfold; and this, after having given half his goods to the poor? Many, too many by far, now think if they give to a church or a churchly object, they have made amends for a life of extortion. But you know full well that no such doctrine was ever taught by Him, who taught as man had not taught, when He was a resident of earth in the body.

What, then, shall we say to the seeker after truth? and what to him who refuses to receive it? To the latter we say, it would be better had you never been born; that you had always remained in the state of dweller in Paradise. To the former, that the seeker shall find, and to him who asks shall be given abundantly, good measure, heaped up and overflowing. For God, who loveth a cheerful giver, is one Himself who is unexcelled in generosity. And the only man who can want, while God has good things to bestow, is he who, filled with pride and conceit, cries out, Begone, all ye workers of iniquity! begone, all ye deceivers! I will have none of your new-fangled dreams, none of your new philosophies, none of your spiritual guides! I have one guide, the Bible, as the Jews had one father, Moses, and as he was sufficient for them, so is it for me! But let me say, I will trust in God, and He shall be the portion of every soul that desires to bring forth good fruit and work righteousness.

We have, thus far, looked only at the reason of the matter and placed our reliance on your willingness to be convinced. Because a man can close his mind to reason, we must also address his feelings. We shall appeal to his humanity, to ask whether he could be happy if his children were in hell and he in heaven? Could God be happy if His creatures suffered eternally and without hope? No man can, from his heart, say Yes. The very beasts would deride him and devils would gloat over his superior heartlessness. There is in every man a repugnance to believe as much of this kind of doctrine as churches, generally, make them say they do. Look at the Roman Church, declaring that only its members can, by any possibility, be saved. Do its officers and subjects act upon such a belief? Oh, no! or, if they do, it is only some deep scheme of deception, for they are not shocked and melancholy at the indifference with which the greater part of the world is going along the path of eternal damnation!

Is the English Episcopal Church any better in this respect? They, too, require membership by baptism to insure salvation. They, too, feel that it is not true; for they comfort their Quaker friends, who lost a lovely infant, by the declaration that it is in Jesus' bosom, or, God has taken the lovely blossom to Himself to save it from rude winds and storms of this wandering life. Have they the boldness to declare, My friends, your child went to hell because you did not have it baptized, as God has appointed, Christ preached, and we practice? No; never would they be so impolite, so heartless, so inhuman, so devilish. That is a proof that it is not true that the child did go to hell; for if you had believed it to have been really in danger of such a fate, rather would you have baptized it by stealth than have it suffer eternal hell-fire. But yet, you have other friends, not belonging to Episcopal churches, whose children have not been baptized; why do you not rush to them and plead, beg, and entreat—weep, wail, and lament—until you persuade the parents, at least, to let you baptize the innocent babe, the unconscious infant, that must be otherwise condemned for its progenitors' fall from grace and happiness? Alas, I might go through with the creed of every church, until I have manifested what is apparent to ordinary observation, if exercised, that they, none of them, believe what they profess and all have erected an impossible standard of faith; while they almost equally discard and reject practicing the legitimate consequences of that faith. While they nearly all require faith as the one thing needful, not one has it. How, then, shall any be saved? God's mercy is infinite. Blessed be God, who will have sinners to be saved. And they shall be saved, because He wills it!

CHAPTER XI

THE SOUL OF MAN IN THE FUTURE STATE

There is a soul in every man, immortal, unchangeable in quality or essence, invisible to bodily eyes, but breathing and living as separately and as permanently separate from God, as the body is from earth, so long as He chooses to maintain its organization and individuality. Why this soul is hampered and restrained, why it is oppressed by animal passion and darkened by error, is explained in the previous pages. But how it is to be freed and released from the stains imparted to it by its union and alliance with the body, is a deeper and more complex subject which must be treated with more exactness and logical method, to carry conviction to impartial and willing seekers of truth. I shall, therefore, divide this part of my subject into three chapters, or sections. First, what man should do while in the body. Second, what change death makes in his relation to God and his fellow-men. And, third, what he must do to be saved with an eternal salvation.

Section one: What must be done in the body.

Man, being born to trouble, as sparks that fly upward are prone to descend; man, being placed here for the express purpose of receiving instruction in the knowledge of good and evil by experience, has no escape from temptation. He must suffer it. He can not escape from the consequences of his condition any more than he can escape from his condition. To be sure, some think that by suicide they may escape from this condition of existence. But such are mistaken. They only fulfill the appointed term of their bodily life and fall, by the last and greatest temptation, into the next stage of being where they soon perceive what opportunities they have wasted, what mercy rejected, what folly exercised, and what useless—and worse than useless—sin they have committed, by taking into their own hands the prerogative of the Deity and closing an existence He designed should then end. But, if they must have then died, why did they not wait for God Himself, by His laws, to dissolve their connection with the body? Because they could not, or would not, resist the temptation which evil thoughts, rebellious designs, and selfish considerations placed before them. They fell, and left mourning friends, disgraced relations, fond children or parents, or weeping brothers or sisters; or, at least, they fell into condemnation for violating the law of God written on the heart, which is the unpardonable sin.

Let us proceed to see what should be the duty of men thus tempted, as this is the sorest temptation that besets men and is, also, one of the most common. My medium has been beset by it, and I know that few men have lived to maturity, or, at the farthest, to old age, without having been beset by it frequently. It offers such a ready solution of the most overwhelming difficulties, such a sudden relief from care and anxiety, that sometimes its almost unending consequences are overlooked and the suffering that others must experience from the rash act is disregarded. It is the highest, or the extreme, manifestation of selfishness. The heart of man is never so desperately wicked as when it resolves to disregard all law, human and divine; to overlook every obligation of friendship, love, and the debts of nature and of business; and, placing the means of destruction where his life must be extinguished by them, he rushes violently, with robber-like audacity, toward the kingdom of heaven.

The next greatest sin is the sin of living for yourselves. To die for yourselves is suicide; to live for yourselves is murder. Yes! murder is only its highest, or extreme manifestation. Theft, robbery, arson, murder, are only steps in the ladder of crime. All end in one object, the serving of self; all are liable to and receive, one condemnation, that of the withdrawal of God's mercy for a time. He leaves man to do his own will until he is willing to accept other guidance. When man finds that all these selfish gratifications result in bitterness and death to enjoyment, he sometimes, even in this life, will reform. That is, he will open his heart to God's influence, who, through His agents—the ministers, the servants, the angels, the messengers, the spirits, or the demons, as they are variously denominated in the translation called the Bible—will visit and help him unto salvation.

We will now take up the second section of our subject—*What of the change which death of the body makes in the relations of man to his Creator.*

Section two: *What change death makes in his relation to God.*

Man, being never released from the dread of death, except through suffering, seldom deliberately resolves to seek it. He often contemplates suicide but seldom braves it. When, at last, death presents himself to the sick, suffering has worn the body and tried the soul of man. He has, by God's mercy, lain for days in a favorable position for seeing the vanity and nothingness of every occupation that had only self in view, and feels sensitively the need he has of God's mercy and loving kindness to raise him from the deep pit in which he finds himself sunk. Much work, efficient work, is often done on these beds of suffering; and the soul, by a day of suffering, gains years, and sometimes myriads of years, advance in the great work of reconciling itself to God. Albeit, many good resolutions are broken by temptations succeeding an unexpected recovery; although the last state of that man is worse than the first—he having ejected the presence of God and the communion of His spirits, or saints—he was nevertheless sincere, and, therefore, actually at that time, was as much reconciled to God as he appeared to be.

But let no one put off the work till tomorrow, much less to the death-bed, for you know not what a day may bring forth. Ye know not that time and sense will be left to you for repentance then. Work now, while the day lasts, for the night cometh when no man can work. That night is the next state of existence, in which no man can work out his salvation—but he is worked upon as he is willing. He is acted upon by God's higher spirits—persuaded by example, taught by precept, instructed by his memory of his earthly experience. He is never forced to receive the good, but he is never allowed to pursue the evil. He may be inactive and unprogressive, but every step he does take is one reducing his distance from God—one leaving error and sin behind. He then marches forward, every step being easier than the preceding one, every step reducing his toil while increasing his enjoyment, every step enlightening his pathway with good actions and wishes; and thus with meek regrets and kind wishes to others and for others, he proceeds on his way rejoicing more and more, until at last having reached the circle in which he can see God, he bursts into praise. He declares himself His servant, begs to be allowed to view Him always, to serve Him forever, in any position it may please God to place him; and, in harmony with God and the servants of God, he proceeds at an accelerated pace.

The mighty roll of time, ceaselessly beating on the shores of eternity, is faintly heard in these lofty and distant halls, or mansions, of bliss. The echoes of the past are no longer reverberating through the ears of the former inhabitants of the earth. Spirits of the great and good, as measured by God, commingle. They harmonize most perfectly: they have one will and one law, one power and one wish, one hope, one love. And all these being common and joint, are equally common to and joined with God. They harmonize with God; they fall down, as it were, at His feet, united to Him by the closest bonds of love; also united by a resolution to have no will but His, to exercise no power but His, to feel no desire He does not implant, to have no motive of action, no action, no feeling, no love but God's. Being in this state one with Him, they are sons of God; joint heirs with Christ; seated, with Him, on the right hand of God, from whence they shall come with Christ—as they are one with Him, being themselves Christ—to judge the quick and the dead; to enter into every soul that is willing to receive them and to lead that soul to God, even as they were led.

Some will find it hard to believe that they may be and shall be, as Christ; that they shall be really Christ. But it is because they mistake the nature of Christ. Christ is the power of God unto salvation. It is the love of God unending; which is, and was, and shall be evermore. It is the Son of God, born of purity, led into suffering, raised into power, and seated on the right hand of God, where, in the wisdom and power of God it rules the world, the whole creation, by its oneness with God. And its oneness makes it God. For things which are one and the same are not separate, or unlike; not separate but, yet, not merged in God. Possessing still an individuality, this Christ is made of all the good and great in righteousness. All ages of the world, and all celestial, or planetary, bodies in the universe contribute to its formation and fullness. Jesus of Nazareth, a man approved of God, was the first of the inhabitants of earth who was raised by God to this high and holy and immaculate position. Jesus is our great exemplar; and His precepts, preached as they have been for eighteen hundred years, are now to be preached again in a new form but not in new substance.

And this brings Me to my third section; namely, *The duties of man,* or the way in which he is to work out his own salvation with fear and trembling.

Let us proceed in order, and by order show how orderly God is in what He requires of man. First, in childhood; second, in maturity; and third in old age.

Section three; Part One: Childhood

When a child is born, the soul is received into its body at its first inspiration. Its first sensation is pain; and this is wise, for thereby it is prepared for pleasure and enjoyment which requires previous pain for its perfection. When in the course of nature, as it is called, the child grows and begins to show signs of intelligence, the activity of the soul commences. We must remember the soul, sometimes erroneously called spirit, has heretofore been in a passive state. That activity has to be learned, as well as everything else that God does not implant. Now, when it has further advanced, the signs of intelligence become stronger and language is learned.

Language in the beginning was an inspiration of God. Without His inspiration its perfection could never have been attained, for all the efforts of the learned and powerful can not produce a perfectly original language. But the language once given has been variously modified and diversified, and though all come from one original source, the lapse of many ages of years, the many transitions through ignorant mediums, the diversity of situation and of race, as varieties of men are termed, have so changed language, that it is now infinite in variety. Or at least, it appears so to men for, besides the thousands of languages, there are to each an almost unending variation in different individuals—some so great as to be called dialects, and others only the remains of a variety, or form, formerly existing.

Having acquired language, the child exercises his newly-acquired faculty like a toy. He asks questions for the pleasure it affords him to talk, rather than with a desire for knowledge. But, in the course of time, some tire of this fun and become rather taciturn; others linger in the pursuit of this pleasure all their lives and talk for the pleasure of hearing themselves talk. So it is with the various faculties of our bodily nature; they are connected with our souls and they have their antitypes in the soul, which the type brings into use. Thus, reverence for parents is the type of reverence for our Heavenly Father; veneration for great men is the type of veneration for God. And so on through the list. Now, when the soul has thus been educated by the body, it has partaken of the fruit of the tree of knowledge of good and evil, and it must then pass through the opening called death into the spirit-world, where it finds its eternal home.

Whatever qualities the father, or mother, would find in the character of their offspring, must be implanted by some one. They are not fixed nor natural, but acquired. The soul of the child is first pure. The tablet of its mind is blank—unwritten upon. Its memory of the past is merged in the future with its memory of earth and are both taken as one. The passiveness of the first period of existence leaves almost no impression upon the soul. But God further obscures that memory while in the body by the body's constitution, which does not take cognizance of anything that was not experienced in itself. So that memory of the first state is impossible while in the body and only returns to the spirit in the next world by slow degrees. First, as a part of its earthly existence but, afterward, it separates itself as the spirit advances and as its new relation to God requires it to possess the knowledge of its former condition, in order that His perfect love, wisdom, and benevolence may be fully manifest. The child then is pure and innocent until passion or example has led it to sin. How careful, then, parents ought to be to set examples of patience, long-suffering, goodness of every kind before their children, from the very first dawn of their intellect. This period commences very early, a few days, at the farthest, from their birth.

Having shown what examples ought to be set before the children, we will suppose every care exercised to secure their continuance in a state of innocence and purity. Then the passions are to be controlled and regulated by the child itself, for the parent can never fully, and scarcely even imperfectly, restrain the child's passion. The child's education should proceed on the plan of forming his character, so that he will, of himself, walk in the paths of virtue and resist temptations. Then, the work is done for life. The child is the father to the man. Train up a child in the way he should go and when he is old he will not depart from it, was the advice of a wise man; and wise men will ever bear in mind, as is the tree, so is its fruit.

When youth begins, the child has already implanted the seeds of good and evil. Careful culture may yet weed out much of the latter, and cause the remainder to be overshadowed by the former. But in a great measure, the die is cast as to the happiness or misery of the youth. He must follow the impulses of nature, if he has not been taught to restrain his desires and modify his actions by love of others. How, then, shall those parents be excused for their neglect, who have failed so to train their children? Surely their sins will be visited upon their children, even to the third and fourth generation. And so it is that such small advances are made by all the wise and prudent of this world. God will not let them be the ruin of more than one generation except in a few instances. He restrains their progress in sin by cutting off their children, or by throwing the children into such circumstances as to relieve them from the example and precept of the neglectful, or incompetent, parent.

What God can do to restrain evil, without infringing the free-will of man, and without debarring him from the experience necessary to the soul's enjoyment of eternal happiness, He does. But man, ungrateful ever, often mourns and repines at the dispensations by which this good, or saving from evil, is effected. Who will be willing to submit all things to God's pleasure, in perfect confidence that His will and governnment are wise and good? None now living, I think I may safely say, and yet, one such lived among men. One example has been given. And, how was it that He attained to such superior excellence, and remained ever in His original purity? Because His father and His mother listened to God, who restrained every evil influence, every impure desire, and led them in peace and purity from city to city, from Judea to Egypt. There, dwelling in quiet solitude, they trained the holy infant and led innocent lives. They returned, filled with God's holy influence, to Jerusalem and settled in Nazareth. Here, other sons and daughters were born to them; but the circumstances of their position were so different that these children were not remarkable, or distinguished readily from their fellow-citizens. True, they received, late and reluctantly, the precepts and example, the life and death of their extraordinary brother, as evidence of His divine character, as being the Messiah so long promised to their nation. But they, after all, were not emancipated from the traditions against which He preached, and never ceased to observe the ceremonies from which Jesus desired to emancipate them. Their descendants, too, continued to differ from their Christian brethren, until at last they were scattered and overthrown in the rebellions and tumults that destroyed their nation and almost exterminated their church.

CHAPTER XII

THE SOUL OF MAN HERE ON EARTH

Section three: Part two: Maturity.

Having shown how the child should be trained, let us now proceed to consider how the man should act. Virtue is its own reward; because the consequences of virtue are happiness, because nothing but purity can result from its practice, because the virtuous man is holy and dependent upon God. For, without God's help he would not have practiced virtue, and without His continued assistance he would not refrain from evil. It is, then, to God we must appeal; entreating Him to aid us, to enlighten us, to be our ever-present guide and powerful helper. If God be on our side, who shall overcome us? Not the prince of this world, nor all the temptations that may be compared to the gates of hell. Never can anything prevail against God.

CHAPTER XIII.

THE SOUL OF MAN CONSIDERED IN ITS EVENTUAL RELATION TO GOD

What, then, shall we say? Shall we say that God does all? Not so. He only helps the man to do what the man wills to do, if the desire, or will, be good; or lets him alone, or restrains him, if it be evil. He, Himself, does nothing, except through His agents, the spirits before described. But even they do nothing but help in the man's will and keep themselves on the alert to aid every good thought, every lofty desire, every pure aspiration. They ever watch the movement of his will, and the instant the man opens his heart's door, by a willingness to receive God's help, they rush in and embrace him, as it were, with tears of joy. But, too often are they as suddenly thrust out. But if man, in the body, was told he must forgive his brother, who had offended him four hundred and ninety times, shall not God's mercy and patience extend as far beyond that as infinity exceeds man's finite being? Well, then, again and again do God's angels ask for admittance. Again and again do they crowd into man's heart, and commence the work of purification. While he remains willing and passive, they stay. But when evil desires or resolutions invade the sanctuary, they can not stay; for good and evil are antagonistic and repel each other. They may linger long enough to warn, to entreat, to sigh or to weep for the misfortunes which man will receive and suffer from, in consequence of his evil. But the will of man is free and uncontrolled. Without that, where would be responsibility? No one but God can control man's will and He will not do it. He designed man for this very state of checkered existence, and when His creation and laws were completed, He pronounced them good.

CHAPTER XIV

THE MEANS BY WHICH GOD'S MERCY IS EXERCISED

Section three, Part three: Old age.

Now, for our last section, or part, on old age. Man has conjectured, that if the laws of health were duly attended to, he might prolong his life almost indefinitely. This is not so. Threescore years and ten is the appointed time to which health should bring a man. His usefulness is then, generally, at an end and if his life be prolonged, it is one of passiveness, or repining. This, too, like all God's laws, is wisdom. The man who passes his seventieth year has lived long enough to experience all that is necessary to his future enjoyment. To continue to encumber the earth would be a waste of eternity. But some will say that life might still be a blessing if health and strength were maintained. So it might. It is a kind of blessing, too, when it is not. For dispensations are blessings. But it is better, as Paul said, to be with God than to remain. And if so, why desire to remain? Perhaps you say, he might reform, or progress here. Alas! by that time the tree is dry and hard. It will no more yield to guidance. As the branches had been bent in youth, so they stand in age, only more stiff and gnarled. The storms that bent it once can bend it no more. It may break, but not yield. It can die, but it can not bring forth fruit. Poor old creature, it is better, far better, to be with God, than to remain in the strife and turmoil of an active and vigorous generation, that knew you not in your prime, among people who have changed their fashions, modified their laws, progressed with inventions, leaving you sole landmark for past spirits, sole remnant of the olden time.

And why was it that men in former days lived nearly a thousand years? Because then mankind was so new to existence, that more time was required to pass through the same experience. The earth was unpeopled, and longer lives insured a numerous offspring. The experience of the past was valuable, because it had originated in revelation and was handed down by tradition. Old men were then the lights of the world; now, it is in the breasts of young men that are found the springs of progress. Then, the struggle was rather to retain than progress; but now, retention is left to take care of itself and progress is the one idea. But, you will say, the change was sudden; the reasons given should only have made it gradual. True; when the world had been peopled completely, it pleased God to destroy the race of man, with a few exceptions. But the reasons for this are foreign to our subject and are, to this generation, unimportant. Suffice it, that men formerly lived longer, because they were needed on earth; but that the necessity having passed away they are released from the bondage of the body. So far from repining at this change, men should rejoice; for, though eternity does not measure time, nor time eternity, yet a thousand years are a long time to be away from God and to be committing sins to be atoned for.

There is no repentance beyond the grave. There, atonement is required; not that of one for another, but each for himself. Being, then, derived from the past and proceeding to the future, the soul longs to reach its goal. But the fetters of sin are strong. The will of the spirit is weak. Its struggles, therefore, are tedious to watch over and pray over. But all these things we convert into pleasures by doing them cheerfully. By desiring to act only as pleases God we have a reward, the praises of Well done, good and faithful servant; thou hast been faithful in small things, I will make thee ruler over much. So we proceed, helping others and blessing or being blessed by it, till we arrive at the Sonship, when, indeed, we become the ruler over much. For to the Son are all things given.

The last enemy is death. Not the outward death of the body but the inward death of the spirit of man. This is death to God, until it repents, or atones for, its sins. Repentance is an act of will; atonement is a punishment received. Choose now, O man, whether you will serve yourself, or God. If God, repent and sin no more. If self, sin and be punished. Your punishment shall not be eternal nor so dreadful as hell-fire; but it may extend over myriads of years and be greater than you would now think you could bear, could you foresee and understand it. Let the spirit, then, persuade you, by the mercies of God, by the love of His Son, by the tears of your brethren, by the woes of your children, by the despair of the fallen, by the hope of the raised to put your trust in God; to lean on His protecting arms; to commune with His saints; to be received up into glory—incorruptible and unspeakable.

But I foresee that many will reject the counsel of the spirits because they come not with power. Power is theirs to make signs and wonders and miracles, that should, if possible, deceive the very elect. Why, then, do they not exercise it? you say. Raise the dead for me, show my brother the hole in the side; let me experience heaven on earth, let him be damned to eternal torture—then I will believe, you say. Alas! My friend, you ask too much. To sit on My right hand, or on My left, is not Mine to give, but it will be given to them for whom it was designed from the foundation of the world. How, then, you say, can I escape condemnation and suffering, if election was made so long ago as to the fate of every man? Thou fool! neither is God, or man, to be confounded by thy impertinence

Need I lay bare the mystery of iniquity that exists in the churches of the East and West, the Catholic and the Protestant, and all others. Alas! all are gone astray. There is none good; no, not one. They shall all perish before My face, saith God; and their place shall be found of them no more. They shall vanish as a scroll, and fire from heaven shall consume them. Alas! what ingenuity has been wasted, what agony suffered, what torture inflicted in the name of the meek and lowly Jesus of Nazareth! He drove from His Father's temple those who sold doves and made a trade of religion; and so will He, in effect, do again. For He cannot abide in their evil hearts. He must go out of them, leaving them to receive hereafter the punishment of their sins, which will be as nearly eternal as mortals can be able to understand.

Turn ye, O Christian people, in every land under the sun; turn ye to God; He will be found of you, and you shall have comfort, aid, succor, living water, heavenly manna, daily bread, the wine that maketh glad the heart. You shall have peace. Peace everlasting with the Father and with the Son and with the Holy Spirit. These are, and were, and will be evermore, One—one God, the Father of all, and Jesus Christ whom He hath sent—one faith, one baptism. The faith and the baptism are of the Spirit which is the brother and soul of Christ; which is the power of God unto salvation; which is the Son of God, the Messiah the Holy One of Israel, so long promised to and looked for by the Jews. The Messiah is, indeed, one with God, the Father; because God, the Father, has placed Him at His right hand and united Him in the bonds of love and perfect harmony to His Divinity; so that they have one will, one thought, one action; and but one motive to action now pervades them and that is to save sinners. O sinner, what an array of names there is against you! But in the great name of God alone is strength. Is not that enough to make you fear overthrow? Do you dare to say He loves you and yet continue to reject Him? If you do, you are a bad man indeed. Gratitude for all His mercies and all the pleasures you have enjoyed ought at least to impel you to love Him; and if you once loved Him you are saved. You are saved for the time. You can fall again, if you listen to temptation. If you take sin to your affection, you cannot love God. But love Him, and you cannot sin so long as you love Him.

My friend, I love you. I want to save you. I ask you, what you would prefer—to be with God in Heaven, or with the lowest spirit out of the body, engaged, for perhaps unaccountable years, in making atonement for your passionate departure from the love of God? Methinks I hear you say, I would be with God; but I don't see how I can bear to give up my will to do His work. You say, the yoke is easy, the burden is light, but whenever I have tried to do right, I have found it very difficult! My friend, I will help you. Only be willing to let us both try and I will guarantee success. Breathe for Me, or for God, for that is the true term, one single prayer; make but one single, even ideal or mental ejaculation, and I am already with you. Tell Me what you want to do. If good,

I will assist. If evil, I shall have to leave you; but only for a time —only until you ask Me again. Fear not to tire Me by your fickleness, but fear those who can kill the body and cast the soul into Gehenna, where shall be weeping and wailing and gnashing of teeth. Who are those who can kill the body? The Lord has appointed unto all men to die, but He has placed governments among men to cut off evil doers. They can kill the body for its crimes, and then the soul, as a consequence, is cast into suffering and obliged to atone by it for its sins. Disease, too, kills the body; and often, far more often than men have ever imagined even, have diseases been the consequence of sins, and so the sinful soul has been separated from the body. Then is the saying equally true, as if the body were cut off by the powers that be.

The end of the matter is this. All that will come, may come, and partake of the waters of life—freely, without money and without price. *This book shall be published by the proceeds of a bad debt; and the proceeds of the book, at a trifling price, shall again be devoted to its further dissemination. If any seek its pages who have better use for the money, My agents shall deliver it without money and without price. But yet, that is only an outward performance of the promise, and not the one that its first utterer had in view. The true interpretation is, that God will teach men Himself; that no man need go to his brother to inquire where is Christ, for, behold He is in you, except ye be reprobate. And, if you are reprobate, then repent and live, repent and receive Christ.

CHAPTER XV

THE CORRECT VIEW OF CREATION

In the beginning God made all that is made, as we have declared. But how He made it, we will declare.

In the beginning was the Word, and the Word was God. It was of God and was God. What was the Word? The Word was the power or wisdom or will of God. For these three are one. The Word took flesh, and we beheld His glory—the glory of the only begotten Son of God. This only begotten Son of God was also Son of man. He is Jesus Christ, who was of Nazareth, and He now sitteth, or existeth, at the right hand of God—in His power, will and glory. The Word of God is quick and powerful, even to the dividing asunder of the joints and the marrow of man. It is a sword of division between good and evil. It is a sword that is two-edged, and sharper than any steel or metal sword that ever was made. It is God! But did God take flesh? Oh, no, thou outward-viewing men! The Word of God is God, and the Word of God took flesh. Then God took flesh, you say, by every rule of logic and reason? Wait; and I will explain to you the difference between the two sayings.

The Word of God is quick and powerful, sharper than any two-

*This refers to the first edition, published in 1851.

edged sword, says Paul. The Word of God took flesh, and we beheld His glory, says John. Now, which is right? Both I say; and I will show how they are to be understood and that they do agree in fact, though not in words.

First, God is not the Word, though the Word is God. How can that be? again exclaims the logician. The same as before, I say, and so I must explain to you again, that in God are all things. All things are from Him, and without Him there is nothing. This is, I believe, admitted by all who admit a God to exist. For, if God were not all, then something would be independent of all else, and therefore independent of Him. For out of God came all, or else it must have come from nothing. If it came from nothing, it came by His command, or by another's. If by another's, whose? Either some other immortal being, independent of Him; or else an independent God; or else a part of Him who made all things but Himself. For He, Himself, having always existed, was never made. Now, if God made all things, He made the Word, or else the Word was a part of Him. If a part of Him, it was not He though He might be the Word. For though a part can not be a whole, a whole can be, or comprise, a part. Thus it is that the Word is God, because it is a part of God. But God is not the Word, because He is more than that and can not be comprised in it.

There is in every man a Word of God, a power of God, provided he is willing to have it. But, in Jesus of Nazareth the Word of God was in abundance, bringing forth fruit unto righteousness, purity, and love. How, then, shall we make it appear that He was not the only begotten Son of God, born of the Virgin Mary? By attentive consideration of what I shall declare, you will, I think, perceive that, of a truth, God was in Him but that He was not God. That He was the only begotten Son of God, but that He was not God. That He was the only Savior given to man, by whom man could be saved; and yet He was not God. That He was the sure and steadfast promised Messiah; so long looked for, so often prophesied of; yet He was not God. That He was King of Kings, and Lord of Lords; yet He was not God. Now all these things are believed of Him, with the addition that He is God, by many churches, comprising, by far, the greatest part of mankind calling themselves Christians.

The ways of God are mysterious and incomprehensible to men. But the ways of men are plain in the sight of God, however men may strive to conceal them. Often men try to conceal their motives from themselves and succeed almost as well as when they try only to deceive their brethren. But God, seeing all things and knowing every motive, sees that a time has come when men are prepared to receive the truth; and He will have it preached unto them, raising up for that purpose such instruments as shall give Him the glory, honor, and praise of all that is accomplished through them, and being willing to receive from Him their equal penny with other laborers who may have done far less. But the race is not to the swift, nor the battle to the strong, and when the contest is over,

God will be All in all. But He will have victory through humble and submissive means, or mediums of His will, who will be satisfied with the rewards of a good conscience.

CHAPTER XVI

THE HISTORY OF JESUS OF NAZARETH

Now, in order that the History of All Things may be complete, I shall proceed to give the History of Jesus of Nazareth, a man approved by God. He was the son of Mary, virgin of Jerusalem. She was betrothed to Joseph, the son of Jacob, and he was the reputed father of Jesus. But he was not His father, in any way, except that his wife bore him that child. Still, he was born in lawful wedlock and so was not subject to any reproach. But, should any one reproach Mary with bearing a child that was not her husband's, she was prepared by God to give an answer; that her husband was satisfied. But the birth of Jesus was in this wise. The child was the result of the will of God operating upon the powers of Mary, who conceived, without desire of a man, a child. The child grew, in the usual course of gestation. It was born in due time; and at its first inspiration received the spirit of Christ, the Son and Sent of God. But how was the spirit of this child miraculous, as was the body? How was the spirit so different from other men? He was born with a different motive. Other men left Paradise because they desired change; they desired knowledge of good and evil. But He left because He desired to do good; and to show to the Father and Creator His gratitude for the happiness He had enjoyed in Paradise. He desired to serve others. God, in His infinite wisdom, selected this spirit for the Messiah, so long promised to the Jews; and who, Daniel had been informed, should be born at this precise time. This spirit, thus selected, was placed in the body, so prepared as to be pure and free from all sensuality. For there had been no sensual excitement in Mary's experience; and consequently, she impressed no trace of it on her offspring.

This is the history of the miraculous birth of Jesus. He was, thus, the only begotten Son of God; because God had, by His will, which is His power, begotten, or caused His conception by Mary, without any sensation on her part of the act; because in no other instance has it pleased God so to cause a being to be produced on earth; because, being thus chosen, He was qualified to become the Son of God while in the body, a state which no other child of God has been able to reach until he has left the body, and been numbered with transgressors in the way that Jesus was; not only in the way that He was numbered, but also by being transgressors themselves, each for himself. Having thus opened this subject, let us pause and reflect that God has caused this event to be described with great particularity by two evangelists, and that Paul, also, refers plainly to the manner and form of it; and yet out of those accounts men have managed to build a blasphemous theory, which has no foundation there or elsewhere. This is the more strange, as it originated at a very early period of the church, while there were some standing in the body who had themselves seen the Holy One of Israel.

CHAPTER XVII

THE LIFE OF CHRIST ON EARTH

The whole world was expecting a great event. The shepherds that watched their flocks, the priests that worshipped in temple and fane, the king on the throne, and the student in the closet—all were expecting some one to rise who would declare the will of God, and be armed with power and authority to teach, to rule, and to condemn. This was produced by the will of God, in various ways we will not stop to describe. But of all the watchers, only the shepherds and the magi were observant of the signs of His coming. They proceeded to visit Him in His humble, His lowly abode, and undaunted by the poverty of His parents they adored His manifestation of the presence of Deity, which shone miraculously, as they thought, in His countenance.

But the child grew and found favor with God and man. Pilate was not yet governor of Judea, but Herod was as willing to do a bloody act as Pilate was afterward. He caused the child to be sought for, that he might put Him to death. But God directed Joseph and Mary to flee. He sustained their health and strength and blessed their exertions to procure food and raiment, so that they were abundantly provided for during their sojourn among a hostile people. For, though Jews were well received in some parts of Egypt, in others they were abhorred. They resided in pagan darkness, but a bright light ever shown from the countenance of the child, which charmed and captivated every beholder. The time arrived when Herod was no more and Judea was again a safe residence for the King of Kings. His parents, with their, as yet, only child, returned and visited Jerusalem and the great feast, or gathering of the Jews from all parts of the world. There He was distinguished, too, by His extraordinary countenance, and He attracted the attention of the dignitaries of the nation, who found the beautiful face was only an index of a lovely disposition and a powerful mind. They were confounded by His answers then, and puzzled by His questions, as they were afterward when He had entered upon His great work of declaring that God was the Father of all, and that the kingdom of Heaven was within men.

So much, or nearly so much, we have from history, which has come down to the present time among men. But there were lives of Him written that were more full, and that described what I shall describe at some future time with particularity and precision. These books would not allow the Homoousian doctrine to be established. Power and presumption destroyed and proscribed them; and at last they were lost forever, unless God pleases, hereafter, to reveal their contents through some humble medium. One that will be willing to give this sacrifice, for the sacrifice of the heart is not enough, unless father and mother, wife and child, soul and body, are laid at the feet of a suffering humanity, a pleased God. That is, God will be pleased when the sacrifice is offered, and humanity must suffer till it is offered upon God's altar of mercy and love. The holy medium must be, in the body, a willing son of God. Such will arise, ready and willing to serve, and to die, for the love of God and men. God will accept the sacrifice, and they shall be blessed forevermore. Peace on earth, and good-will to men, will be their salutation.

This is now My salutation, as I commence the History of the Ministry and Sacrifices of Jesus. He was a carpenter by trade and worked at His trade in Nazareth for years before He began to preach. He was thirty-one years old when He first left His residence, to follow the direction of God in His ministry, but during His earlier years He had wrought miracles and had been regarded as a Divine personage by His mother and some devout old men. But at thirty-one years of age His public ministry commenced.

His first act was to conform to the new light, or form, which John, the last of the Jewish prophets and His own forerunner, was declaring to the people would help to purify them and prepare them for the kingdom of Heaven. When John preached baptism, it was not sprinkling but immersion that he used and enforced. How did these help to prepare men for heaven? some will ask, who now call it unessential. Because it was a type of regeneration. It signified that the recipient had taken the pledge, that he would henceforth try to serve God and watch for His Son and Sent. Then John taught them that that Son and Sent of God would teach and practice a different kind of baptism, and that the kind he practiced would cease. He must increase, but I must decrease, said John the Baptizer. But when Jesus came, He was baptized; and His disciples baptized. Yes; Jesus fulfilled every custom of the Mosaic dispensation and baptism was common in their ritual. We find it now recorded that Naaman, the leper, was directed to baptize three times in the Jordan and he was indignant, that so common a proceeding should be the only prescription the prophet of God would give him.

The last enemy that man encounters is death. But Christ triumphs over death, and the Son of Man triumphed over death, even in so horrible a form as that of the cross. But when He cried out, My God, why hast Thou forsaken Me? He had not experienced that God was with Him to remain forever. He was, momentarily, at a loss for the heavenly consolation He had so constantly experienced. He had turned within Himself, as usual, to have the counsel of God, and, to His astonishment, found no responsive spirit. Christ was withdrawn from Jesus. The man suffered, the spirit was withdrawn to God. Not the spirit of the man, but the spirit of Christ. How in this, you say; was not Jesus and Christ one person? Was he not Jesus Christ? Yes, He was Jesus Christ, and Jesus and Christ were united as one. But they were two persons. Thus, Jesus was the name, among men, of that body and spirit of which Mary was delivered. But the spirit of Christ was the Sent of God that had so pleased God as to be called His beloved Son. But the spirit of the man, the soul as it is generally called, was also the Sent of God, the Messiah long promised, the glorious Son of God, the only begotten Son of God. And this Jesus, too, possessed the Christ that is also sent to every man. But to Jesus it was sent more abundantly; for to him that hath, much shall be given, while to him that hath little, or none, what he has shall be taken from him. This Jesus, then, being so filled with the Christ or the Sent Spirit, Son of God, was properly called Jesus Christ.

Yet, Christ had another signification in which it was really used and applied to Him. He was called Jesus Messiah and Messiah is rendered in the Greek, Christos, in the English, Christ. But, then, what is the difference between the Christ that Jesus Christ had, and the Christ that Jesus had? It is this. Jesus was the Christ, the Son of God, because He was chosen by God to be the Shiloh, or Prince of Peace, the Messiah, or the Sent of God to the Jews, as He had promised their forefathers many times. But He, being chosen (because in Paradise, He had desired to be of service to others, to do God's will, and be His servant) was sent. He entered the body God had caused Mary to bring forth. His birth was miraculous, or contrary to the general order of generation. He was the only example of such a procedure.

Then, being born as He was, being thus pure and holy, the son of God as well as the son of man, He was worthy to be the Son of God, because of His innocence, His purity, and His benevolence. He was also passive to the influence of God upon Him. He strove to do His Father's will by bringing, or keeping, His own will in entire submission to God's. Whoever does this, will receive a Son of God, a Christ, into his heart or mind or soul, as it would be variously termed by different persons under different constitutions of faith or learning. Having received this last Christ, He was armed with a double armor. He had put on the whole armor of God, and nothing earthly, nor even heavenly, could prevail against Him. Because His will was in unison with heaven, there could be no collision between them; because earth was powerless against heaven, there could be no contest there. So there was no contest, but He had the victory. He overcame without fighting. Resistance would have been impossible. His will was law. It was God's will. God's, by His will being submissive to God's will; and so, of course, all others must yield to God's will.

Yet, Christ was deserted by this second Christ, when Jesus the Messiah, or first Christ, was on the cross. Then immediately the man cried out, as in the original, Eli, Eli, lama Sabachthani? My God, why hast Thou forsaken me? This cry of pain showed that the two Christs were not inseparable. What was the cause of this separation, think you? It was the desire to save Himself from death upon the cross, that had been so reluctantly yielded in the garden of Gethsemane and had not triumphed over Him in the severity of

50

His pain and suffering. Alas! that one rebellion against God's will was a sin for which He had to atone. He descended into hell, or the place of departed spirits, for this sin. He soon rose again—purified, sanctified, glorified. He ascended into heaven, and there He dwells in power and majesty and dominion of God.

He is united to God by perfect submission to God's will. He can never fall from His grace; neither can any other son of God when in the spirit form, for then none can go backward, all go forward after having been so united to the Father? Yes, indeed; all creation is progressive, and as Jesus Christ, the high and holy Son of God, King of Kings, and Lord of Lords, becomes nearer and nearer to a perfect God, He loses, little by little, the imperfection of His nature. At last He will be almost like God, almost God Himself in all His attributes. Still since but one can be God, and only God is perfect, so then perfection is approached at every step, yet it can never be attained. For if mathematicians say truly, no object can ever be reached if every progressive movement toward it lessens only a part of the distance; so, though Jesus of Nazareth is now the highest Spirit among the sons of God, next indeed to God Himself, He can never reach perfection so long as He only becomes more and more nearly perfect—that is, less and less imperfect. Amen.

The holy medium says,

Amen.

Blessed be God, evermore, for all His mercies,

and for all His promises.

AMEN,

saith the Spirit,

and, God will bless those who believe, as thou hast done,

WITHOUT HAVING SEEN.

History Of The Origin Of All Things

PART II

The Chronological, Geological,
Geographical, and Social History
of the World, with the Story
of Divine Influence.

INTRODUCTION

Let all the people praise Thee, O God! for all Thy mighty works and for all Thy loving promises. Let every nation, kindred, tongue and people, praise Thy holy name; for Thou art greatly to be loved, and separation from Thee greatly to be feared. Then, O God! let us experience Thy mercy and loving kindness, in this our day of probation, whilst we are left free to choose the good and reject the evil; or to choose the evil and reject the good. And so, O God! lead us to advancement in Thy great chain of existence; which, link by link extends to the lowest particle, or atom, of Thy works, even to their most attenuated, or unformed, state. Let us who read this book, O God! receive the truths and revelations it contains as truths and revelations, and let not our pride, our prejudices, our education, or our passions, separate us from the truth, or divide our affections, which we desire to place on Thee, O Most Holy, Most Kind, Most Loving, and Merciful God and Savior. Amen.

The deep instruction and the lofty truths contained in this book will, in many instances, be pearls cast before swine, who will desire to turn and rend My holy medium. But, though he is resolved to bear with patience any persecution, he shall not be found suffering from it. This land of America is free, and however some may desire to make men's opinions in religious matters a test of fitness for business or political office, they never have succeeded and never will succeed in overthrowing any servant of Mine who acted in My will. When mediums act in their own wills, they may often receive such opposition as to confound them and destroy their work. But this only shows that he who would proceed rightly must rely on God and proceed no faster than He directs. Such a one must be careful not to go too fast as he is to keep up to what is required of him. Submission of will—a surrender of man's free will—is required in order to have God's sure help. He will assuredly save those who rely wholly upon Him when, to all human reason, salvation is impossible.

Read Daniel's account of Shadrach, Meshach, and Abednego's being cast into the fiery furnace, and of his own salvation from the hungry lions; be assured, first, that it is literally true, having occurred precisely according to the simple relation of it; and, second, that God is able to save now as He was then. And, if necessary now, any one of the world's rulers could be made to eat grass like an ox, and his kingdom be taken from him to be restored no more, or to be restored at the end of seven years, as easily as Nebuchadnezzar was turned out of his palace for exalting himself in the midst of extraordinary grandeur and unlimited power over men. Remember God's power, and remember the advice of him who, 1800 years ago, learned and pious, though not convinced of the truth of Christianity, had warned the Sanhedrin to let the preachers of new doctrines alone; for if the thing were of God it would prosper and they might be found fighting against Him, whilst if it were not of

God it would come to naught. Let every man, then, look carefully about him to see on what foundation he stands, and let him who thinketh he is already on a sure foundation, take heed lest he fall. For this day is the word of the Lord fulfilled—that your old men shall dream dreams, and your young men see visions, and the Spirit of God is poured out in the land. Let the earth rejoice and all the sons of the earth give joyful thanks. Let the floods clap their hands and all the people shout for joy. For unto us a Son is given! Who will declare His generation? Who will show His forthcoming? Let every medium attend well to what I shall declare through him, and let every one who has believed himself inspired by Me or by God Himself or by His Holy Spirit, for all are one, let them, I say, attend to their impressions; for if they are willing to receive the unmixed truth, I will impress upon them the conviction of the truth of this book, and of the verity of all My sayings through this holy medium.

And you, Mediums! and you, O Inspired Receivers! do you declare publicly of My impressions, as it were upon or from the housetop, what I give to you in a corner or in your own hearts or minds. Do this and live. Smother it, and you shall die! Die to My communion, to My impressions, to My communications. Attend, O People of America! and prepare for your great destiny by submission to instruction, and by being willing to come under the authority and guidance of the King of Kings.

PREFACE

Part II of this book is the higher manifestation promised in Part I. It is published sooner than might have been expected, because the need is great and the holy medium was ready, passive and submissive. It has been written, as the first part was, by the direct revelation of the Son of God, Jesus Christ, formerly of Nazareth, now the First Spirit of all the sons of Earth who have reached the seventh circle of the seventh sphere. He is now the only son of man there from the earth. But others are in the sphere below Him, advancing steadily, and with greater and greater proportionate rapidity, in that chain of degrees of existence which extends from God to man in the body, in this earth and in the other globes of matter. It extends thus far spiritually and even beyond it, one step, to the spirits in Paradise. It also extends by infinitely small links, or degrees of graduation, to the lowest forms, or manifestations, of matter, though this may seem below the dignity of My subject, as God is above Spirit.

Man may control matter even as God controls spirit. But men are controlled by the laws of matter, and God by the laws of spirit. The difference is only that God made all laws, including those by which He Himself is governed. Let us read with care, and with high and pure motives and earnest endeavor to find the truth, and be assured, O son of Earth! you will rise from the perusal of this revelation from God, a wiser and better man. But if you read to find flaws and faults, I have enough to satisfy you, and to excuse yourself, to yourself, for your contempt of the knowledge of the hidden things of God here offered to mankind. Things which many have desired with a great desire to obtain a knowledge of, but have not: things which can give no satisfaction when known unless received with submission and obedience to the light they display and the precepts they inculcate. That you may be benefitted, I have made an earnest prayer to God in your behalf, which you will find near the close. Read it when you come to it in regular progression, and if you desire to receive the greatest possible benefit from this book, read in the order in which you find it printed, and with constant asking of the Everlasting Father, and Prince of Peace, that He will help you to understand its high and eternal truths. The errors are trifling and will do you no harm, if believed, or acted upon. So read and receive with confidence, for the Holy One of Israel, whom you have so often asked for knowledge with your lips, now offers it to you, and it only needs the heart's prayer and work to enable you now to obtain it, through a humble but correct holy medium.

Let Us Pray

Almighty God! who dost, from Thy throne, behold all the nations of the earth, all the hearts of men, and all the Creation of Thy Will, look down, I pray thee, upon this intended reader of Thy revelation. Sanctify to him its precepts; bless to him its knowledge; purify in him his nature; subdue in him his will, by leading his reason to see the beauty of this revelation of Thy Will and Purpose, the knowledge of which has heretofore been hidden from men in the body. Let all who read understand, and all who understand, be wise; and all who are wise will praise and honor Thee, the Everlasting and Ever-Loving God. Amen.

Reader, reader, be wise, understand, and be profited. For the riches of God's kingdom are greater than those of Golconda, and the glory of God beyond the glory of this world, so far that men cannot appreciate it, whilst in the body.

Let all the people praise the Lord;
Yea, let all the people praise Him,
For His mighty works,
And for His noble revelations;
For His great mercies,
And for His loving kindness.

CHAPTER XVIII

The Chronology of Mankind conformed to the real Chronology, as ascertained by Spirits after their Ascension to the right hand of God.

In the beginning was the Word, and the Word was with God. The Word was with God in the beginning. But God has no beginning. In this case, beginning must be taken to mean all eternity, or else we must believe the Word was created by God. For God was, always, and ever will be. But nothing but God is eternal: what then is the Word? Not God! for He is one, and He Himself never took flesh. Not a part of God always separate because then two existences must have been eternal. But there was a time, or period, when the Word was in God, united, unseparated. Then there came a time when God separated the Word from Himself, but subservient to Him and ever having but one will with Himself. Inasmuch as the Word always acted in accordance with God's will, it was always equally endowed with power to become a son of God, equal in power to God. Equal in power, because to him who does God's will is given God's power. But the least departure from God's will destroys that unity with God on which this power depends, and Word or being becomes powerless; unless God has also given it power of its own, allowing it to exercise this power within certain limits, according to its own free will. How, then, and when, was the Word created? Long ages before the world was created, long ages before the command went forth: Let there be light, the Word was created, that is, separated from God and made a separate, but depedent, existence. Shall I attempt to declare His generation, and number the years of His age? No: finite beings are not possessed of the capacity, in the body, to conceive of the length of time in the period that existed after the Word was separated into an existence separate from God, until the world was spoken into existence. But before matter existed at all in the creation of God, when all was void and all was God, then the Word was brought forth from God by His will and power and made His servant. By it, the worlds were made; and through it, man was brought forth.

But how were the worlds made by this Word? By the operation of God's will, through the Word, the laws of progression were established. God spake, and it was done. He commanded, and it stood fast. So it is recorded He did, and so it was. Well then, how did the Word assist? The Word received the command and as God's servant executed it. The Word was obedient and made the worlds. As they now are, so they were made to be, by the Word. The same Word, that was in the beginning with God, is yet with Him in the eternal existence. Such as man can comprehend I am permitted to unfold. But there are speedy limits to man's comprehension.

It is enough for man to know that there were laws, or rules of proceeding, established as the foundations of the universe, and that these laws, or rules, still enable the Word to maintain the universe of matter in its place, and to be the means of its progress towards perfection, which it can never reach. What then is the purpose of the laws of progress being spoken of, if they cannot be explained? What is the use of revealing any part, without telling all? some will ask. I say, that some are glad to make an addition to their knowledge without asking for the whole counsel of God. Sufficient it is for them, that God makes them rulers over a little. These shall, however, receive the more for being satisfied. The others shall be confounded by the utterance of strange voices who will make them doubt that they know anything.

This is the end of the matter. God made the laws, and the Word made the worlds. Without the Word was not any thing made that was made. What then is the Word? Is it a gigantic, powerful, lusty, and hard-working assistant of God? Oh no! God needs no help. He could as well have proceeded without the Word. Why then did He make the Word to make the worlds, when He could have made them without it? Because it was His will to make the Word first, and to have the worlds made by it. Because the Word has other duties to perform, besides making, guiding, and preserving the worlds of matter. The least of the Word's duties are comprised in its relations to matter. It is to spirit that the Word is most faithful, or constant in attention. Spirit, then, is under its rule! Yes, by it all things were made that were made and without it was not any thing made that was made.

John spoke not of himself but by revelation. He was a holy medium, such a one as I am using. A man in the body, not wholly free from sin, but desirous to do the will of the Deity, and to be passive in His hands, and in the hands of His spirit. The Word is Spirit. Then matter was made by spirit? Certainly, you cannot doubt that, if you believe God is a spirit and that God made, or caused to be made, all things. Well, the Word was with God, and the Word was God! How is this true? In the first part of this book to which I have alluded before, this is explained. I do not choose to do the work twice. Read that. If you have read it once, or twice, read it again; and if you do not see more in it than you did before, set Me down as an unfaithful guide. For I know that whoever shall read that book twenty times, shall each and every time derive new instruction from it and, if a sincere inquirer after truth, shall be advanced in his pursuit.

CHAPTER XIX

THE WORD

The History of the Word, continued, and carried to the present time.

The Word is eternal, as a part of the Deity. But, by itself, it is finite in its powers, and terminable in its existence. But will God terminate the existence of His Word? Not so long as men have a separate existence, for the Word has the care of men. The Word took flesh, and John, and others, beheld His glory as of the only begotten son of God. This glory was undoubtedly a great and a surpassing one. It was seen, however, in the person of Jesus Christ, the only begotten son of God, as described in the first part of this book. Then the Word was beheld by men. It was not beheld by the bodily eyes of John and others. Its glory was spiritual and consisted in Its superiority in morals, works, and love. It was the only begotten Son because It had pure desires, and because It was the promised Prince of Peace. It was a body, endowed with a high and holy spirit from Paradise, that had entered this world to benefit mankind. He had no narrow views of saving from sin and misery a family, or a nation; but all the inhabitants of the world, being equally God's children, were equally intended for the receipt of His love, manifested in His proclaiming the great truths relating to man's acceptance with God, and to man's conduct—socially, politically, and morally.

But having taught the sublime doctrines He did, how came it that He was disregarded by so large a portion of His hearers? For, at the time of His crucifixion, scarcely twenty believers could be found. And even after He had risen from the dead and ascended before eyes of men into the clouds of glory, from whence He shall come again in clouds of glory, how many believed besides the apostles? Few, indeed; perhaps not twenty in all. For all the mighty works, the stupendous doctrines, the all-pervading love, would not, nay, could not, bring men from their self-will, and make them have faith, and submit themselves to God's will. It was expedient that He should go away, or the Comforter would not come. He declared the Comforter should lead them into all peace. It is the Comforter that has ever since given men peace, when they have had it. And the Comforter is the Word of God. The same Word that took flesh, and the same Word that is so described by Paul as being quick and powerful, sharper than a two-edged sword, to the dividing asunder of the joints and marrow, and discerning the thoughts and intentions of men.

What then is the Word? It is the Power, the Will, of God. It is the Great Harmonizer of man—the Intercessor, the Mediator, the Redeemer. But you thought Jesus of Nazareth was all this! So He was, as far as He was one with the Power, or Word, or Will, of God. He had no power except from the Father. None of His works are done of Himself. The works that I do, He declared, are not Mine, but My Father's who is in heaven. Alas! that man should have been unwilling to take the testimony of Jesus Himself, as to what and who He was. But the world is ever ready to construe itself by itself. Man is ever ready to help God, if God will let him help in such a way as pleases the man. But God wants no help. He wants sacrifice. Sacrifice of man's will, and nothing but that will he receive as the acceptable offering.

The love of God never tires of being neglected by men. It continues to be offered up on the Cross of Christ to this day. And who shall suffer now a martyr's death? No one, for God has established a government here, in this political body, that will not execute the sentence of ecclesiastical bodies. If it could be brought to do it, think you that My holy mediums, various and contradictory in appearance as they are, would be allowed to live in peace and quiet, doing My will? Not for a day. Anathema, Maranatha, would be hurled upon their devoted heads by every organized church known in Christendom. Why then has God allowed these churches to grow up under the supervision of the Word (for, undoubtedly, they have all, at times, had sincerely inspired men within their communion, or pale), and why has not the Word shown them the iniquity of their association and the destruction that impends upon them? Because the laws of God promulgated at the creation or formation of matter would not permit the Word to proceed in his own will, nor to proceed out of time. A time, a period was established, when the light should shine into the darkness and be comprehended. Heretofore, it shone into darkness and the darkness comprehended it not. What, then, is the time when the Word will act upon men? When will the light and the day dawn, that is so often spoken of in the Bible? A day of glory, eternal, unfading, and more lovely than old Jerusalem, more heavenly than the New Jerusalem. It is now dawning. The Word operates now in the Will of God and in accordance with the old enduring laws. The Word will cause Itself to become known, and Himself to be heard and listened to. The Word will be the light of men and at last man shall know God. Yea, all, even from the least to the greatest. And the last shall be first; and the first, last.

Now that the Word is about to be declared present amongst men, whither shall we turn to know how to distinguish Him from others, who will be desirous to assume His powers, and declare the duty of men? Try the spirits and see whether they be of God, was of old a direction to men. It remains as the only guide and test ever given by which spirits may be judged. Try Me, then, and try other spirits, or pretended spirits, by this rule. Every spirit which confesseth not that Jesus Christ is come in the flesh, is not of God. Beware of evil spirits. Beware of deceivers, that would, if possible, deceive the very elect. But it is not possible to deceive the elect, for the elect are those who choose God for their portion, who trust in Him. The elect are those who have elected Him to be their Ruler, their Guide, their Counsellor, their King. They are those who, when trials surround and troubles beset, trust in God. They are those who do not their own wills, but God's. They are those who pray earnestly and sincerely from their hearts: Thy will be done on the earth, as it is in heaven. Can these be deceived? No; it is not possible. God is in them and no man, nor devil, nor deceiver, can eject Him. Who then is in danger of being deceived? The inquirers who are looking here, there, and everywhere for something to confirm their pre-existent ideas, to help their creeds, so cherished; and their doctrines, so interwoven into society that they fear society would fall to pieces without their cohesive qualities.

Alas! could society be relieved of them, it would appear more as it was intended to be, more as it was in the beginning of man's sojourn upon this planet.

CHAPTER XX

DECLINE OF KNOWLEDGE

The causes of the decline of man's Knowledge of God, as first possessed by Him in this world.

The world was fair to look upon when men first roamed over its hills and vales; first gazed upon its mountains, its rocks, its rivers, and its seas. As it is now, so it was then. Nature is ever changing, yet ever repeating herself. Man, too, was then what he is now; a being sentient, but not wise; prudent, but not foreknowing; active, but not realizing. He was blessed by God with the pronouncing of a declaration that he was good. So was all God's creation. Then let no one seek to alter what God declared good. God implanted in man the desire to extend his species, and to advance in knowledge. But designing men contrived to obscure the desire for knowledge, and stored up in their own order all that was known, by God's revelation of Himself, and of man's duties towards Him. Having so possessed themselves of the keys of heaven (as it seemed to them) they allowed mankind to fall into deeper and deeper ignorance, until, being unable to distinguish between the Creator and the created, they ceased to worship the former except through the latter. God was not so much offended at this as some would suppose. He is not jealous of His dignity and He fears no rival. He pitied men, but He did not desire to revenge upon them the wrongs of themselves. They had been wronged by priests, and kings, but priests and kings may be pitied too. For they knew not that what they did would cause the loss of the tradition of God's action towards men, which they had received from generation to generation, even from the earliest of their appearance on earth. Gradually it was obscured; gradually it disappeared. At last it was no longer distinguishable as truth.

But Moses was educated in all the learning of the Egyptians. He aspired to make the people of Egypt, in general, acquainted with the truths hidden from them by the priestly order. But he, though the adopted son of Pharoah's daughter and the legal heir to the crown, was not powerful enough. He was compelled to flee for his life. For the time had not come when God's Providence, or Word, was ready to act, or to have Moses act efficiently. Forty years afterward, Moses, in the will of God, entered Egypt and preached the knowledge of God. He led forth, from the tents and cities of Egypt, an immense multitude whom no man could number. They went forth, not as Hebrews nor the children of Abraham, but as the believers in the God of Abraham, Isaac, and Jacob. God sustained their courage by mighty deliverances. He fed them by miracles and preserved them by His power. He prepared the land of Canaan for their reception by desolation of sword, pestilence, and famine. He made His Word, or Power, to go before them by day and by night, until, having placed them in the promised land, He allowed them to exterminate the inhabitants and to apportion it amongst themselves.

But the institutions with which he furnished them, however plain and pure, were soon perverted by men desirous of ruling and the nation soon sank into its former dark ignorance, until they again worshipped the idols of stone and wood. Again and again God called them by His holy mediums, or prophets. Again and again did He deliver them, by mighty miracles, from their enemies and punish them for transgression, or reward them for obedience. But, at last, He had them all deported to Assyria. Here a purer religion than that of Egypt prevailed, and a long captivity purchased their restoration to their country, confirming and strengthening their desire to keep the statutes of Moses and obey the counsels of God, as declared through His holy mediums, in those days called prophets. Daniel was allowed to declare the very year when the Messiah, the Prince, should come and be cut off. Yet when He came, the children rejected the knowledge of their fathers, and would not believe the prophet of their own time, John the Baptizer. This John plainly declared Jesus to be the Messiah, and though the Jews believed Him to be a prophet, and a seer, they still rejected the Messiah.

Why do I tell you this? Is it not written mostly in the Bible? Oh, yes. But do you draw from it this instruction—that the ways of God are past finding out except as He chooses to reveal them? Can man by seeking find out God? asked My servant Job, three thousand three hundred years ago. No; never can reason bring Him down, or strength raise itself to Him. Be then patient, passive, and willing to be God's servant. Then will you be invited to stand still and see the glory of God.

CHAPTER XXI

CAUSES OF THE CRUCIFIXION

The Causes that required the Crucifixion of Jesus

The time when Jesus preached was a time similar to the present, when all enlightened and inquiring minds are seeking and expecting a better knowledge of God. A higher sentiment than reason impels man to prepare himself for futurity. This sentiment is the Word of God, operating through His agents. These agents are spirits, who, having found salvation through the power of the Word and the mercy of God, are desirous of helping, or at least of being partici-pators, in the work of the redemption of their fellow men from the bondage of sin and death, once suffered by them and now being suffered by most of those in the body even in this favored land.

America! thou art to the present time what the Roman Empire

was to the time of Jesus. Large as are thy bounds, they shall still be enlarged. Strong as are the bonds of the Union of thy States, they shall be stronger. Dissolution shall not take place, until the work is done for which I brought thee into existence. Let the dead bury their dead. Let the contentious wrangle and the envious aspire; but, oh, ye sons, or servants who do My will, be ye steadfast, immovable, undoubting, unfearing. Resist not evil. Let the heathen rage, for they shall be confounded, and all the gates of hell shall never cause any unhappiness to him, or them, whose mind is stayed on God. Be then of good cheer, if you have overcome the world, for God has appointed America, or more properly the United States, to stand as the tolerant receptacle of His holy mediums, the great and constantly extending area for the operations of His spirits.

Do you read again the prophecies of Daniel, and of Isaiah and see if you cannot find that a kingdom would succeed the fourth great kingdom, the Roman Empire, which indeed still exists in its last phase. Cromwell's Fifth Monarchy men had an inkling of the truth and they cheerfully abandoned the country that rejected them. They fled to the wilderness, where a great eagle has sustain'd them. There they have been preserved from the Dragon, the seven-headed. monster of Rome; and the False Prophet, the Reformed churches, so called, of Europe, that would have bound the Woman if they could and would now undertake to destroy her if they dared. But the foundations laid by God's laws are not to be overthrown until the superstructure has been finished, and the purposes for which it was built, accomplished.

What the uses are to which America is to be applied, when the superstructure shall have been erected, of which the foundation is now laid, shall be explained by or through this holy medium, when he shall have finished some other work I have in store for him after this book shall have been completed. But the last of his works will be his resignation to God. For he does not yet submit as fully as I desire to have men submit, nor as fully as men have submitted in former days. He is, however, the best medium I now have and, being such, through him will the higher revelations be given to men until another shall excel him, or he shall be taken from works here to works in the spirit world. As he can now view death without apprehension, he is in a good state to progress, and he will continue to progress in submission, I believe, for a long period.

Having now opened the subject, I will explain that Jesus of Nazareth must necessarily have suffered a violent death unless God had withdrawn Him before His time. For the days in which He appeared were those of ignorance, and though by a constant miracle God could have maintained Jesus' existence upon the earth, it would have led only to idolatry of Him. He would have appeared to be God and would have been worshipped as such a being. The Greeks and other Gentiles would have been confirmed in their previous belief in many gods that had, as they believed, lived and acted amongst men. Why, then, was He required to suffer so painful a death as that of the cross? Because His example was required to sustain His followers in the persecutions they were required to undergo, and to endure to the extremity of torture. Many martyrs suffered more horrible and torturing deaths than Jesus. But none suffered much, for he whose mind is stayed on God and who trusts in God as his Savior, Redeemer, and Preserver can never feel the pain that others may attempt to inflict, or feel the pangs of death.

The true life is inward. Fear not those who kill the body, but those who cast body and soul into hell fire—the fire of evil. Fear the fire that rages in man's will, that feeds upon the man and leads his spirit into the outward darkness of a departure from God's light, that feels itself to be suffering from the indulgence of its own propensities and evil desires, and leaves itself in the outer, or outward, darkness, where it weeps and wails and gnashes its teeth with vain imprecations upon God and men, upon itself, and God's creation in general. What greater hell can be conceived of than this? Can material fire burn the body worse than the remorse a man who rejects the counsels of wisdom and sins against knowledge, must feel when he finds that the tempting apples of desire and lust, of self-gratification of every kind, are indeed dust and ashes? Bitterly and severely will he wail and weep and lash his passions with impotent fury, when he finds he has destroyed his happiness and separated himself from the love of God by pursuing a vain world's transitory and fleeting pleasures, instead of laying up treasure where moth and rust do not destroy nor corrupt; neither do thieves break in and steal the treasures of heaven which he has thus placed in a safe garner, with a safe and powerful keeper.

But, some will say, God might have overruled the wrath of men and brought them to believe in Jesus as the Messiah! Could the world have been brought to believe in Him as such, they would have been, by the mighty works that were done in Galilee. For the three full years He spent in His ministry were a daily preaching and working of miracles. No; the eyes of men were not open to the light of God's love. The light shone into the darkness, and the darkness comprehended it not.

Surely God could have forced men to believe! No, my friend, not without violating His own purpose of leaving man a free agent to choose good or evil. God works upon men as they are willing to be worked upon. He pleads with them, condescends even to reason with them, but He never forces their will into subjection to His. That would take from man his individuality, his responsibility, his distinctive nature. God therefore leaves man to hear, to accept, or to reject; to see, to believe, or to reject; to feel and know but yet to be able to reject the evidence of his senses, the convictions of his reason, and the hard taskmaster of his own cruel will casts him down into the pit of ignorance and despair because he consults, after all this work, after all these demonstrations, the will of himself, the traditions of his fathers, and the tears or entreaties of his brethren, rather than to cast his burden upon God. Oh, men, why will ye die? Leave the ways of self-will! Be passive to

God's holy and divine influence; to God's love and power and will. He will lead you to living waters, from drinking which ye shall be refreshed to thirst no more. He will give to you bread from Heaven of which those who eat shall never die.

Come then, O people of America! You are free, intelligent, and independent above all others. Why will you reject reason, sensation, revelation! Why will you refuse the gifts of God, receivable only by one sacrifice on your part—the sacrifice of your heart to God, of submitting your will to Him. Seek, and ye shall find; knock, and it shall be opened unto you. Be then willing servants in the days of the manifestation of His power. The extraordinary demonstration of God's spirits showing to men outward miracles will not much longer continue. They were given for a time 1800 years ago. Then they were withdrawn. They were given again occasionally, during the time that has since elapsed, to see if men would be persuaded by them. But now they are proceeding with unwonted power. Now is the accepted time; now is the day of salvation. If you reject Christ now, you are lost to God for the present time. If you believe, you shall be saved with an everlasting salvation. Not that you may not reject after having received God's presence in you, but, unless you reject Him, you shall not fall from grace.

Come then, O people of sincere desires for truth and righteousness! Lay aside prejudice and tradition; fall down before God in earnest, private supplication. Continue to do it. Pray without ceasing to God that He will enlighten your understandings, that He will make the crooked paths straight, that He will lead you to the fountains of living waters, that He will impress upon you a knowledge of your duty, that He will raise you to a knowledge of the deep things of God, and that He will show you how to serve Him and how it pleases Him to be served. Walk humbly. Be cheerful. Be content with your wages. God will hear prayer. God will answer prayer. He even answers and grants prayers of men made in their own will, though it results later in their own suffering. Why? Because He answers fools according to their folly; because when you ask Him for a stone, He will give you what you ask for and not the bread that you did not ask for. How, then, you say, shall I know what to pray for! I will tell you. I will write a prayer for you to make in sincerity and from the depth of your heart. Make it in private—standing, sitting, lying in bed, or walking in the street. Make it audibly or mentally. Only make it with sincerity as your own prayer—and it will be answered.

Prayer

O Thou eternal, incomprehensible, almighty, and ever-loving Father and Friend! Oh, listen to the humble supplication of Thy deeply desiring servant; or, if not Thy servant, O God! make me Thy servant. Grant, O most loving, kind, and powerful Father and Friend! that I may have wisdom from Thee to see what way

I should take, to feel what I ought to feel, to love what I ought to love. Be thou, O most kind Parent; my Helper, my Savior, my Intercessor, my Redeemer, my Friend. I know, O God, that Thou art all these; but yet, O kind Parent! make me feel its surety more. Let me know the peace that the world cannot give, or take away. Be Thou, O Father! my helper in this world's affairs and my savior in spiritual matters. O God! I desire to serve Thee, and to do Thy will. May it please Thee to help me do it. Help me, O Father! to walk as Thou wouldst have me and to pray acceptably to Thee. Help me, O God! to say at all times and under every dispensation, when troubles surround me and trials depress me, then, O God! help me more and more, till I can say truly, sincerely and with perfect reliance on Thy goodness, mercy, and loving kindness, all, like Thyself, infinite; to say then, O Lord God Almighty! not my will, but Thine, O Heavenly Father, be done! Amen.

Can you make this prayer now? If so, do it. If you cannot, try. Repeating it over and over will not make it yours, but repeating it with a desire to make it yours will enable you, in time, to make it as yours. Try, O son or daughter! I say always son, and use the masculine gender, but remember that all are one in Christ. There is no distinction of sex or color or condition before, or with, God. All are His children and all are equally loved, if equally obedient. Be then earnest in seeking, fervent in asking, constant in desire, immovable in faith, unmistakable in your position. Fear not the world nor men nor devils. There is One God, the Father of All, the Creator of All, the Preserver of All. He can save. Through Christ He chooses to do it, and you cannot be saved in any other way than that. What Christ is, I have explained in Part 1, of this Book. Read that, attentively, many times if you wish to progress in the knowledge and love of God.

CHAPTER XXII

THE WORLD

Where Were The Foundations of the World Laid?

When the morning stars sang together and all the sons of God shouted for joy, then the earth and its attendant planets existed. But who were the sons of God before men had left the body? There were other bodies in the universe to which, ages before, innumerable and incomprehensible to man's understanding, God had given inhabitants, many of whom had become sons of God. All these united in shouting for joy, that another creation had appeared, and other beings had been created to participate in the heavenly bliss enjoyed by them. No envious spirit dwells in heaven. No hater of his brother can ever reach there, whilst he is such. But the last of God's creation, so far as starry globes extend, has not yet taken place. New heavens and new earths are being created. Heavens are spiritual; earths are material. But that was not first which is spiritual but that which is natural. After the natural comes that which is spiritual. How is this? Have I not given a different explanation in Part I of this book? Look and see; I am consistent. Be ye understanding. Be ye desirous to find Me right and you will not find Me wrong. But, if you desire the contrary, you will obtain your desire for I have explained to you that God gives stones when they are asked for. Be seekers of the truth and not seekers after discord. For seek, and you shall find; knock, and it shall be opened unto you. What is good, ask for and receive. What is evil, avoid, and pray for preservation from it.

The cause that really produced and made necessary the crucifixion of Jesus was the hard-heartedness and ignorance of mankind. Produced by man's self will, it resulted in a resolution to reject and overcome all that should oppose his will, and thus tend to relieve his fellow men from the rule of priestly order in Judea. The civil power chose to sustain the ecclesiastical that it might the more easily rule the turbulent Jewish nation. Now if God had by His power overruled their intentions, He would only have transferred the scene of operations to another spot of the same empire or to some other empire. No other spot had all the advantage of proceeding from the only nation or people that, as such, believed in God.

True, the Jews were a despised people amongst the Greeks and Romans. But the Christian religion discarded at once the very causes that produced this aversion and contempt. The Christian religion, in reality, had no greater obstacles to overcome then than now. Then, as now, many were interested pecuniarily in resisting revelation. Then, as now, many were ambitious of swaying the church, whenever a considerable body of believers were gathered. Then, as now, the lust of the body and the pride or vanity of the mind made fearful inroads upon the ranks of those who were almost persuaded to be Christians. But, for all that, for all these, the truth did become manifest to most in a distorted form, perhaps, yet, here and there, in purity and in strength.

Now the world is better prepared because education is more generally diffused, independence of thought and action is more general amongst men, and the rule of pontiffs and of kings is maintaind with great difficulty only by the most cruel and stringent policy. But the long-suffering of God is near to its end. The fifth kingdom is established on its foundation. The corner stone is laid. The rock is Christ, the Son and Sent of God. The forerunner of Christ was then an outward dispensation by Moses, and an outward sign was given to John the Baptizer. The forerunner then disappeared and was forgotten in the splendor of the following displays of Divine Love and Power. Then, the last of the old prophets saw and rejoiced that the New Jerusalem was descending to take the place of that outward city and temple, in which had centered the hopes and affections of believers in God. Then the last prophet of the old welcomed the first and greatest prophet of the new.

Now, the old prophets or teachers in the assemblies, or professed churches of God, resist the new prophets; and instead of pointing the people to them, they excommunicate those who may dare to follow the new revelation. They would have God to stand still and see their glory. Wait and see them compass sea and land to make one convert who, when made, is two-fold more the child of hell than before. But now, as then, God calls on His servants to stand still and see His glory. To wait for Him to convert the unbelieving and to lift up the lame, the halt, and the blind. The last prophets shall yet acknowledge their errors, and the new prophets shall yet acknowledge the glory of God, and His mighty works, to have succeeded in making men believe them.

There is in every man a Christ, a spirit of God; as I showed in Part I. But the Christ, or spirit, that resided most surely and constantly with Jesus of Nazareth, was the spirit or soul of a being whose bodily existence had been passed on another planet. It was the planet Saturn that had borne the body of this spirit. There, He had been regarded as a superior inhabitant during His life, and divine honors were paid to Him after His bodily death. But this result did not change His position with God. The being, or individual, had done His duty, though others, His companions, had departed from or excelled theirs. They ought to have regarded His teachings as divine, and Himself as a servant of God. But how did He so excel all others of that race as to be deemed worthy of worship by them, and worthy to be the Christ of God to Jesus of Nazareth? Because He, like Jesus of Nazareth, had left Paradise from a desire to be of service; to be a servant and helper of God, to be useful to His fellow creatures. How long He lived on Saturn and how long He afterward existed in the spirit state before the advent of Jesus, I will not at present say. Your minds must be gradually prepared for the full effulgence of the revelations I have to make through this holy medium, and I shall have a long course of exercise for him, and for you, before you can believe, implicitly and unhesitatingly, all that God is willing you should know.

The time since Earth's foundations were laid is so long that I could scarcely write on a page of this book the figures that would express the number of years. Ten thousand times a million, ten thousand times repeated, would still be shorter than this period. But, for myriads and thousands of years, the earth was in its foundation, without form and void. Then God said, Let there be light, and there was light. Then the starry globes, the sun, and the more perfectly and earlier formed planets, appeared. But the chaos was not brought into its present result at once. Myriads and thousands of myriads of years rolled by, with the earth becoming formed gradually. Its processes have been guessed at by the geologists, and some of them have dared to believe that the result was an inherent property of matter instead of a glorious manifestation of God's power and will. He spake, and it was done. He commanded, and it stood fast. So it was. He spake, and the law was promulgated by which the earth and its inhabitants were formed, and established in progress to their present and future state. Their present state you know. Their past I will unfold in this book and their future in another, but not the next book this holy medium will receive.

All that God wills to let man know of Himself is now to be unfolded. He shall possess all the knowledge that spirits of the fourth sphere have. Not that I am limited to that sphere or knowledge, but that I am not authorized to unfold more than that.

There are seven spheres and seven circles in each sphere. The perfection of God is above all spheres. Jesus Christ, formerly of Nazareth, is of the seventh circle of the seventh sphere. There, He has as yet no associate from this planet, but He has associates from Jupiter, Saturn, and other globes in this and other solar systems and other combinations. Not many, compared with the innumerable worlds of matter in God's creation; not many, compared with the numbers in other circles of the same sphere; still less are they many, compared with circles in lower spheres; but vast, incomprehensible to man, are their numbers.

You may think it derogatory to Jesus that He should be only the equal of so many other beings; that God should choose to have so many sons sitting, or existing, in right hand nearness to Him. But how much more derogatory then will you consider it that you, too, shall hereafter be His companion in that same seventh circle of the seventh sphere. This is literally true. For all the beings that have emanated from God shall at last arrive at this superior position, and all shall be sons of God, equal to Christ in glory, honor, power, and love. All shall be one with the Father, even as Jesus was, and is, One with Him. They shall even be thus One with the Father before they arrive at the seventh circle. In the fourth circle the spirits see God, in His glory and honor and power. They cannot see God till they have so submitted themselves to Him as to have no will but His. When they have done this, they are united to Him so intimately as to know His will so far as He makes it known to spirits in any circle. They are left in ignorance of the time and manner of its execution, but they know His will. Then in the fifth circle, they know His power. His power executes His will, and, as they know of the execution of His will, they know its time and manner. But the sixth circle is distinguished from the fifth by knowing the form of the passing law; the present form of His intention respecting the future. The seventh circle is so perfectly one with Him that those in it know all that God knows. But they are separate and below God in that they cannot cause anything but an execution of His will. God causes. God proceeds to know what will be. The highest spirits cannot do this. They can only know what God does know; and God knows all He has done, all He has resolved to do, and how and when it shall, or will, be done. But He Himself cannot be said to know an intention He has not formed; though He has, of course, the power to form any intention not contrary to His nature. He cannot form such an intention, because He is Himself; because He will not be led into inconsistency by any cause. He cannot destroy Himself, and inconsistency would destroy Him. Then what is contrary to God's nature is impossible for this reason. And though, therefore, some things may be said to be impossible to Him, it is because such things are not only undesirable but destructive of all good.

70

CHAPTER XXIII

The Reason and Truth of Revelation, asserted and maintained

That revelation is true is evident, if it be a revelation from God. God is, in His nature, true, and nothing from Him can be inconsistent with His nature, as we have just shown. Revelation from God, then, is always true. If always true, it must always be consistent with all His other gifts; because God is a unit and all that proceeds from Him is, in like manner, a partaker of His unity. Do you say, if man, as I have declared him to be, is endowed with the power of opposing God or God's manifestations, that therefore, here is an inconsistency? I say, man is only an emanation from God spiritually. Bodily, he is a creation. Not a gift, nor a proceeding, but a creation; primarily from nothing but secondarily from the matter of the earth. Creation is harmonious as a whole, and man, as a whole, is harmonious. As seven shades unite to produce white, so all the varieties of men's manifestations unite to produce harmony with themselves and with God. As two discordant musical notes are united and harmonized by a third, equally discordant with each of the others, so do these inharmonious men, and the inharmonious actions of men, unite in a whole tuneful manifestation of God's will and power. The truth is, however, one that man is conscious of and may accept without reasoning, that all God's gifts are harmonious, and apparent discords are the production of His creatures and not of Himself.

Let us, then, proceed to inquire, what are God's gifts to men? First, He gave man existence. Second, consciousness. Third, individuality. Fourth, wisdom. Fifth, reason. Sixth, judgment. Seventh, love. These are all God's gifts, but they are not all of God's gifts. The next seven, being of another sphere, I will name. First, love of God. Second, love of man. Third, love of self. Fourth, love of family. Fifth, sexual love. Sixth, heavenly desire of God's love. Seventh, the love of existence. These make a sphere, the seventh sphere of man's gifts from God. But all of the other first named six gifts comprise seven circles. I will name them:

First Sphere

First Circle. Existence, as man in spirit, in Paradise.

Second Circle. Existence in the body, in infancy and childhood.

Third Circle. Existence in the body in youth, as connected with courtship or sexual love.

Fourth Circle. Existence in the body in its maturity as properly developed in the conjugal relation and in parental love.

Fifth Circle. Existence in the body in its decline as properly manifested in grandparents, who live over again the duties of parents.

Sixth Circle. Existence in the body in sickness and death.

Seventh Circle. Spiritual existence in continual progress towards perfection.

Second Sphere

First Circle. Consciousness of existence.
Second Circle. Consciousness of love.
Third Circle. Consciousness of providential care.
Fourth Circle. Consciousness of God's love for man.
Fifth Circle. Consciousness of God's resolution to save him from his errors. (This has been almost lost but it was not the less a gift.)
Sixth Circle. Consciousness that he has free-will.
Seventh Circle. Consciousness that God must be served.

Third Sphere

First Circle. Separation from God.
Second Circle. Separation from others in like nature.
Third Circle. Separation from the evil of despair.
Fourth Circle. Separation from the evil of hate.
Fifth Circle. Separation from all the evils of doubt.
Sixth Circle. Separation from self, consciousness of the past.
Seventh Circle. Separation from the body, and unity with God.

Fourth Sphere

First Circle. Wisdom by reasoning powers, or intellect.
Second Circle. Wisdom by reasoning of the mental faculty of the spirit, which is sometimes called conscience.
Third Circle. Wisdom by intuition, or by God's impression.
Fourth Circle. Wisdom by intuition, or the instinct of animals.
Fifth Circle. Wisdom by the laws of men, or educational wisdom.
Sixth Circle. Wisdom by the laws of progress, or self-education, or experience, or memory.
Seventh Circle. Wisdom by spiritual communication.

Fifth Sphere

First Circle. Reason by intuition, as in animals.
Second Circle. Reason by logical form, as by education.
Third Circle. Reason by religious sentiment, or conscience, developing the reasoning faculty to higher objects.
Fourth Circle. Reason by power of will. (This is often manifested in so-called psychological experiments.)
Fifth Circle. Reason by power of love. (This is often manifested by compliance of belief in consequence of a tender relation.)
Sixth Circle. Reason by power of God, manifested by yielding to spiritual influences.
Seventh Circle. Reason manifested by intuition of the spiritual or Divine essence of man's soul or spirit.

Sixth Sphere

First Circle. Judgment by intuition, or instinctive action. (This is sometimes called presence of mind.)
Second Circle. Judgment by the effort of reason. (This is sometimes called a conclusion.)

72

Third Circle. Judgment by the power of love. (This is sometimes called passion.)

Fourth Circle. Judgment by the power of wisdom. (This is sometimes called revelation. It is, however, only from within the individual himself. It was highly manifest in Socrates and Plato, and is the cause of the general harmony of their writings or recorded sayings with Divine revelation.)

Fifth Circle. Judgment by the Divine influence upon the mind, or intellect.

Sixth Circle. Judgment by the Divine impression upon the spiritual intellectual power. (This spiritual intellectual power is resident in the soul or spirit of man, whilst the intellect is a combination of spiritual and material organizations.)

Seventh Circle. Judgment by love of God. This is a surrender of the will to God's influence and will, the highest manifestation of which, to Earth's inhabitants, was in Jesus of Nazareth. Another high manifestation was John the Baptizer. Another was Moses after he was eighty years of age. Another was Daniel after he was taken to Babylon. Another was Luther, though he soon yielded again to the influence of the lower instinctive judgment. Modern times have not produced any manifestations equal to any of the preceding, though Bush, in America, and Swedenborg, in Europe, may be placed in the highest list; Channing and Wilberforce in the next; George Fox and Charles Wesley in a third; and so on. I might name many to gratify a curiosity morbid though honest, innocent though unprofitable.

Having given a very full list of God's gifts to man, do you perceive any inharmonious one excepting his FREE-WILL? This occupies, to man and these gifts, the same position that God does to the seven spheres of spiritual existences. Free-will is man's deity.

The only similarity wanting, to make man God, is infinity. If he had coexistence with God but was not infinite, he could not be God. Infinity is the distinctive nature of God. All spirits are finite. No spirit is omniscient or omnipresent. No spirit is all pervading in its love, except through God. No spirit is all-powerful except through God. No spirit is omniscient nor can any spirit ever arrive at this quality or nature, for then the spirit would be God. And the existence of two Gods is as impossible as the existence of two infinities, which is a contradiction in itself, to state which is to bring into exercise man's judgment by the spiritual nature or spiritual intuition and make the hearer or reader declare it false. Even the heathen world when enlightened believed in a supreme God to whom all others were subordinate. See Homer's picture of the threatened rebellion during the siege of Troy, when Jove declared that all the gods might try together to pull him down from his seat but that he would, by the same chain, lift them and earth together, and that he could hurl them all to the depths beneath the solid face of the earth and bind them into eternal bondage, if he pleased.

The life of man in the body is short but the spirit never dies.

Is it then eternal? Yes. Is not that being infinite in existence? Yes. But eternity of existence is not infinity of existence. It is eternity of existence that man possesses as being a part of God and not a being apart from God. God separated a portion of Himself to be man. But, as I explained in Part I of this book, a part cannot be or comprise the whole, though the whole, necessarily, is or comprises a part.

There is so much speculation now as to the future and so much striving to throw light upon the past, by deep research into all the existing remains of former ages, that God grants the prayer of man to be possessed of more knowledge, because the prayer is a general one, and is from a good and pure motive in most of those who make it. For it is founded in a desire to make proof outwardly of the truth and authenticity of His revelations made in former ages to a less intelligent, but a more refined people! Yes. The ancient world was refined. It was more spiritual than the present. The material never had so high and perfect a manifestation as now. Railroads, steamboats, airplanes, telegraphs, and radio never existed in the former ages. That they exist now, is because God desired to have progress take place, and man, making no spiritual progress, was glad to cultivate the next lower faculty and to advance materially, whilst the prejudices of education restrained his spiritual speculations and aspirations. The good of these material advances is manifest to all; but not all their good, nor their chief good, which is to bring the material nature and development of man nearer to Deity's spiritual nature and manifestations. This is accomplished by the union of electric and magnetic forces; by the union of all the purest material natures with a combination of grosser materials, guided and governed by man's highest reasoning power, and controlled and impelled by his highest judgment.

The seventh sphere will next come into action more powerfully and then Love will become the ruling principle, and all the Sons of God will again shout for joy. For then will be the reign of Christ upon earth, in the flesh. Then the millennium will be. All men who have admitted Christ .to reign over them will be harmonious; and neither nation nor individual will any longer learn war, any longer have strife; and the lion of passion and the lamb of innocence will lie down together, and even a little child shall lead them into his own nature and mind. I can scarcely refrain from declaring more of the glories of that period, so soon to advance rapidly upon the earth, but I know that men are not yet prepared for the revelation and I will refrain. *Prepare yourselves by faith, and you shall know all.*

The revelation of the past agrees with that which I give, and with all that God's spirits give. There may be discordant communications caused by the imperfection of the holy mediums, few of whom in the present day have sacrificed their wills, even in part, to God. But the outward manifestations are connected with imperfect revelation, because the minds of men could be so reached and so brought to listen to the counsels of God, given through higher, or more perfect, mediums. Had I waited for this medium to be in his present state, before commencing to give him revelations, I should have waited in vain. He would still have been a seeker who had not found. But God had revealed that the seeker should find, and he did. For when he first placed his hand with a pen or pencil to paper, I moved it. I did not wait for him to ask in submission, but I soon required submission of him. Long and strenuously did he resist. Long and perseveringly did he require an outward sign—rapping, a vision, or prophecy—to persuade him and to excuse to his reason, submission. But after many struggles during which he sometimes ordered Me to leave him, sometimes prayed to God to save him from evil spirits, which can do evil only in the body, sometimes resolved to submit for a time only, he at last succumbed to My influence. He did this because he had found Me consistent with his spiritual advancement, and because he had found that all things worked together for good, though some appeared, when taken separately, to be evil or retarding. Then his reason was shown thus:—first, that God is good only; second, that all God does is good; third, that all good must come from a good source; fourth, that all good must come from God, who only is good; fifth, that man is from God, unless he originated himself, which, after all, is unbelievable; sixth, that man, being from God, must be good and must be at last united to God in harmony; seventh, that man being united to God in the world, or state to come, can arrive at unity with God only by progress, and that progress must be a gift of God.

Then having found that man is naturally good, he desired to get back, and his next effort was to get forward. For this he asked God's help, and received it.

Having now become willing to be indebted to God for all his progress, he soon became willing to believe God had done him all the good that he had received at all times; and that even what appeared evil must, if it had been the work or gift of God, have been good. Then, having found that God was Lord of Lords and King of Kings, he was willing to be subject, because he saw he could not be free, in reality, from sin and death till he was in submission to, and in unity with, God. But to sacrifice his free-will was the last and greatest trial. To give up the guidance of reason, to withstand the pleadings of affection, the threats of the world, the censures of the church, the universal skepticism of his associated society, was severely trying to him. But, finding that passiveness was the great requirement here and the greatest glory hereafter, he yielded his will as a sacrifice acceptable to God. He withstood the trials of being made a fool for Christ's sake, and of being led into the wilderness of desertion by the spirit of God. There he was assailed by the enemy, the devil as he is called, which is the spirit of man's will, arising from its overthrow and rebelling against God. Having resisted this and having measurably resisted Satan, or the desire to accuse his brethren of short-coming, I have accepted him as the best medium who has yet offered to do My work. For I force no man to work. It must all be agreeable to the man, or it will not be done.

My holy medium did not know that he would not be used until he was willing, and his severest trials arose from the apprehension of being called to do just what he is now called to do. He is now rejoiced that he is found worthy to be so used. But it was not until this very day that he fully sacrificed his best remains of will to God's will. He has now declared himself willing to work even in any way; even in the way formerly most dreaded—I might say abhorred, by him. But I am not yet ready to use him in speaking directly to men. For the present I shall continue to address men through him in writing, which sometimes he will read, and sometimes print. But to him I still speak directly, as I have since he first began to make sacrifices to Me, or to God's will, which last is the same as Me, for I am in perfect unity with it. I am a high spirit, but I shall not declare how high, except to him and to such as he chooses to tell it to in conversation. I am the son of God. So are all who love and serve God, in perfect subjection of their will to His. But the last shall be first; and the first, last. And the last and the first, and the first and the last, are all equally Sons and Sent, or Christs, of God.

Let us pray

O God! who art the giver of every good and perfect gift; who art the eternal, the everlasting Redeemer and Savior of men; by whom the worlds and the whole creation was made, may it please Thee to look with Thy ever-untiring mercy and love upon this people, who are desirous to know Thee better, and to love Thee more. O God! may it please Thee to give us such knowledge as we need of Thy loving kindnesses, and such faith in Thy ever-loving nature as will impel us most heartily to love Thee; most fervently we desire to see Thy rule established in the world of men and to make our submission to Thee in the right way. But, O God, be Thou merciful, for we are foolish before Thee. We are now, O Lord! assembled for hearing Thy word proclaimed in this way through this medium. Bless the medium with passiveness, so that he may fearlessly and unhesitatingly declare whatever it pleases Thee to reveal; and be Thou, O God! our Savior, our Redeemer, our Intercessor, our ever-kind ever-loving God. Almighty Father! Thou canst impress us with faith in him, and in Thy loving kindness. May it please Thee so to do, to our enlightenment and to our advancement in the knowledge and love of Thee. What we want, O God, Thou knowest better than we know; and if Thou, O God! will be pleased to confer upon us Thy love, we shall not want; Thy kindness will feed us and Thy arm will strengthen us to resist evil. O God! we do not know how to pray to Thee but we do know that Thou are worthy of all honor, praise, and glorification, though we cannot give it because we know not how to make it acceptable to Thee.

Then, O Almighty Father! give us new hearts and wills, submissive to Thine, so that all old things shall be done away and all new things appear in their places. Save us, as it may best please Thee, O God! and let us be thankful and obedient to Thy will on earth, as the spirits before Thy throne are in heaven. Amen.

I am now going to write for you a chronological table, beginning at the foundation of Saturn and leaving out the outside planets, because I am not ready to declare prophesy or unknown scientific facts through this holy medium. But yet there will be unsettled questions of science determined by My announcement.

First, Saturn separated into a continuous ring, revolving around the central body of the solar system, now so called by men of earth. This occurred when the contractile effort of matter, by its law of progressive contraction, had overcome the cohesion of the particles which connected what is now called Saturn with what is now called Jupiter.

The ages of centuries, myriads of years, rolled by during which the contraction of the central matter continued till Jupiter also separated. About the same time Saturn fell into fragments of itself, by the ring form becoming so attenuated as to be incapable of maintaining equally its relative motion around the Sun, or center. But this disruption was not sudden or gradual; first one part separated, and, the contraction being continued, it separated further and further at that place, until a rotary motion was required for it to maintain its equilibrium. This rotary motion commenced in this way.

The ends having been separated as far as nearly one fourth of its orbit, or first length of circumference, the end, which may be called the forward one, rolled, or was gradually doubled under, towards the center of the system by the retarding force of the fluid called the aura, in which all the planets move. Then this rolling continued to proceed with acceleration, because as the whole mass necessarily retained its center of gravity in the same position that it would have maintained had this doubling under not taken place, the outside had to move faster than before. This disproportion and the resistance of aura continuing to increase, the mass more and more rapidly assumed its present globular form and arrived at its present period of axial revolution.

Saturn, being now a planet in form, had as yet no attendant bodies. But the contraction of its body continued, because of the existence and action of the same law before referred to, and it soon separated into rings—first one, then another, about the time the first began to separate, preparatory to a folding or rolling up into a moon. And thus it continued to progress until it had reached its present state of rings and moons, which will in time be further modified by the rings' becoming moons and new rings being formed. A change of this kind will take place very soon, but the particular time will not be declared, either through this medium or any other, to Earth's inhabitants.

The same process continued to proceed in the solar and other systems until they arrived at their present form. Mars has no moon because the contraction has not yet reached a degree that will separate a ring from the central body. The Earth has one. For Mars and Earth were both separated from the central body about the same time, just as now more than one ring exists with Saturn.

The asteroids, as men call the small planets between Mars and Jupiter, are the result of several very narrow rings, which existed at nearly the same time, and were also similar to the narrow and near rings of Saturn. The Earth's moon was separated at the time the deluge occurred, for such an event did take place. The Earth was inhabited before that time even for many myriads of ages. Man, though, was not placed in an earthly body till six thousand years before the deluge.

This does not agree with the Bible, you say, and yet I have said revelation should agree with itself. Well then, let me explain that the imperfect chronology of the Bible is not revelation but history, written by men who were often inspired but not necessarily always so. Further I will also say that history cannot be truly called revelation unless it be written by spirit. Now the Bible does not give its chronology as revelation but as history. Then what the Bible itself does not affirm to be revelation should not be understood as such. But very little of the Bible is said to be inspiration, you say. Look again, and you will find that much more than you think was declared by the writers to have been received by them from God's spirit. And much more, too, than the greater part of Christians, so-called, fully believe. How is this, you say, do not Christians believe the whole of the Bible when it is one of the articles of faith of most churches that it must be implicitly received? and when it is made the separating line, to determine whether the professor is worthy of salvation by the church's efforts?

There is a want of true faith. Profession has taken its place. Men cannot reconcile the dark passages of the Bible with the light within them and, as the church will not let them receive from the light within an explanation of the difficulties named, their faith suffers deterioration and is often turned into complete skepticism. Still, as belonging to churches is honorable, few are willing to declare their unbelief. Few are willing even to confess doubts. They want at least to stand well with the church, though they cannot reconcile themselves with God or with the Bible.

Let us proceed. You will find I believe the Bible, for I intend to explain its most difficult and puzzling passages in the course of this book. And I have just given you a solution of the cause of the deluge that you never thought of. A theory was once promulgated by a scientific and pious man, that the earth once had two moons and the collision or combination of one of them with the earth caused the deluge. But you can easily perceive that nature does not go backward, or the solar system would fall into disorder. God does not rule and guide like men, imperfectly. But His will sustains all in continual progress, and He never makes any mistakes.

Adam, then, existed about 6000 years before the deluge, and for that length of time the antediluvians populated and cultivated the earth. Empires rose and fell, but their names and languages have not been recorded. Neither would it be interesting to report them now. Continents, islands, oceans, seas, lakes, rivers, mountains, plains, then existed on the earth but not in the same relative or absolute form as now. The old surface of the earth was entirely broken up and the fountains of the great deep were opened. The windows of heaven, too, were opened. What are the windows of heaven, and how do they affect the accumulation of the waters upon the face of the earth? is a natural and interesting question, which I will now answer.

The windows of heaven have puzzled scientific pious men, more than almost any other passage, for it is either a mistake of the writer, a mistranslation, or else the term is used in a metaphorical sense. If metaphorical, there seems no evident and plain type of which it may be the antitype; if a mistake of the writer, either from ignorance or other cause, it must cause us to distrust the remainder of his writings; if a mistranslation, it must also cause us to distrust the remainder, for there may be many others as far wrong.

The windows of heaven are the portals of God's mercy. And what are these portals like? They are like the passage from death to life, or life to death, as it is generally called. Not that the mere change of condition is equivalent to salvation, but that the life of man in the world to come is mercifully ordained to be a recipient of love and mercy, whilst it is no longer possible for the spirit to diverge from God. The people of the antediluvian world were so deeply sunk in error, so stupendously imbued in sin, so darkly resolved on scaling heaven in their own way, so outrageously disregardful of decency, or propriety of action, or love of self prevailed so unobstructedly, that no man could be brought nearer to God in that life. Nearly all had departed from piety.

One family yet remained, and God directed the head of this family how to save himself and his race from destruction. For, by the laws of progress, the time for another creation of man's earthly body had gone by, and the earth must have been left uninhabited, or a new law must have been promulgated, to bring into being more men upon the earth. This man, called in the Bible Noah, which was in fact the name of his nation instead of an individual name, became the progenitor of all the men who have since dwelt upon the earth. Every nation retains some tradition of his escape from the confusion of the land and water. Every nation or race possesses the individual marks of this progenitor by having a caudal extremity, by having a depression in the throat above the breast, by having five fingers and five toes upon each hand or foot, for the thumb is also a finger. Through these marks, which distinguished Noah from his fellow men, all men are known by us to be descendants of Noah; while those who had tails, six fingers, six toes, and a full straight neck like a baboon were, or are, of antediluvian existence in the body. To be sure, occasional manifestations of six fingers or toes occur down to the present time and are generally conjoined with great stature, showing that at times the influence of Noah's progenitors extends itself beyond his. So too, sometimes, the full necks are seen in very sensual men, and even approaches to tails are seen in some parts of the earth but these are among the lowest of the race in development.

79

Such revelations do not admit of outward proof, so most men will be incredulous. But I reveal to you what you have desired to know and if you are not satisfied, blame yourselves, not Me. It is true. My assertion will satisfy My medium, if no other. And one man's happiness, or pleasure, or gratification, is thought worthy of regard by the highest of God's spirits.

How long then did the moon continue to revolve in a ring about the earth? A thousand years is a long time for man to reckon, and subtracted from the last of Europe's history would leave little worthy of the pursuit of the present race of men. But a thousand times a thousand would no more than embrace the period during which the moon revolved around the earth in a ring. Then its gradual rolling up commenced, and this process required thousands of years. Then the earth is very old, you think! Yes, it is as old as the other planets, but its matter has, like theirs, assumed various shapes. Then the earth has had other revolutions? Yes, thousands of changes, such as being demolished and reconstructed, have taken place with its matter; but God can again, and again, cause these changes and the matter will never be worn out. But then, God must have made mistakes to make over His works so often! He never makes them over. He always brings forth new works. Old things are done away with and new things appear. Such a change is now impending. What! is the earth to be destroyed, or reformed, so as to dissolve all these works of man upon it? Yes, my anxious reader, yes. But not in your time, so you will have full opportunity to prepare to meet God in the usual course of nature, or by His plan.

We will now return to the creation, or formation of the earth in its present shape, after the moon was disrupted from it. In that primeval time, Noah found a rugged home in the central table land of Asia. Indeed, he himself founded the empire still existing under the name of China, though its seat of power has been removed to the western coast of the continent. The Chinese truly record their great antiquity. The learned fools who endeavor to show that it was forged at a late period waste their time and pains. All Chinese history, and all Chinese art, bears the impress of the truth of their chronology. For thousands of years their empire has been stationary in art, language, and form of government. How did this stamp of permanence become so peculiar to them? Not by any reason other than that God revealed to their founder a system of government which would maintain itself. The head of the nation is theoretically its Father; the priests work, and are not maintained in idleness; all work, none are idle. The king and the beggar, in China, alike are required to maintain the respect that it is proper should be paid to labor.

Three hundred and sixty thousand years did the empire continue uncontrolled by foreign influence, under the peaceful sway of lineally descended princes. Father and son were harmonious, for the rule was patriarchal. The people dwelt in peaceful happiness and each practiced his father's trade or profession. So art became fixed and limits were set to progress. It came to be regarded as sacrilegious to be wiser than one's father, or to attempt to excel him in skill or contrivance. At last China was conquered by barbarians from the south, and, though these were absorbed into its ample population, they infused into it some spirit of change, slight indeed, but perceptible. Another irruption followed from the east, by the Thibetan hordes, or ramblers of the deserts and great plateaus of the district where the Chinese nation itself was founded. This irruption corrupted the religion of China, which until then had been very pure, and almost perfectly retained from its first promulgation by Noah. Again a southern irruption occurred from India, and so the refluent waves of population which had originally separated from Noah's family or descendants in early time, began to roll back upon the great primeval nation, until its ancient barriers and powerful armies were broken and destroyed. Wars, famines, pestilences, and all their attendant evils in the corruption of the people, nearly depopulated China. But the principles upon which it was founded had become so implanted in the nature of her people that they could no more be eradicatd than their oblique eyes or enormous ears.

But have the Chinese a history or chronology for more than a million of years? Yes. It is obscured now by fables interwoven by ignorance and doubt. It is perplexed by transposition and errors. But in its general outline it is true. It calls a race or dynasty, a man. It calls a man a race, perhaps. But that such was its history, its civilization and its religion, is truly stated. How, then, did men come to depart so widely from the original stock, in form, features,' and color; in habits, laws and government? First, it pleased God to separate the spirits who desired to leave Paradise into classes. Those whose object was travel or wandering, principally, were placed in one class which would make, when formed in bodies, nomadic nations. Those who were inclined to be patient were placed into another, the impatient into a third, and so on. But if existence is so passive there how could these qualities be developed? They were not developed, but God could foresee that such development would take place. Again, how came it that these beings, all emanating from God, were so different in character. Because God willed not to produce sameness, but dissimilarity. So that no two were alike, though there are a number greater than the grains of sand containable in a globe like the earth. And why then did not the former inhabitants earlier attack and overcome the Chinese! Because the greater part of the world's population was under that rule, so paternal in its character, so Divine in its execution, that no rebellion would take place.

Such is the history of the earth chronologically. But you have not given us the years or dates of any events! No, I give no outward evidence through this holy medium. I reserve him for the highest revelations I have to make, and these revelations are spiritual. The outward too much obscures the inward, and the material too much controls the spiritual in most mediums, and I should obscure his spiritual nature, without really advancing your faith, if the revealments of the character you desire, and this holy medium at first desired, were given. But other holy mediums will arise who will be competent to take the revealment of the outward, though not passive enough for the purely inward. Through them I will cause lower spirits to manifest signs and tokens which shall satisfy all reasonable seekers, and I will allow them to declare the chronology by years and dates, of all the great empires which have existed of late upon the earth; that is, of all in which men by their traditions feel an interest.

Having disposed of chronology, let us proceed to discuss theology, as founded upon the History of All Things, chronologically considered.

CHAPTER XXIV

THE HISTORY OF CHRONOLOGICAL THEOLOGY

The Bible is composed of various writings, by various authors. It is a compilation from different nations too, but of course, the oldest nation being the Chinese, all that is authentic of the earliest period must have come from their records. Language was by them first reduced to writing. Ideas can be pictured, but language must be written, to be preserved in purity or vigor. When the first records of this kind were made, it was found that tradition differed as to the past and an attempt was made, at that early period, to force an agreement between different traditions. In this way the time and order of God's creation was erroneously recorded in the first record. But the tradition as to the formation of the earth was further obscured by men, endeavoring to bring the account into such a shape that they could comprehend it. Their knowledge of science and of astronomy was less than now, and they could not conceive of the truth. Therefore the surety of God's power was found, by their wise men, to be best understood by the people when it was placed in familiar images, and days were used for periods of time. Evenings and mornings were used for the ends of one period, and the beginnings of another. So the book of Genesis was commenced.

But Adam was placed in Paradise and its locality described as being in that central region, south of the Caspian sea, where a delightful climate and beautiful scenery ever rejoices the eye and from whence the celestial nation, as they still love to call themselves, were forever excluded by their love of stationary habits and their distant position. But it says, the Lord God planted a garden eastward in Eden, and Eden is westward from China, if it be south of the Caspian! Well, this too I will explain satisfactorily to candid minds.

But first I will proceed with My other details. There was also a call heard in Eden, Adam, where art thou? This was symbolic of the progress of souls, or rather spirits, of men in Paradisical state, and shows that even in Eden the tradition held that God appointed man to be led by Him and to rely on His parental care and oversight. It is by such calls on spirits, indeed, that they are aroused from their monotonous and passive state of enjoyment, and led to ask themselves the question that God has been represented to ask in proper person Himself. But the voice of God is heard in each of the Adams, or spirits of men, in Paradise, before they leave it, asking of them where they are, and, Why hast thou found the desire for change? That is, Why hast thou dared to eat of the tree of knowledge of good and evil? The evil and the good are to be tasted only on earth, but the desire for them is formed in Paradise, and then, as now, men reap as they have so⸱ n. Various are the motives and extraordinary the excuses given. ut none is more common than that the parital companion having tasted of the fruit, or of the knowledge, the remaining half desires to follow the example. The account therefore is beautifully symbolic, and the glories of Paradise are obscured only by being taken too literally in the present day.

When the creation account reaches the expulsion from Paradise, it becomes plainer. The first offspring of Adam and Eve, the parital man, was Cain, that is, evil; the second was Abel, that is, peace. But the second was overcome by the first, and slain by his own altar, which altar was in Adam's heart. But then Seth was born, and Seth was a man; from him and other sons and daughters of Adam were derived the succeeding races as they are termed, or race, as I would say, of men. The Cain principle left the immediate vicinity, which was under Adam's wise management, and retired to the land of Nod, where it associated itself with a companion and had a large progeny. It was there that evil brought forth good by seeking to teach men arts and sciences, manufactures and life in cities.

But where is the land of Nod and how came it to be peopled, as it appears from the account that it was! It was still in Adam. The first man, or pair, having brought forth evil, first overcame it and held it in subjection by Abel, or good, or peace. But, in a moment of fury, evil again prevailed and slew Adam's peace. Then the enormity of his offence appeared to Adam and, again, he brought this rebellious son into subjection. Cain's heart again became purified and the duty of providing for his now numerous descendants, from that revealed store of knowledge which God had given him, became his chief desire. He became convinced that he was his brother's keeper. Then he strove, by imparting to his descendants the knowledge respecting the workings of metals, the construction of implements of husbandry and of mechanical and manufacturing labor, to perform the duties which God had called him to do, and from which heretofore the evil principle, or desire, had restrained him.

Here then was the land of Nod, the place, or work, or theater, of duty. Here he found his companion, Industry. Her name was not given in the Bible, but it may be inferred from the change which took place in the manifestation of Cain, or evil desire, when placed under the control again of the man, or Adam and Eve, and united to Industry. So, to the present day it is a proverb: Let a man be idle and the enemy, which is the evil of man's will left free, will find him employment. Idleness begets sin. Industry begets good works. Faithfulness begets reformation, and Industry brings forth works meet for repentance. These works are being useful to our fellow men and purifying our hearts from rebellious desires against the happiness of others, or the desire to avoid, or shrink from our own duties. Cain and Abel then represent principles, and the peace principle having been slain, its blood, or memory, cries aloud from the earth, or grossness of the heart, for revenge; that is, for restoration to life, for resuscitation in the heart, and for having its sacrifices again acceptable to man's highest nature, so that evil shall be disregarded, and peace alone be found worthy.

This outline is sufficient to enable the wise to trace out more instruction from it and from the Mosaical, as it is called, but the traditional, as I call it, account of the first proceedings of men on the earth. Seth, and the other natural children of Adam and Eve, are spoken of as sons and daughters and the age of Adam is given at Seth's birth. But the age is speculative. It was not long after his entrance into life that Eve became prolific; and, though then the race had long lives, they bore children as rapidly as now, and by such means the world became so fully populated in 6000 years that men were with difficulty restrained from eating each other.

The account of the descendants of Adam and Eve is continued to Noah. But we should regard it as only a similar, or like statement of truth, made by wise men from the traditions existing at the invention of writing, which was not antediluvian. They selected such as would seem most in accordance with their experience; their motives were good and their course was good. It was history they were writing and not revelation. Noah, the wisest of his generation, was the link between the old and new, and was a predicted prophet raised up by God to warn and convict a wicked world of men, that destruction was consequent on disobedience. They would not listen to him. But he listened to God and by God's direction built a vessel called the Ark, which was the wonder of those who saw it. It was so superior to all other vessels which the antediluvians had seen that they were willing to worship Noah. They did make him king over a vast multitude and it was by his power in that position that he procured provisions and stored them; that he collected all the various animals which accompanied him in the Ark.

So, having sufficiently sketched the antediluvian history, let us proceed with the postdiluvian. The vigor of constitution, the long life inherent to the antediluvians, continued to decline under the sons and descendants of Noah. But this was gradual, and was owing to the modification of the atmosphere consequent upon the disruption of the earth's moon. Shem, Ham, and Japhet, or rather Japhet, Ham, and Shem, for such was their order of birth, became the rulers of their descendants because of the patriarchal character of the government of Noah, who himself continued to be the supreme ruler of the race of men until his death, which was 600 years after the flood.

This you say differs from the Genesis account, because in that is a transposition—his 600 years before the flood, as in that account, should be after it. Of this you can have no proof, except that I am revealing to you the knowledge possessed by the fourth sphere of spirits. The first circle spirits of that sphere are attentive to the openings of divine harmony and receive with implicit faith these historical truths. Till they have done so, they cannot progress to the second circle. In the second circle they find the history of other planets; in the third circle, that of other systems; in the fourth circle, the history of the universe to which the solar system of the earth belongs; in the fifth circle, the history of the previous condition of the vast circle of universes, which revolve in harmony around a vast circuit of the illimitable space in which God's creation is expanded and unfolded. Myriads of myriads of universes compose this coelum or heavenly association of associated universes revolving about their common center. Each universe is attended by, or composed of, myriads of myriads of myriads of suns, each having its planets, moons, comets. and invisible, or unprogressed, satellites.

So proceeds the Order of God's creation. So it is governed by one law. For God does not amplify His will into a long extended fiat. He speaks and it is done. But beyond this coelum, or heavenly association of associated universes, comes, or succeeds, another, vaster' and more illimitable, more incomprehensible to man; an association of coelums, or associated associations of universes, the history of which becomes revealed to spirits who reach the sixth circle of the fourth sphere. Again, the great whole of God's creation bears a still higher relation to this circle, or association of associations, in than its association of associations of associations is far more numerous, far more illimitable than the lower ones. In fact, each upward ascent gives increased numbers of principal associated bodies, so that this last may well be called infinite, though it is not infinite. God only is infinite. It is then the knowledge of the history of all the created bodies of God's creation becomes revealed to the spirits of men, either from the earth or any other inhabited globe, (and all are inhabited)—I say, all this knowledge, so nearly infinite as it is, and as you can perceive it must be, is revealed to spirits of men in the seventh circle of the fourth sphere.

What then remains for higher circles, when so much is received in passing through the fourth sphere? God will be able to provide novelties more stupendous even than these, for He is unlimited by any thing but His own will. But moreover He has provided already for those who are in the fifth sphere, in the following manner.

First Circle. Knowledge of Law, in its manifestations of love.

Second Circle. Knowledge of Love, in its manifestations in universal progress.

Third Circle. Knowledge of the Infinite nature of God's attributes of Love, and Mercy, and every good man's fate.

Fourth Circle. The infinite knowledge of infinity of worlds and men, of their relations to God, and to each other. By this foresight is highly developed. Prophets may be inspired by this circle, and though they may sometimes fail, it will seldom be so.

Fifth Circle. Prophecy, by knowledge of God's revealed purposes. Not all the revelations that God makes to spirits can, however, give them a knowledge of what a man will do. For men are free agents. But the laws of being are so understood, that man's course can be conjectured or judged of with reasonable certainty. Then those in this circle have such knowledge as will assist them materially, by being above all the lower and baser affections of men's spirits, which are not entirely got rid of until elevated into this circle. The last circle was purified of all but love of power and this advance completed the purification.

Sixth Circle. Advance manifested by the power of spirits to discern the intentions of men by looking into future minds, yet in Paradise, as well as by judging from what can be seen in present minds, in bodies.

Seventh Circle of the fifth sphere. Where can we place any more extension to knowledge of men? For the knowledge obtained in this sphere is a knowledge of man. In the fourth, it is a knowledge of material or outward history. In the fifth, of spiritual or inward history. What then remains for the seventh circle? It is a knowledge of spirit revelation from the beginning. Of all that God has revealed in all ages of creation, in all time of His existence. But this existence is eternal. Yes, and the knowledge which spirits add to their former stock by passing through this circle is almost infinite. But not infinite, because infinity cannot be received by a part of God. And man's soul, or self, is a part of God, as I showed in Part I of this book.

What then remains for the sixth sphere? Abundance. God has not exhausted His resources of employing and forwarding in progress spirits who know all that I have described. In the sixth sphere, spirits are employed in calmly waiting upon God's benevolent merciful works. They are the Servants of His Will, the Word of His Power, The Honor of His Nature. They stand before the throne of His mercy, ever praising God for His loving kindness to all men. And, when they would change the song: Lord, have mercy upon us miserable sinners, which men so delight to declare with their lips, they sing: Great and marvelous are Thy works; just and true are all Thy Ways, Thou King of saints. Well, but is this peculiar to the sixth sphere? Oh, no! All spirits do this who know God. And they who have reached the third sphere, may be said to know Him, though necessarily it is an imperfect knowledge of His attributes and actions. But what are the distinctive features of the progressive march of spirits through the circles of this high sphere? They are Sons of God, now on His right hand. Not fully entered into His glory, but yet judging the world from which they have escaped. They are elevated to power. Power is the distinguishing element of advance in this sphere. Action becomes their duty, instead of mere reception of knowledge, as in the fifth sphere; of mere learning history, as in the fourth sphere; or of mere learning of memory, as in the third sphere; or of mere reconciliation with God, as in the second sphere; or of mere experience of good and evil as in the first sphere.

Now we will proceed to give you a list of the employments, or advances, of the various circles of the sixth sphere, and of the first three circles of the seventh, the last four circles of that sphere having been described first of all.

First Circle, sixth sphere. Power of Prophecy respecting future events, as manifested in Daniel, Ezekiel and other prophets. The Angel Gabriel, as the spirits of this sphere were called, was a manifestation in the outward of this circle. The Jews had received the names of several circles from the traditions of older nations, but only Gabriel has been handed down to us as a high spirit of God, in the record called the Bible. What then are the names of the other circles? you may inquire. Not profitably though, for though the names are significant, they are only so to knowledge of their signification. Even Gabriel is now misunderstood. Its true meaning is what I have declared, but do you find commentators arriving at this conclusion? By no means. What then are the ways and preachings of the other circles of this sphere? In the second circle, the power of Love of God to reform the souls of men enters into the progressing spirit. Next, in the third circle, it receives the power of Will. That is, the power to will the accomplishment of objects through the manifestation of the physical.

It is to spirits of this circle that the rappings in their origin may be referred. Not that they rap, but they by their will cause outward demonstration to be made through lower spirits, none of whom are above the second sphere, nor above the fourth circle of that sphere. But then, many have declared their communications to be higher than that! Yes; but these spirits either spoke in the name of the higher spirit, or they assumed a position, or the name of a position, they had not arrived at. But can these spirits be permitted to mislead men so! Have they misled them? When they have declared themselves belonging to the higher spheres, did you believe them? No: you scouted that idea. You resolved to have low ones, those who had been, of late, associated in the body in the family circle. And if they deceived you into a belief of their advancement, it was because you willed to have them higher, and because you did not submit to be led by them, but asked, not for the will of God to be done, but for the will you possessed to be done. And the will you had was a will of your own, which was controlling to the spirit.

For the spirit, or spirits, can communicate to you only in some will, and in accordance with that will. If it be your will which is acted in, or under, it will be variable, or constant, as that will may happen to be. If it be the will of a higher spirit, then the communication will be truthful, so long as you submit and are passive. But when you contend with the spirit, either by doubting or disputing the ideas, or disclaiming its agency, either you stop the communion, or bring the spirit again under your power. For whatever man chooses, he does. If he chooses to have his own will gratified, he may have it, at least to his own confusion. But if he chooses God's will for his ruler, or communicator, or for his interpreter, he will obtain truth if he gets revelation.

How is that? Does not the man get an answer always when he acts in God's will? When he acts in God's will, he does not ask for information, or communications. He leaves it to God. He receives what God gives, without question, or answer, without doubt, and without rebellion. But few, my friends, very few have been willing, even at times, to receive so. Nearly all have a purpose to obtain, other than doing God's work. My holy medium first acted in his own will. I tried his willingness to obey Me. I found an inclination, but not a controlling one. Then I commenced trying him by doubts, and false impressions, by fooling him as he called it, and by various trials, during which he sometimes rebelled most thoroughly, and at other times submitted most perfectly. But he was unstable. I could not depend on him. I left him. Then I caused him to read, and reason, and at last he saw his error. He saw that passiveness and submission is the great requirement, that his will was the sacrifice demanded. Even now, he fears that he does not keep himself passive enough. But I do not allow him to write erroneously, because I have other objects than his own improvement. I have now to promote the increase of general knowledge, the advancement of sound doctrine, the true knowledge of God, and the everlasting welfare of mankind. With such results affectable by My errors, or My holy medium's errors, I shall not allow any to escape notice, either now or hereafter.

Be no longer fearful, then, or unbelieving, but submissive, and be taught as a little child. For unless ye become as little children, ye can in no wise enter the kingdom of heaven. It is now true, as it was when Jesus walked the streets of Jerusalem, that men must become as little children. This is, they must receive the Divine teachings with childlike faith and confidence, or they cannot make any progress. It is in every man's power to resist all progress of belief in his mind, or faith in his heart. It is also in his power to yield either, or both. Many say they cannot help their belief. But this is an error. We can control our belief. Free-will is the deity of man, as I have before stated and shown. But free-will can cause no man to believe, or to refuse to entertain the truth. If men would believe the truth, they must ask God sincerely in their private hours and in every act of their lives to help them to know what to believe. They must learn to be independent of man, whose breath is in his nostrils, and rely on the help of God, which will probably be given through the medium of men, whose breath is the power and will of God; that is, spirits of the third society, or circle, of the sixth sphere.

The fourth society or circle has the power of the Will of man.

That is, it has the power to influence the will of man by externals, such as miracles. Such were performed for Jesus of Nazareth, for Gideon, for Samson, and for others of the past time. But in the present time they have acted through lower spirits, such as the fifth circle of the second sphere, sending forward the influence of the higher spirit, manifest to men as their work. That is, the moving of material objects; disturbing the atmosphere, by which sounds are produced, proceeds from the circle below, as I have explained; but the moving of solid materials is the province, or peculiar sphere of power, of those who have attained to this circle. The high circles always have the power of lower ones in addition to those peculiar to them. Then why do these spirits not act directly upon man? Because man desires lower spirits to act, or else he does not render himself fit, by subjection and patience, to be associated directly with the higher.

But the sixth circle of the fourth sphere has also the power of knowledge of God's intentions respecting man's government, and through it some of the miracles are performed. Miracles, we call them, because that name conveys to you the proper meaning. They are proceedings or operations in the Will of God, yet beyond, or apart from the ordinary law, or rule governing material substances. They are departures from what are properly called laws of nature, or the laws which subject matter to relations of materiality with regard to men.

The fifth circle of this fourth sphere, devoted to the acquisition of knowledge of the past in the universe of God, pertaining most, intimately to their original earth or globe of matter, has also the power to affect the matter of their own planet, or globe. The sixth circle of that sphere has power over the matter of the system of which they were inhabitants; the seventh, over the matter of which its universe is composed.

Then, proceeding upwards, this power is extended step by step, in each circle of the fifth sphere, until it arrives at general power in all created bodies, in the third circle. Then commences another manifestation of power, in the fourth circle of the fifth sphere. That is, the power of Knowledge of the Laws of Matter, in their various relations. First, confined in this circle to their own planet or globe, and so extending, step by step, as knowledge of history extended; a step to each circle of advancement, until all is attained in this branch of knowledge, in the sixth sphere third circle. The fourth circle of the sixth sphere then takes cognizance of the Laws of Spiritual Relations, first, in their own globe or planet, whether it be primary, secondary, or central. In a similar way as with the knowledge of material laws, now they proceed step by step and circle by circle, until, having received knowledge of all the spiritual laws of government, in the third circle of the seventh sphere, the Mind, or Spirit of man, is ready to act in the fourth circle of the seventh sphere, in the Will of God in any part of His creation. This power is exercised, in general, through lower spirits, each having its proper working sphere. Then the higher circles are employed in accordance with the position before described in this book.

Now the sixth circle of the fifth sphere has no knowledge of the duties of the seventh circle. For the knowledge of a part is not a knowledge of another part in God's creation. Spirits cannot infer from analogy, as men do. This is because the laws of God, though concise and general, are so influenced or modified in their application, without there being, however, any change of law, that they produce infinite variety. As well might a man undertake to describe a Malay because he had seen a Hollander, as to describe Saturn because his spirit was acquainted with Earth; or the solar system of Sirius because the full knowledge of this solar system had been revealed to him. And so it is, through all God's Works, Laws, and Will.

We will now return to the explanation of the extension of spirit capacity for action in the sixth sphere. The fifth circle extends a step beyond the fourth, having the power of the knowledge of the will of men, throughout the system to which he originally, or more properly, in the body, belonged. This power extends to influencing the will of all these beings in the body, which is a higher power than influencing beings in spirit, by miracles or disturbances of matter in a novel or unaccustomed manner. And thus the extension of the sphere of operations extends upwards, step by step and circle by circle, until the fourth circle of the seventh sphere is reached, where all knowledge of Will is received, including the whole expressed Will of God. All in the fifth circle know His Power. Power, therefore, is a higher form of Knowledge than Will, and commences its development in spirits, or its revelation bestowed upon them, in all cases, and each form of manifestation, at one step above Will. So too, Love is higher than Power, and Action higher than Love, and so each of these forms of God's nature becomes impressed on the advancing spirits of men, at a step above and preceding, and continues always in the same precise relation to all the others, in the spirit, or mind, of man, either in or out of the body.

The digression of explaining the nature of the different circles seems to have been required for the satisfaction of such minds as have been very curious about the future state of their souls, and have doubted whether they could continue to learn to all eternity, supposing they could soon arrive at perfect knowledge, if they had the opportunity. Now they will perceive that advancement is made slowly, and that of necessity the progression of spirits is never ended, as shown in Part I. But, then, Jesus of Nazareth was said to have reached the high station of Son of God and Ruler of the Earth very soon after the death of the body. For He declared: All power is given Me both in Heaven, and on Earth. This power, though, was given Him under the influence and by the direction of the same spirit that aided Him as His Christ on Earth, when He was in the body. By continuing His perfect submission and passiveness, His progress was rapid, and He arrived very soon, comparatively, at the seventh circle of the seventh sphere. And yet it is only lately that He arrived in it, and it is by His Action, which could commence only when He arrived in that sphere, and could be perfect only when in the seventh circle, that He causes, or is allowed by God to cause, the manifestations which are now awakening mankind "from ignorance, fear, and torturing doubt." (See title page of Part I.)

THE HISTORY OF THE DESCENDANTS OF NOAH

The sons of Noah, Shem, Ham and Japhet, or, as I have said, more properly Japhet, Ham and Shem, left their father Noah to rule the Chinese nation. They divided the earth between them, and as they had long lives, each lived to found a powerful kingdom and to rule in patriarchal simplicity, and unlimited parental power over extensive countries.

Shem founded first the nation afterwards dwelling in India. They first explored the valleys of the Ganges and its branches, and there the seat of their power and principal cities continued ever to remain. Japhet was the second to leave the parent stock with a colony. He led them to Greece and there founded his empire or kingdom. The mountains and valleys of this region interposed obstacles to the retention of power long in his proper line. Soon his descendants branched themselves into numerous colonies, of which one, the most important, led his followers backward towards China until they reached the fertile plains of Mesopotamia. There Nimrod, a great grandson of Asshur, or grandson of Noah, grew so powerful as to excite the jealousy of the poorer branches of Japhet's stock, who desired to share in the wealth and luxury which already began to prevail.

How could the world so soon become so populous? By long life yet inherent in mankind and by great vigor and health, which also characterized the race, as the remains of still greater manifestations of both these, formerly existing anterior to the deluge. So rapid was the increase, that vast works were undertaken. The pyramids of Egypt are only built after the earlier models of the works of Japhet's descendants, settled in Assyria and Mesopotamia. The first great disturbance of harmony arose from an immense structure undertaken by an early ruler of Babylon, called in the Bible the Tower of Babel. A general rebellion took place, and under various patriarchal leaders, the people separated into various nations, or tribes of people; some eastward, some northward, some westward again into Palestine, Phoenicia, and Ionia. Others penetrated into Egypt, and there founded another durable kingdom, which long retained in great purity the primeval religion.

There still remained a vast multitude who, though exasperated with their prince, who had been such a severe task-master, again submitted to him on his promise of amendment, and he relinquished his burdensome proceedings. When all were again quiet, he took such measures as enabled him to make his rule more sure and independent. He maintained, for the first time, an army trained to obedience, and urged by various motives to adhere to him rather than to his nation. Then, having an army, he found it convenient to use it against other nations, and gradually be acquired the rule again over a large portion of those who had fled from his burdens. But Egypt for a long period retained its independence, distinguished not as an aggressive, but as a resistant people.

The last great branch or colony that left the Chinese family, under the conduct of a son of Noah, was Ham. He led his followers westward, and, finding all the most fertile valleys already occupied by his brother's descendants, he passed on in a peaceful and easy way until he reached the Nile. He ascended that oceanic river to its sources, where he established his seat of empire. This was long maintained as a splendid, wealthy, and powerful empire until centuries after the downfall of Troy. The last of the kingdoms of the early world after the deluge was founded in Italy. A few dissatisfied, law-breaking, vagabondic individuals of Greece, as since called, were led by Pyrrhus to Rome, or to where Rome afterward came into existence. This is a scheme which bears a resemblance to history, sacred and profane, as men choose to designate it! Now, I will give a brief, but true sketch of it, as it really was.

Noah was the first ruler of China, or the nation since called Chinese. Its seat was then central Asia or Thibet. His sons were at first his only subjects, but the prolific nature of the antediluvians, and their fondness for the multiplication of the species, promoted the fulfillment of God's first command after the deluge, to increase and multiply and replenish the earth. Their long lives, nearly all passed in maturity, their vigor and their powerful frames, enabled them to subdue nature, and behold their descendants thickly settled around them. But none of them ever left the parent stock, or dispersed themselves. Long centuries afterward, individuals scattered themselves abroad and pushed beyond the outskirts or boundaries of the general population. Then, sometimes, they established a separate government, for at that time government rested upon the will and consent of the governed. These scattered people also progressed in population and, being of a wandering disposition, transmitted the more of such inclination to their descendants, who thus became more and more roving in their disposition and habits, until they abandoned settled homes and roamed at large upon the vast plains that abound with rich pasturage, and yield an easy, roving form of living for nations to this day. They gladly left the old stock, which as gladly spared them.

These offshoots carried with them their family heads, who became their rulers and, as chance directed, led their followers to the various parts of the earth. The names of the sons of Noah, and of their descendants, given in the book of Genesis, are but the names of nations, or tribes of people, who scattered themselves gradually over the face of the earth. Ship building was as well known then by Chinese as it is now, and seas formed no obstacle to their extension. America was settled early, and Africa became a nursery of nations. India was an early government, which retained much of the customs of the primeval or antediluvian world. Egypt was a later one, and though a greater difference was perceptible here, it was still a near relationship. Very soon three great empires became prominent: Egypt, Assyria, and India. These were the ones which bore the names of Ham, Japhet, and Shem. The parent stock of Noah tended to its present seaboard location, and no emigration proceeded from it after the first offshoots were established. I have stated, the people were stationary by habit, custom, and inclination. The restless natures of Paradise were born into other nations, and by that arrangement quiet was preserved in one corner of the earth, whilst turbulence and progression appeared in other parts.

But this turbulence led to a decline of knowledge and civilization, and in the lapse of ages many tribes became barbarous, and afterward savage. Large portions of Earth became a wilderness of forest, and men gathered about the few civilized and central empires, watching for spoil when not wandering as shepherds or hunters. The whole of Asia was sometimes under the nominal sway of the parent stock, but the parent stock never secured permanent sway beyond their original boundaries and settlements, as established by gradual increase of their population. The last time their sway was extended over Asia was ten thousand years ago, when the learned men of the other nations, conceiving that such a rule would increase the general happiness of all, persuaded the rulers of the western nations to submit their differences to arbitration, and themselves to the headship of the prince, or sovereign, or patriarch, of China. But the virtue of the nobles was not equal to the wisdom of the priests, and they soon broke up the confederacy, as it was, rather than an empire. Names and dates in these distant periods cannot have an interest to men so unlearned as those in the body, unless they refer to those of whom tradition has preserved some record. And as I am not prepared to give an outward proof, or test, through this medium, I shall not trench upon the authentic, or partially authentic, historical records. But I will briefly describe the relation that existed between the religious and military aristocracy in the early empires, and the course of descent by which the father of the Jewish nation came.

The military aristocracy was founded later than the priestly. For when Noah left the ark he built an altar, and made sacrifice thereon to God. The patriarch continued to be the chief, or high priest, and the sovereign continued to call assistance from the wisest and most venerable in after times in the performance of religious worship, which was addressed to the One Almighty Creator, whom they knew by revelation originally made to the first man, the tradition of which had descended in the antediluvian population with great purity in one branch, that to which Noah belonged. Noah, too, was a spiritually minded man, and a prophet and seer. He received corrected ideas of the nature of God and of his relationship to Him directly from the spirit world. God continued to raise up spiritually-minded men amongst the early inhabitants of the world, and continued thus to receive a pure worship for a long period.

At last the high priests established rules promoting their dignity and the dignity of their order. This was in their own will, separated then from God and from spiritual communications. The priesthood should be the most humble of mankind, for if they best know God, it should make them most desirous of serving Him and other men than themselves. This desire to serve others will prevent them from assuming to rule, where they need only to serve. It is men's invention, by which they undertake to benefit others by force, or by assuming the leadership, or government. The last exhibition of this change of relationship which occurred in the ancient world, was in Assyria, where, until a very late period, the worship was pure, and the priests humble. Egypt and India departed sooner, and the smaller tribes or peoples scattered abroad over the globe very soon allowed their revelations to become corrupted by love of power. The priests bear rule, for the people will have it so, says the Scripture, and the temptation to the priest to yield is very strong, when he conducts himself with sincere desires for truth and good. But priests, like other men, are fallible. Even holy mediums, like other men, are fallible. The only way they keep themselves in union with truth is by a constant sacrifice of their own wills upon the altar of God's love, and in submission, entire and unwavering, to His slightest spiritual impression, or revelation.

The difference between impression and revelation is this: the former is felt within a man; the latter is manifest to his senses. The one exists merely in ideas and feelings; the other is known by words and actions. The one is the kind generally experienced; the other is extraordinary, and for special purposes. The one will always be found within man, if he will attend to it when he seeks for it; the other will only be found when God has a work for the man to perform towards others. The one, is for the man himself; the other, is for the man's fellow men. The one may lead a man to declare his impressions, but the influence of the man's mind must pervade them; the other will probably lead a man to declare his knowledge, but he will not do it without command, or permission. He will not declare either, if obedient, except in the will of God, and as far as possible in the words given him.

This book is revelation. Most sermons are impressions. The one can greatly instruct the individual, and lead him to benefit, and teach others; the other leads the medium to serve God and others, by specific acts, the form, manner, and time of whose performance God, or His spirit, makes known to the holy medium, who cannot disobey without condemnation. The general life of the one must be more consistent than the other necessarily is, because the one acts generally, and much by his intellect. The other acts specifically and without much, if any, exertion or use of intellect. What, then, is the relation they bear to each other? They are often conjoined. But when separate, the one declares; the other conforms. The one reveals; the other receives, and judges. The one is direct revelation; the other is secondary revelation, or rather the reception of revelation. The one is a holy medium of transmission; the other is a holy medium of reception.

The one may not be benefitted personally; the other must necessarily be. So it is better to receive, than to deliver, when the thing transmitted is from God. But the things of earth are the opposite of heavenly, and it is better to give than to receive from men. The change I make in designating the two kinds of revealment is instructive. In the first part, I call the revelation, *the other*. In the last part, I call the revelation, *the one*. But in the first part, I speak of it as it has been; and in the last part, I allude to it as it will be. Hereafter, direct revelation will be more common than formerly, and the days of general reception of revelation are at hand. These will be happy days for men, when they can receive constant and reliable directions respecting everything in which their temporal or eternal welfare is concerned: when advice and aid will be freely rendered to all who can serve and obey God, or, what is the same, His holy spirits.

But now let us return to our subject, the chronology of mankind after the flood.

The Jews are descended from Abraham. Abraham was a Chaldean, or Assyrian, or a descendant of citizens of the primeval empire of the great and fertile valley of Mesopotamia. For all these names may with propriety be bestowed upon the region in which Terah and his ancestors resided. Abraham left that country, impressed by God with the belief that he should found a mighty nation, and, having settled in Canaan, he cultivated the most friendly relations with its wisest princes. His existence in the body was real, for he was a man, and not a nation, as was Heber, and Terah, and others whose names are mentioned in the Genesis account. Abraham lived many hundred years earlier than chronology generally reckons him to have done. But yet his life was comparatively recent. Egypt records a long line of kings who reigned before Abraham visited that country, and yet, when Abraham was there, reverence for God, as one God, existed in full force, as may be seen by the allusions of Pharoah, or the High Priest Sovereign of Egypt, as recorded in Genesis. He feared God, and feared to do evil to Abraham or his wife because he believed that God required him to dispense justice instead of gratifying his passions. Few absolute kings behave better in these Christian times.

Four hundred and eighty years after the death of Abraham, the descendants of Jacob left Egypt, under the leadership of Moses, as I have already specified, not as a nation, but as a party in favor of the restoration to the people of revelation and religious knowledge. Even now, the pyramids contain the records of the revelation of former ages. The traditions of the Noahic family of man are nowhere else so well preserved as there. Moses knew them all. He was educated in all their learning, and, like every other heir apparent to the Pharaonic throne, was educated as the future High Priest of Egypt. The chief portions of them he embodied in the Book of Genesis. But the disorders of the early Jewish condition, during which they were often subject to the surrounding nations, and oftener plunged as a nation into shameful and odious idolatry and superstition, caused the loss of their fullness, so that the beautiful consistent account he recorded has thus been reduced to a few fragments. My medium is not yet passive enough to let Me write this account, as I would, for restoration, but the time is not distant when it will be discovered in the Pyramid of Ghizeh, known as the Great Pyramid*. Why do I not tell you just where to look? some will say. Because, as I have said, I give through this holy medium no outward proof. Why do You tell them, then, that it will be found soon in a certain pyramid, if You give no outward proof? Because I know you will not take it as any proof when found, for you will attribute the coincidence to chance, or a bold guess. Let us proceed.

Did Joshua march his men about Jericho for seven days, till the wall fell at the sound of his trumpets? Yes. But meanwhile his armies had underworked the walls, and his attentive enemy had watched only his outward maneuvers. Did the sun and moon stand still at his command, or prayer, so that the daylight and moonlight were prolonged? Yes. The sun and the moon were upon the banners of the Canaanites, and by his prayer to God they were brought to a standstill upon Gibeon, and Ajalon. Then the light of day was prolonged by a peculiar kind of zodiacal light, sometimes seen in those regions. The wonderful destruction thus caused in the enemy's army was long remembered, and was connected with the manner of its accomplishment in such a way as most naturally to lead to considering it a stupendous miracle. It would have been more than a miracle, because it would have required a suspension of God's laws of movement in all the space of creation, or an exception to have been established for the earth and its solar system, in which case it would have been equivalent to a new law of God. The record of the creation would then have been incomplete. God could not have been in a state of rest, after man was made a living body and soul. For, by Joshua's request He must have made another law, which would have been, to God, the same as making another creation. For matter was spoken into existence and order by a law.

No, the author of the Pentateuch did not regard it as a miracle of that kind. For he does not mention it as a very extraordinary thing, as it would have been had it involved a new creation of law by God. A miracle is not a departure from God's laws. It is a manifestation of a law of God unknown to man. When Christ healed the sick, restored the lame, the blind, the palsied, and the lunatic, He did not violate, He but exercised, God's laws. He did not use new laws, but applied old ones. He did not fall down in wonder, nor ask those who saw Him to do so. He did them as simple acts of benevolence. And though the same kind of works were often performed by the apostles, before and after His death upon the cross, and by others of the primitive Christian church,

*Cf. David Davidson, of Leeds, England, distinguished engineer, archeologist, and Egyptologist—Ed.

yet none of the workers or witnesses of them thought they saw God's laws, which are His will's manifestations, violated.

Then was not the raising of Lazarus a miracle? Yes, a miracle, but not a violation of God's law. The extraordinary and isolated character of this manifestation of knowledge of God's law leads Me to dwell longer upon this subject, and to relate its circumstances more fully. When Jesus started for Bethany, He did so by a Divine intimation that Lazarus was sick. When He arrived there, the family were weeping for his loss. Jesus asked, Where have you laid him? They conducted Him to the tomb, and He ordered the stone to be removed. His sisters tried to persuaded Jesus that their brother was already corrupted by the commencement of decomposition of the body. Jesus knew better than they did, for He perceived that Lazarus had been buried in a trance. He then called him, saying, Lazarus, come forth, and he came forth, bound hand and foot, and the napkin, as usual with corpses in those days, tied over his head and face. Loose him, and let him go, said Jesus. His life was saved, not restored. He was not dead, but in a trance. How then did he come forth bound hand and foot. The angels of God, the spirits that once dwelt in bodies as men, attended with pleasure to the wants and wishes of Jesus, for Jesus served His Father, and brought His whole life and ministry into entire subjection to, and passiveness before God. He acted always, then, in God's will, and in God's pleasure. Not that God did the will of Jesus, but that Jesus did the will of God. God, then, being willing to have Lazarus continue longer upon the earth in a bodily condition, preserved his life when threatened by disease. He allowed the appearance of death to take place, and still kept him alive, even in the tomb, for three days. Then Jesus came, sent there by God. Then He caused Jesus to pray, and give to God the glory of saving the life of Lazarus. Then God caused the knowledge of this will to be known to spirits, who, delighting to do God's will, brought Lazarus forth by their invisible bodies and strength, and placed him upon his feet, where they sustained him till the attendants in the body obeyed Jesus' command to loose him and let him go.

To all appearance, the dead was raised. But yet there was no such violation of God's order, and law. For God Himself does not for Himself, contradict His own law, or set aside His own resolves. In Him is no shadow of turning. But He foresees all, and provides for all, and for every contingency and emergency. As this was easily foreseen, there could have been no need to violate His own law, much less to allow it to be violated, for the sake of a body more or less in the world. No, it was a miracle, and not a violation of God's laws. It was a manifestation of God's provident care for all His servants, and all His creatures. Not a sparrow falls to the ground without His notice. How then should Lazarus' danger escape His observation? Well, then, miracles are not miracles after all? you ask. No, my friend, I say not so. I say, miracles are actions under, or manifestations of, God's laws, the existence of which is not generally understood, and sometimes not even understood by those who are agents in their performance. So when Jesus declared He could, by prayer, have the aid of more than twelve legions of angels, He knew that God would answer His prayer if it was a consistent one, and that a much larger number than that, (72,000 men would be the usual complement) were constantly about Him, ready and desirous of doing the will of God either by making an outward demonstration, or assisting in a spiritual manner by operations upon the hearts of such men as opened their hearts at the words of encouragement, or warning, spoken by the Holy Jesus.

CHAPTER XXV

JESUS CHRIST

The Continuation of History taken Chronologically upon which Theology has been built

The last points I shall notice in this part of My subject will be the death and resurrection of Jesus, His ascension, and His legacies.

His history in general I gave in Part I. When He knew that He was to be crucified, He informed His disciples of it, and it was then that Peter was rebuked as Satan, the Enemy, or Accuser of his brother. For in accusing Jesus, as he virtually did, of acting in His own will, he derided His inspiration. But Jesus resisted the temptation of shrinking from the horrible death He was directed to, and went on His way peacefully, endeavoring to be useful to His brethren to the last. But in the garden of Gethsemane He departed from that perfect resignation which had previously possessed Him, and in praying for the Father to change His determination, brought upon Himself condemnation, for which He atoned by a descent into the place of departed spirits instead of an ascension to the brightness of a redeemed Son of God. Still, though this unpardonable sin of disobeying, or declining for a time to obey, a known law of God, that is, any expressed will of His, had to be atoned for, yet the atonement was slight compared with the offence, for God knew the great trial it was, and felt that the sacrifice of youthful life and vigor in the body was a painful one. And He, at last, made the sacrifice with dignity, propriety, and resignation. And was not that atonement enough? Not in this case, for though He had performed so many mighty works, and led such a useful and blameless life, yet He had been most highly favored by the aid of another Christ, or Son and Sent, of God, who had inspired and led Him to a knowledge of the power and love of God, and the duties all men owe to Him. And of one to whom much is given, much is required. This is the great reason. His advantages for having willingness to be obedient to God's inspiration, or revelation within Him, were greater than any other. He had started in life in the body with a spirit actuated by a desire to serve God, and be useful to men. He had been born free from bodily lust, and was thus secured from one of the most powerful temptations that assail men in the body. He had been carefully trained by a pious father, and an affectionate mother, who devoted the first twelve years of their union entirely to His service. He was filled with a high and powerful Christ, operating upon His mind from His childhood—that Christ, whose parentage or birthplace I have explained, and who was then already a glorified Son of God, existing at His right hand, and elevated to the seventh circle of the seventh sphere. With these advantages, His fall was more reprehensible than it would have been in another man. Yet God's justice was tempered by mercy, and the sacrifice that Jesus made atoned with God for His sin.

God raised Him from the place of departed spirits and, placing Him in a glorified body, elevated Him above the common laws of matter. Not that the body of Jesus did not corrupt. For all bodies of men are of one flesh, and that flesh is of grass. But there are bodies celestial, and bodies terrestrial. When the terrestrial body was deposited in the tomb of Joseph of Arimathea, it was already dead. But the celestial body partook of its semblance and form, and was a living and a sentient body, composed of spiritual, or highly refined matter. It was, indeed, the same body that the spirit of Jesus had worn upon His soul or spirit, and was changed by the power of God from the corruptible earthy nature to the incorruptible celestial nature. There was then no remaining earthly body. It was changed by God's power, under the operation of laws previously existing, to an incorruptible body, which was essentially spiritual, and was worn by Him afterwards during a considerable period upon the earth, mingling in the sight of men with His disciples, after which it ascended from their sight, and was dispersed in the atmosphere like a cloud.

What then is the reason that other men have never by chance received such a body, under this constantly existing law? Because they have never one of them been so purified from gross desires, and imagined themselves able to bear the change from life to death in accordance with the laws of such a transformation. But Peter is said to have suffered in a similar manner to his great exemplar, whose precepts and example he steadily made his guide, after the ascension of Jesus' purified body. Yes, Peter was crucified at Rome, but not in the manner that Jesus was. He was not so led by the spirit, and so favored by the Christ, or Sent of God. And that he was not so favored was because he would not fully resign himself to its guidance, but continued often to act in his one will, whilst it was always Jesus' will to do the will of His Heavenly Father, except for the one time already mentioned. And before He could receive the purified and glorified celestial body, He had to atone for that departure from God's will. Happily, the atonement was brief; the return to perfect obedience was almost immediate; and when the spirit sought the body again, it had not begun to corrupt. It was still warm with the lately departed life, still capable of receiving the spirit into its ramifications and, being pervaded again by it, was, in accordance with the law still existing, changed from death to life: from corruptible to incorruptible; from earthly to heavenly; from terrestrial to celestial; and from the flesh and blood, derived from the grass, to the heavenly or ethereal particles which form grass, and all other matter which men see, feel, or in any way take cognizance of. All are, in their ultimate particles, invisible. For though chemists still put down in their analysis of bodies of men, or of vegetables, that there is a residuum of earthy matter, irresolvable by them into gaseous compound, or unities, this is because they are unable to carry the analysis to its full extent. They are equally changeable into invisible particles, as if the most solid matter were water, or oil, or pure carbon.

Jesus then was crucified, was dead, was buried. Thirty-eight hours afterwards He rose from the dead, and assumed His celestial body. With that, He journeyed from Jerusalem to Emmaus, visited His assembled disciples in the evening, and displayed His wounded side and hands to the incredulous Thomas, a week afterward. He was afterwards seen as related by John and Paul, and at last was transfigured, from a celestial to a spiritual body, before the eyes of a selected number of followers. What was this transfiguration? The celestial refined particles of terrestrial materials which composed His purified and glorified body, were dispersed in the atmosphere, and assumed the form of a cloud. His spiritual body then remained and by the laws constantly, or often, used by spirits, was made visible to those present. The same spiritual body afterwards was seen by Paul, and its appearance converted him from an opponent to a supporter of the precepts that Jesus had preached when in the body of earth. This spiritual body shone like the sun, for its brightness was commensurate with the exceeding purity and love of His nature. It was a long time before the apostle recovered his vision, though the three did not lose theirs. How was this so different? Because during the interval of time, or eternity, Jesus, in His spiritual form, or body, had progressed rapidly, in ascending, in heaven. He ascended into heaven immediately, but He then ascended *in* heaven, reaching higher and higher in circles and spheres, until, lately, He has arrived at the seventh circle of the seventh sphere.

But then He declared long ago that all power was given Him in heaven and earth! Yes, and He had all power to guide and assist men; to send the Comforter to those who wanted Him; to elevate the thoughts, actions and aspirations of His followers; and to aid them by many outward manifestations. But this you say was not all power. It was all the power He required. It was all the power He desired. For He was still as devoted a servant, as perfect a son of God, as before. He did, not His will, but His Father's. Doing His Father's will, He had His Father's power with Him. He was not having God for His servant, but He was the high, the faithful, the devoted, the ever obedient, ever deeply humble, son of God, and His equally faithful, obedient, humble servant.

In what then is He superior to other spirits? By His obedience in the body, He was endowed with a spirit of progress, which advanced Him while in the body, so that He was qualified for a very high position in the order of spiritual degrees immediately after, or upon, entering the spirit world. He continued thus imbued with this spirit, or habit of progress, so that His advance has been rapid. Being now arrived at the highest circle of the highest sphere, His unity with God is such that He participates in God's action. He shares in His counsel, or reflections. He does this in common with other spirits in the same circle, as I have before explained, in this book. But having now all knowledge, all love, all power, and all thought, or action, He becomes the director of all the spirits, of all the circles, and spheres, as far as the execution of God's will, power, love, and thought, are extended.

He now directs a new, or rather, more constant and visible, proceeding from spirits to men, or, more properly speaking, from spiritual bodies to those yet in earthly bodies, which is designed, first, to awaken men of earth to a knowledge and sure consciousness of the fact that the spirit of man is immortal, that it exists in another state, conscious of its former existence on earth, and retaining its individuality, affections, and character; somewhat modified to be sure, but not, at first, essentially different from its manifestation in the body. Second, the way in which spirits progress in the world to come, from a low state to a higher one, thus giving to man the hope of salvation by an eternal and general law. Third, the particular manner of this progress and what it depends upon. This I am now unfolding through this holy medium, to incite men to virtue and good works. For a belief that salvation is inevitable does lessen a man's care of his efforts, and attention to his duties. And yet happiness results from the performance of duties more than any other act, or acts. But God is pleased to make known to men not only that they shall be saved, but that they shall be saved by works, as well as by mercy. The last is indispensable, but the first is useful, since a speedier arrival at bliss, and elevated circles, depends on them.

This speedy arrival does not shorten enjoyment. Eternity is not lessened because it is sooner entered upon. Neither does a man's spirit have any more enjoyment in the highest, for having dwelt longer in lower circles, because the existence in the body furnishes the state of comparison, not the lower circles of the second sphere. But these lower circles of the second sphere long hold men within them. It is there they most obstinately resist the influences of God's spirits, acting in His will. Then the will they had indulged most on earth continues most active, and its manifestation leads to the exhibition of such representations as Swedenborg witnessed when he was in the spiritual state, except that he mistook some movements as downward which were not so. For there is no retrogression beyond the grave. No repentance, no retrogression. They must

101

either be stationary, or submit their wills so much as to desire to be better, to be improved, to have higher spirits instruct them. Sooner or later all will have this desire. But there are spirits of antediluvians now in the lowest, or first, circle of the second sphere. And yet, eternity is long enough to carry them through the whole remaining forty-one circles, before it ends. It is unending, and at last every spirit will be equally the son of God, and the sharer of His will, power, thought, love and action! This will be the time referred to in the text which God asserted to be as true as that He lives: My spirit shall not always strive with man.

This then must come to pass, and when strife shall have ceased and all shall be united to God, what then? Then they will continue to enjoy all the pleasure which harmony with God and memories of good works can bring. Then it will yet appear that eye hath not seen, ear hath not heard, neither hath it entered into the heart of man to conceive, the bliss that God hath prepared for those that love Him, and do His pleasure. Let us all then, O ye people! love God, and serve Him, seeing that our reward will be so great and so sure. Has the first sphere any temptation to offer that reason can affirm is equal to drawing you away from God? God asks of you and of every man, one great sacrifice. That one sacrifice will reconcile you to Him. That will entitle you to communion with His spirit. That will allow Him to shower upon you blessings unnumbered, innumerable. That will enable you to enjoy the peace the world cannot give, or take away. That will enable you to bear every affliction, every disquietude, that then can approach, with one resigned expression: Not my will, but Thine, O God! be done. After this you can say with Paul, that nothing can separate you from the love of God, and of His Christ, or Sent Spirit. That Sent Spirit will converse with you mentally, even as I converse with this medium. He will help you on all occasions, even as I help this medium on all occasions, either apparently trifling and harmless, or even immoral, yet always affecting the character to the staining or to the purifying of it. It helps on every occasion, too, because the difference between the greatest and the least of men's desires, or actions, is as nothing compared with the employment of influencing God's whole creation, which, with His higher spirits, can be as easily done. God sees the sparrow fall, and has numbered the very hairs of your heads.

> To him, no high, no low, no great, no small,
> He fills, he bounds, connects, and equals, all.*

Pope, of all the poets, arrived most nearly at an appreciation of God's relation to man. His Essay on Man abounds with beauties and truths. You will find it a profitable study, as you advance in your spiritual belief. But yet he places the self of man too much as his motive of action. Not that man has not so acted, but that poesy should not lend its approving numbers to any such low motive as self interest.

*Note by the author—not by the holy medium: This quotation is from Glacomous, who lived in Thrace when I was on earth. Given during the revision, July 25, 1936.

102

Let us pray

O God! let us be guided by Thee. Oh, let us be Thy willing servants, submissive to every intimation of Thy will, every intimation of our work. O God! Thou who knowest all things, grant such of our wants to us as Thy will may not be opposed to, and so far as Thy love may not withhold from gratification for us, to our good. For in Thee, O God! I will trust, and my portion shall be Thy pleasure, and Thy will. O God! Father Almighty! hear the prayer of Thy sensuous and material subject, and raise me to the dignity of being Thy servant in spiritual matters, Thy follower, and Thy son, in all things. O God! let me be taught to praise Thee, and to glorify Thy name, for Thou, O God! art worthy to be praised, and without Thee there is no Savior, or Redeemer. For what is man, O Lord! that Thou art mindful of him, or the son of man, that Thou regardest him? Thou hast created him a little lower than the angels, and hast raised him to glory, and honor! O God! help me to make the only sacrifice that delights Thee, that of the heart, for my heart, O God! is desperately wicked, and there is no health in me except as Thou bestowest on me strength and life. O God! let me be Thy servant, amongst Thy servants, for I am convinced it is better to be a mere doorkeeper in Thy house than to be a guest, and an honored one, of the world's revered or loved idols of flesh. Let me, O God, be united to Thee in the bond of unity, as perfectly as my nature will permit, and let me, O God! be very near to Thy love, and to Thy Son's love, and to union with Thy high and holy sons, and servants, in all Thy heavenly circles. O God! be very good, I pray Thee, and very kind and merciful; for I am a sinner, and there is no power in me to be purified, except through Thy laws, which are the sure and faithful harbingers of Thy mercy, and which lead me surely and truly and inevitably to Thy feet, and leave me there, rejoicing in the supreme happiness of being Thy son. O God! be ever present, and let me not forget Thy presence, but let me think of Thee often, and let my life be devoted to Thy service, and my death be a triumphant passage from works to rewards. And to Thee shall be all praise, honor, glory, and high renown, now and forever, and forever and forever. Amen.

CHAPTER XXVI

EXPLANATIONS OF PROPHECY

Continuation of explanations of prophetical declarations of Hebrew prophets, some of whom wrote from revelation, but most of them from impression.

In Daniel's day, geography was not well understood. The deserts east of Persia or Babylonia, the wild tribes of Scythians on the north, the sea, or salt water, on the south, and Egypt, Greece, and

Thrace on the west, bounded the known parts of the earth. In earlier days some intercourse with China took place, but then, almost none. With India the general relation was hostile, and without much activity. Commerce scarcely existed, with any eastern country, and the confused and contradictory accounts, which from time to time excited a passing emotion of wonder, failed to awaken a desire to know more of countries with which they possessed no common interest, and made no exchange of products.

In Nebuchadnezzar's dream, then, was represented the whole earth that he knew of. And Daniel gave a true interpretation of the dream. Still, there have since been in those regions two or more great empires which were unnoticed in the prophecy, the empires of the Sassanides, and of the Parthians. The irruptions of the Tartars were too transient in their effects to deserve the name of empire. But these latter empires were excluded, as having no relations to Christianity or to Israel. The former was the great coming event to which all prophecy in Jewish annals turned and looked forward as the great crowning, glorious time of joyful reigns, of peaceful and happy kings, of peaceful and happy people, living in the enjoyment of divine favor and intercourse.

This time is yet to come, and it will come, for the prophecy was sure but dark. In general the prophecies were made by impression. By that I mean that the prophet would declare the glories of the future kingdom and people of God, without a knowledge, or impression, of time and place. Still his outward associations would lead him generally to refer to Jerusalem and Judea and the Jewish nation, those impressions of the future manifestations of God's favor and love, which are, indeed, for them only in common with all other nations, kingdoms, tongues, and people, upon the whole circumference of this earth. This time is nearer than it was, as is obviously necessary. But, yet, its full fruition is very distant. Still the preparations are more apparent, the dawnings of its day more evident, the signs of its near approach more visible, than any other age ever even imagined themselves to perceive. In this age its progress will become so evident to all observers that the glorious name of God will be more called on than ever before, since the early religions lost their purity.

The kingdoms of this world will not become the kingdoms of the Lord Jesus Christ now, or in this age. But the fifth monarchy, or kingdom, will be evidently established in it. And though the kingdom will have no king outwardly, it will all the more resemble the Jewish polity, as originally established. It will all the more resemble the reign of the saints of the Most High, of whom it is declared that they shall take the kingdom. And of the power, and dominion, of their successor, there shall be no end.

Who, then, is ready to place himself under the government of the Lord and Savior, Jesus Christ? For in this way shall His kingdom begin, by the adhesion of one man to His government, and the addition of more and more, singly or by scores, until at last all men shall own Him as Lord and as God. How is this? You say. Will He indeed be God? No. I only say He will be recognized as God. And will He not be recognized in His true character? Certainly. He is as God, being one with God in power, will, honor, glory and action. But He is not God, though we may regard Him as God, because He is as God, though He is not God, and though we ought not to call Him God. He is the son of God, in unity with God, to whom God the Father has given all power in heaven and in all creation. He is, however, the servant of God, and His meat and His drink is to do the will of His and our Father. Now He is the highest spirit whose body was of this globe, and none will ever be higher than He is now. He will never be higher than He is now, but others will be as high, and He will welcome them with joy, as joint heirs with Him, to the glory, honor, praise, will, love, and action of God. Glory be to God, and to His Son, the Lord Jesus Christ, for He is worthy to receive glory, honor, praise, thanksgiving, and power, now and forever, world beyond end.

Then, when men regard Him as God, it will be because they will receive His counsel, advice, warning, command, will, power, action, and revelation of these, as being each of God. This is what men should do now, and what they must do, to place themselves under His government, by which they will forward the extension of His power, and raise themselves to His nature and harmony. If we are one with Him, we are also one with God. And unless we are one with Him, we cannot be one with God. And unless we are in union and harmony with Him and God, we are in our own will, and acting independent of Him and of God, to a greater or less extent.

It is this will, man's free-will, which keeps him away from God and Christ. It is that which I ask man to sacrifice to God's Will; one sacrifice, and that only, will reconcile him to God, and to Christ. That sacrifice man will make eventually, and can make here much more easily than in the state to come. There, or here, it will be made. Here you can make it easily; there with difficulty. If made here, it will place you higher there. If made here, your enjoyment here will be far more constant and greater, and will continue in the state, or world to come, to be higher, purer, and more perfect.

This is the condition I urge you to press forward to obtain. Now is the accepted time; now is the day of salvation. Now, declare yourself on the side of God in the world. Now, be willing to be His servant. Now, resolve whilst you feel an impulse in your heart. Fear not a delusion. Fear not the sneers, or the carping, or the taunts of the world. But fear God, who is liable to be separated from you by your own act. Fear yourself, your own will, that may be to you an accuser and an enemy, if you submit it to worldly desire, and dwell in subjection to such desires. Be sure to know your own heart; listen to your own reason. The former will assure you no happiness is found separate from God, and the latter will make you conclude that no one can bestow upon you so much help as God; that God loves to help His creatures; that God will do what He loves to do; and that, therefore, you will receive from Him abundantly, from His inexhaustible store. He leaves you your free-will only because without it you would have no responsibility, no individuality, no separate existence. But He has given you full power to become the Son of God, the joint heir of Christ, and this power you must exercise to become so. He will not force you; He only persuades. He appeals to your reason, to your affections, to your self-love. He asks you to do it, for your own sake, and for the sake of your fellow men. He will be God, and He will be happy, whether you do it or not. But you will not be God's son, neither will you be happy, until you do it. Choose now, O son of man! whether you will now be a son of God. Be wise to-day, for you know not what a day may bring forth. Be wise now, for now is the accepted time. Now, God calls you through this writing.

Now resolve, for if you lay down the book, and neglect My appeal, how will you again arouse in your heart and mind the same earnest desire and hope that you now feel! Say not that you cannot, and that you must take time to consider. Say not that you would, if you could, for all that is needed is that you make the prayer I wrote for you in the twenty-first chapter with heartfelt sincerity, and with an earnest endeavor to make it your own, and to enter into its spirit, and to make the one great sacrifice, that of your FREE-WILL.

CHAPTER XXVII

SALVATION

The Surety of the Salvation of Men

When the foundations of the worlds were laid, man was ordained to be. When he was ordained to be, he was fixed by certain barriers, of which prominent events are a part. God resolved to have certain effects follow in due course, and He established such laws for the government of mankind, and placed in them such constitutions of action, as would secure the accomplishment of His designs. What were these prominent events that God foreordained? They were that man should progress from pure animal to pure spiritual nature. That in the course of that progression, he should grow in the body, die in the body, be saved in the body, be born in the spirit, and be saved in the spirit with an eternal salvation.

All men have sinned, says the Apostle Paul, and come short of the glory of God. This is death in the body, from which even Jesus of Nazareth, the purest and holiest of men, was not exempt, as I have before shown. But are all men saved in the body? This, though not so evidently declared in the scriptures, nor so plain to the observation of mankind, is yet evident on inquiry. The most wicked and ungodly men have their moments of compunction, their periods of remorse, their time of repentance. True, they did not continue and bear fruit, but nevertheless the genuine repentance and remorse and regret was there. It is this which secures salvation, when conjoined with a resolution, however futile, to sacrifice their will to God, or to good works, if they know not God. Why then do not more continue in this state of salvation? Because if they go again into the ways of evil, many are again so strongly tempted that they fall immediately from grace. They depart from the good resolution, and the last state of that man is worse than the first, because his chance or power of reformation is lessened. His resolutions become weaker and weaker, the oftener they are broken.

Look upon mankind, by looking in your own heart, and see if it be not so. And look upon your own broken good resolutions, and then say, if you can, that there lives a single, solitary man who has not also formed at least one resolution, organized at least one effort, to save himself from the dark descent into wickedness and crime, which engulfs so many who might as easily be angels of light, if they would only sacrifice to God what they sacrifice to pride. And that is their free-will. Yes, these workers of iniquity do not enjoy a freedom of will, any more than he who sacrifices his will upon the altar of duty to God. But there is this great difference, he who submits to God secures a helper in every time of need, a friend in every difficulty, a savior in every trial, a protector in every emergency, a guide in every doubtful place, a redeemer from every sin, a loving, kind, affectionate and ever faithful Father in every period of a long life, and a bestower of good gifts upon His children, bountifully, cheerfully, without reproaches or upbraiding. Can any man say that for his companions in the world, for whom he has made sacrifices, for whom he has toiled, or wasted his time? No. And if he has sacrificed to some principle of pride, or ambition, or love of possession of earthly things, has not his reward been earthly? perishable? unsatisfactory? No, my friend, be assured by one who, looking over the past and the present, beholds all the sons of men existing, or having existed, at a glance; who knows all they have hoped, and feared, all they have longed for, possessed, or enjoyed; who, having had the power to select one truly consistent man for a promotion to a high state, or lofty office, has never yet been able to perceive one such in the whole course of man's existence. Then all have sinned, all have repented; all have fallen, and all have been raised; and now, I have already declared to you that all shall be saved, and given you unanswerable reasons for it.

But, how shall I show you that the salvation is eternal? I have shown you that God is good; that all men are equal before Him, and that His justice and mercy combine to save man from destruction, or continuance in evil. I have shown you that all will be saved, and that none will ever be destroyed. Now if God is good, and none are destroyed, and there comes a time when God's spirit does not strive with man, then all must continue in grace, or else must be annihilated; for good and evil cannot dwell together in peace and harmony, neither will your intuition require any argument to sustain this position. But I have told you that the last shall be first, and the first, last; and I have not shown you that the wicked man is not the first man; that is, thus to be the last, as some believe.

Well, then, let me call again to your mind that God has no pleasure in the death of a sinner, and see whether God will re-condemn men who have been once saved, unless they fall again. And if God, only, sustains the created worlds and He alone saves sinners, who else can draw them from Him, and keep Him ever striving with man? For man is necessarily to be saved, or God must continue to strive with sinners. For man will be saved, if God has no pleasure in the death of the sinner, for God's pleasure must be the end of man's creation. God created man for His pleasure, as I declared in Part I, and, having created him, He declared him good. Now evil is not the consequence of good. And, if evil is not the consequence of good, it cannot be more than accident, that good is contemporary with it in the soul of man. Not all the efforts of all men can save one other man, unless he is willing to be saved. Not all the efforts of all spirits of men can save one man, unless he is willing to be saved. Not God, Himself, can save a man, unless he is willing to be saved, unless He establishes a new law. No; willingness to be saved is the first requisite from all men, each for himself. The desire being manifested, God helps; all spirits in unity with God help; and all good men help. But can none of these do anything to make the man want to be saved? Yes, that we can do. That any man may attempt. But the higher the intelligence that directs the effort, the more understandingly and the more effectually will it be made. At last such efforts will succeed, because in the spirit world all that is established in the spirit's favor is retained. Evil no longer approaches or tempts him. The evil he has, works within him, leading him to many absurdities, the greatest of which is his refusal to submit to God's will. But it can never extend its power, and the slightest secession of the perseverance of the spirit in its own will, admits the further advance of good intentions and desires.

That there are spirits confirmed in hardened wickedness, is evident from what I have said of some in the first and second circles of the second sphere. But those of this obstinate kind are few, comparatively speaking, and even these, must, at last, see their own want of more happiness, and that only God can confer that addition. When they have made this discovery, progress will take place. But their progress will be exceedingly slow. Their free-will will combat it in every new position they find themselves in. They will persuade themselves that they are happy enough in each advance, without ever taking another, and they will refuse to go on, until, again and again, new efforts are made to convince and assure them that greater happiness may be attained by another, or further, submission of their still free-will to God. But, then, do they part forever with their free-will by this sacrifice of it? Do they irrecoverably lessen it at every submission? Yes, they do. The highest spirits no longer have free-will. How then is their individuality, and their separativeness, maintained? By the law of God, adapted to their condition, they are separated unto themselves; they are the sons of God, and, being joint heirs of all His power and goodness, they receive power to become the sons of God, and be the sons of men. This power is given in the body, and is a necessary consequence or reward of the sacrifice of their will.

But have they still a responsibility for their acts, when they have thus submitted to God's will! Not the least. God's will is that in which they act. The responsibility is His. He is responsible only to Himself, and that is only that He cannot contradict His own nature. The spirit or the man has no responsibility for God's acts, nor for acts performed entirely in His will. They have submitted to Him, and are His servants, or His sons. But they have still the responsibility of maintaining and completing this submission. What responsibility can that be, when they cannot withdraw the submission already made? It is a responsibility continually lessening as they advance. At last it ceases, because the spirit is entirely submitted to God, and perfectly united to Him. It is indeed impossible for them to withdraw any submission made, but that impossibility arises from their own will's being so far united with God's as to have, that far, no affinity, or desire for evil of the kind or nature surrendered by that submission. The sacrifice of a desire for a particular kind of evil, once made, is made forever, by spirits. By those in the body it has to be repeated, though each sacrifice is lessened by repetition. Your own experience tells you this. Indulgence in evil desires strengthens them. Perseverance in good works makes them easier and easier of performance.

What, then, remains of the independence of man's spirit when the entire sacrifice of his will has been made, and his perfect unity with God secured? There is still his nature derived from God, endowed with memory of its former independence, and all its experience as an individual, through the whole course of its existence from paradise to the seventh circle of the seventh sphere. This furnishes the spirit with an individuality, and is a self-consciousness that it can never lose. It feels and knows that it was itself, as a separate and free-willed being that performed those actions, thought those thoughts, and imagined and wondered and labored in contests of will with God, till God became all in all.

This memory, or consciousness, it never loses. It is that which makes itself, and but for that it would be only a part of God, without any separate thought from God; or itself, as itself, would be God; and all the spirits, united to the God from whom they proceeded originally, would be God—the same God of whom they originally constituted a part. But, as God is infinite, the separation of a part does not lessen His whole. And as God is in all, so are all in God.

This part of our subject demands the very highest exercise of man's reason and inspiration to understand it, for it is the very extreme limit to which he can proceed in his idea, or conception, of infinity. I say, then, that God is all in all. That the spirits are of Him, as they were of Him before separation and placing in paradise, except that they have added to their Divine nature the consciousness, or memory, of all their acts, thoughts, or imaginings experienced, or performed, in that interval of eternity. They then continue to live, sons of God, one with God in power, one in glory, one in honor, one in love, one in will, one in action. But they are separate in knowledge of the past, or memory of their separate existences. This they never lose; neither can they ever lose it without annihilation, which could be only an annihilation of their memory, or knowledge of the past, as separate beings. That annihilation would leave them united to God, as if they had never been separated from Him. But can you suppose God would so confer individuality upon proceedings from, or portions of Himself, to take pleasure at last in demolishing the work of His laws, continued through such an almost infinite period? Can you suppose He would do it if it would be a matter of indifference to Him? And how could it be a matter of indifference to Him that so many myriads of myriads of beings from each of the myriads of myriads of His various orbs, or combinations of orbs already faintly alluded to and shadowed forth, should cease to have a separate existence after they had, by their struggles and repentances, at last been led to partake of His unity of love and power? After they had been led to raise themselves to be His sons, and after He had manifested to them the love of a father? No. We may confidently say, God will never annihilate His spiritual creation, not because He cannot, but because He cannot without a sacrifice to Himself, and we know that that would be contrary to His nature to make, and, in that partial sense, impossible for Him to will it made. He lacks, then, not the absolute power, but the will, or inclination, and without the last, the former is never, and will never be, exercised. It is absurd to suppose the contrary, or any other view.

And reason and revelation will always assert the truth of all I have declared to you, because it is all true, and reason is true, and revelation is true. Now reason is not infallible, because men reason in their own will, but revelation is infallible, because it comes from God. But when revelation is mixed with reason, it is often obscured; and when it is given through men, it is often distorted by the medium through which it passes, unless that medium is one perfectly passive and submissive to the holy encircling influences of God's high and lofty sons. When the holy medium is so submissive and passive to their, or God's, influence, and actuated solely by a desire to do God's will, the revelation will be like its source, and will pass through him perfectly, and in perfection.

This holy medium is the best we have now amongst the sons of Earth; but better could exist, and a better has existed. But yet he is a good medium, and though I may not deem him worthy, or qualified, to transmit all I would willingly make known to men, yet I guard against the transmission of error through him, or his perversion of truth, to speak more exactly, by revealing only what he can perfectly and truthfully transmit. To do this, I am compelled to depart from the orderly arrangement and discussion I would prefer, and which I attempted in the commencement of this book. Yet I have written it so it can be read with profit and instruction, and can be rearranged to the understanding of the reader who gives it repeated perusals. And who will not read often, and carefully, a book written by a son of God, though the holy medium of transmission be but a humble and unheard-of individual. This part, as well as the first, is beyond his material power of mind to compose, even with all the aid that time and preparation, and libraries, and leisure, could bestow. And without these, it has been written in the midst of such disturbances as are generally thought most hostile to careful and correct composition, and, like the first, is submitted to the printer in original draft or copy, with a perfection of preparedness for publication as rare as it is commendable.

CHAPTER XXVIII

REVELATIONS: HOLY MEDIUMS

Explanations of many Bible passages, or portions

The Divine Influx has proceeded at all times and seasons to spirits created or unformed, but set apart from God. In paradise, in the body, in heaven, or in hell, it has proceeded from God, through such holy mediums as He has appointed by His laws, until it has rested upon and vivified, or saved, the spirits of His creation or the emanated spirits, from their own wills, and from the effects of the indulgence of their own wills. It is now this procedure or influx of, which I am about to give a history.

Which of the spirits in Paradise were first chosen to appear on earth? For we will leave the history of the Divine Influx to the beings of other globes, or worlds of matter, for a future revelation, which may or may not be given through this holy medium, whilst he is in the body. I have much for him to do, and he has but a short time to remain. However, he shall perform cheerfully, or not at all, all the work I have to impose upon him, and his reward will be as much as if he had done nothing, for doing all. But, as you may easily understand from what has been written before, he will receive his reward sooner than if he does nothing now. How, then, shall I describe to you what I design to do if he refuses, or neglects, or becomes tired, or from any cause puts his will in opposition to God's! I shall then find some other medium, whom I will undertake to bring to the same point of submission, or even to carry further. I am now trying to obtain such a one, because not only is his life uncertain, but also his will. Yet his nature is such that I am convinced he will not be swerved to the right nor to the left by threats or persuasions of friends or enemies, by urgent remonstrance, or pitiful tears. I know something of him, and of what he will do, by knowing fully what he has done, and what he intends to do. I know how circumstances of his previous experience have affected him, and I know with what desire he has sought for Divine Influx, in various periods most strenuously, but at all times since his maturity, with ardent wishes. My servants, in acting under impressions from Divine Influx, have often assured him publicly and privately that he would be called on to do God's work. He always scouted the idea, and declared his opposition to their forms of expectation.

Now those same servants of Mine will, I doubt not, most generally declare him departed from Me, and acting under delusion. But I will at last impress them, if they will go down into Jordan, or into the humble and life-giving influence of submission to God, with the fact that he declares My will and revelation, and that My will and revelation is Divine, because it is in perfect harmony with the will of God. But all that I shall require of them is obedience to the impressions I give them, and though that may be a hard call for them to comply with, yet they shall not have peace until they do comply with it. So, of him I shall require obedience without reference to results, and he may not see any considerable number of believers in him or in his Divine Authority, yet he must persevere in doing his duty, regardless of the small progress he may perceive to be made, by the manifestations or sacrifices or revelations he may make.

What can be more discreditable than to be despised by your own generation? It is far more discreditable to be despised, or condemned, by God, and His holy spirits. Fear not those who can kill the body, but fear the displeasure or the offending of Him who can cast both body and soul into hell, or hades, or the place of departed spirits.

Well, does not this construction conflict with that given by Me in the first part? Yes, I gave that as an outward signification. I gave that thus, in effect, though perhaps you did not fully understand it: Fear not disease, nor the assassin, but fear the magistrate, who would condemn you for cause, and cast your body into the outward fire of Gehenna, as was formerly the usage with criminals. Yea, fear him, for he would thus punish you for cause, for wickedness, for which you would have to atone in the life, or state, to come.

Now I will explain the same passage spiritually. Fear not those who can kill the body. That is an outward death, that does not affect the spirit. Its existence proceeds harmoniously as ever before, and it assumes in the world to come that position to which its progress towards submission to God entitles it, and qualifies it to receive. Fear those, however, those sins, those acts, those companions, who kill both body and spirit by leading the soul or spirit from God and out of submission, or passiveness to His Divine Influx. Yea, fear them, for they will cast your souls into the next world, or into the place of departed spirits, in lower degree than you might otherwise have attained, where there is weeping, and wailing, and gnashing of teeth; where their worm dieth not. Yea, fear them, for verily I say ye shall not come out of that condition until you have made restitution to God of that submission to which He is entitled. To that power you must at last submit, and until you do submit you will often and again find disappointment in the realization of the very object for which you may have long striven; you will wail over your own success in evil; you will gnash your teeth to find that what had been so hardly obtained was so unsatisfactory; that the love of evil, or of your own will, brings only affliction, and that you are in the outer circle of darkness, as compared with the innermost, or God-loving center, where all is light and glory; where all is peace, and harmony; where the wicked cease from troubling, and the weary of contention are at rest. This is true, the unchangeable interpretation of this text, whether viewed outwardly or inwardly.

The Bible has in this manner two significations, or uses; one outward and evident to observation, the other, only revealed by the spirit, to those who search for it earnestly, and submissively seeking for truth. Those who want only to support a favorite theory cannot arrive at a true interpretation, but the wise of this world shall be confounded by babes and sucklings, for out of their mouths hath God perfected praise. This means in the outward, that God will speak to men through holy mediums they despise for their want of education and intelligence, and that they will declare the perfect things of God. But its inner, or spiritual, signification is that the mouths of the most submissive and passive of the human family shall be used to declare the praises of God, the eternal truths relating to the future world, and the undying, and undieable truths which will become the joy of the higher and nobler natures, in and out of the body.

What, then, shall this man do? What is that to thee? If I will that he continues till I come, what is that to thee? said Jesus to Peter in reference to John. John was the beloved disciple. Peter was the disciple who had thrice in one morning denied his master. Think you Peter was preferred to John? Or was not Peter rebuked, by being set to tend the sheep and lambs of the flock, whilst John was set apart for gospel writing and revelation? Yes; John, the beloved disciple, did write the book of Revelation, as it is called. But it is not a revelation of anything until it is interpreted. This I will endeavor to do, if My holy medium will steer clear of his own will and leave Me free to act in My own. He will try, but only those who have experienced the operation can tell what the difficulty is of having no will, when novelties on most interesting and important subjects are being unfolded; and we fear, slightly, to be sure, yet measurably, that reason may fail to receive the explanation, or the statement, as being in strict accordance with herself. Still, as before, I will proceed carefully and declare what his will submits to whilst it submits.

For man's acts cannot be foreseen, even by God, as far as those acts are acts of will. But as far as they are the effect of circumstances under which the man exists, God and his higher spirits know what man will desire to do, and how far he can obtain his desire. These are the limits of prophecy, which have puzzled many. For man's free-will was plainly expressed, and understood to be a fact. So, too, was the power of prophecy as dependent upon a foreknowledge of God. How to reconcile these equally and fully acknowledged facts, or truths, has been the anxious study of many well-meaning but unreasoning minds. For true reason would have told them that God could make it known, and that reason by man cannot find out God. Now I will commence to explain to you, in as orderly an arrangement as My holy medium's submission will permit, the revelation of John the Divine, as he was called, because he had so large a portion, or influx of the Divine Spirit.

I, John, was in the spirit on the Lord's day; that is, the day set apart by the Lord for My attempting to record His revelation, and I received a direction to write to the seven churches which were in Asia, the churches of Smyrna, Pergamus, Ephesus, Laodicea, etc. These churches were the assemblies of believers, without reference to any professed membership or acknowledged, or known, heads, or organization. The angels, or messengers, were those appointed to express God's will to them, whether or not ordained by men. To these John wrote by My direction and inspiration. To these he made the warnings and promises. To these he declared I would come quickly, or I would bring destruction. To these he declared I would give gifts, or withdraw presence. And were these the churches of those particular cities at that particular time? Oh no! Those warnings were given for all churches at all times, in all parts of Asia and in all parts of the earth, both in the early and in the latter time, or times. Behold, I will come as a thief in the night, and if you would be ready, you must be ever watchful.

The churches in all the world were beginning to show signs of that heresy, or corruption, alluded to under the name of that woman Jezebel. This was Manicheanism. It was the belief that God was not merely the author of good, but the creator of evil. This was a remarkable error for men to fall into whilst they yet had those present who had received the outward preaching of Jesus, and of Paul, and of other divinely inspired men; who had yet amongst them the Divine John, and many of whose members yet received the baptism of the Holy Spirit and of the fire of God's love, which continued to purify and consume the evil of their natures.

The next important error that assailed the church was that of the power of the clergy, or bishops, as they styled themselves. They no longer left the church free to receive or reject them, but imposed themselves on the people, and insisted on being kept in authority as long as they continued to exist in the body, and they assumed that in them resided a superior manifestation of God's power, and a higher knowledge of His will. In some instances this was true, for in many respects the Apostolic churches were weak and faithless. But the bishops established a general rule and forced a general compliance, under the threat of censure by the collective representation of a large body of churches. Another error against which John warned them, without much effect, was that of the laboring compensation being made dependent upon the size of the congregation. For he assured them that the church's candlestick should be removed unless the angel did the first works, and the first works were preaching to the Gentiles in faith and spirit, and without settled salary or compensation. They had their support guaranteed to them by God, and it was their duty to trust to Him for the whole of that support, which they strove to secure by donations of land and settlements of money upon themselves.

The next error, or heresy, was a love of power in the state, to which the angels or bishops already aspired, and which the pagan authority of the state was willing to confer on them for the sake of their assistance and their return of ease of government and collection of taxes, which was thereby ensured to it. The governors of provinces were glad to appoint a Christian bishop prefect of a city, or chief magistrate in a colony, if thereby their revenue and their established authority were secured to them without care on their part. And thus it was that jealousy of the power of the church became the cause of persecution when the governor, on the other hand, did not appoint the bishop to this authority, but required him still to secure the collection of the revenue and the tranquillity of his followers. While in this way the church was corrupted, it suffered numerous persecutions which led to rebellions against the authority of the bishops, which induced them to insist the more strongly on their divine right of governing the church temporally, and directing in all ways the temporalities and spiritualities of the assemblies of believers and the whole body of professed Christians.

There yet remains for me to notice the one error consequent upon these. It was a placing of the outward before the inward; a watching for outward evidences and works from believers, rather than the spiritual or inward evidence of faith, and declarations of creeds commenced in this way. At first in the church as instituted or accepted by the apostles and evangelists, men were only asked to profess a belief that Jesus was the Messiah; that is, the Sent of God prophesied of in olden time, and expected in His own time by the world in general. But it was even now that the Apostles' creed was manufactured as given in Part I, and as it continues to the present day, with a very slight alteration or addition.

Can it be possible, you ask, that this could be an error and be sanctioned by the Apostles and quoted as your faith in this Book? Yes, it is an error for all that. Because, though founded in truth and true in itself, it leads and has led to error and misconstruction. I admit its truth and desire to explain it to you, not that you may continue to use it, but that you may be able to dispense with it, and rely on the higher and nobler manifestation of doctrine and direction that I will afford to every man who opens the door of his heart to Me. I will preach to him personally, spiritually, if he will only hear Me and seek to know the truth; for I am and will be the Comforter to all who receive Me as an emanation from God, or as God's Son and Sent, or as Christ, or as the Holy Spirit. I am not particular about names, and if you only call sincerely for any of these I will answer, and if you follow My directions I will guide you and direct you in every time of doubt or perplexity; in every time of trial I will be your consoler and your counsellor; in every period of difficulty I will bear your burden, and comfort you in every time of sorrow. All this I will do without fail if you will rely on Me. For this is My mission from God. This is the work to which He has appointed Me, and not Me only, but every spirit that shall hereafter be the Son of God in the same manner and degree that I am.

I called Myself John in the beginning of this chapter not because that was My name in the body, but because My servant John acted for Me in writing the Book of Revelation and united with Me in explaining now what he did not fully understand. Besides, he is a high son of God, being in the sixth circle of the sixth sphere. He is a noble spirit who delights to serve God, and who did reveal himself to My clairvoyant spirit Davis when he was submissive to the directions he received as a clairvoyant, and was content to follow them without ambition or sordid desires. But his unity with him ceased when Davis left the control of himself to men of other motives, and it can never be renewed whilst he continues in his present state of rebellion. It is true that I permit him to write many truths, and that I allow spirits in the first, second, and third spheres to influence or direct him, but they are not allowed even to declare all they know of Me to him, because he rebels against My authority and seeks to elevate wisdom above love, and will above action. The only way for him to become a truthful holy medium is to return to his former subjection to the Divine John, and he can do that only by returning to the state from which he departed, when he was kept in subjection to the interior and holy directions he received in his clairvoyant and unconscious state. Whereas since he has been used in the will of those around him, until he was permitted to use himself in his own will. His impressions have been overruled to be a benefit, and a foundation for belief to many. They have been so guided as to be the means of releasing many from bondage to tradition, and from worship of idols of flesh which men have delighted to worship ever since the foundation of the error, or heresy, was laid in the apostolic times referred to in My revelation through John the Divine.

This will surprise many who have almost begun to worship Davis, and others who have honored him as a guide. Many spiritual believers, too, will ask how it can be that he is wrong, when so many spirits have, by outward declarations through rappings and writings, asserted that his works were in the main true, and that believers or inquirers should read them. This was because the works of Davis lead the mind to repose on itself and disencumber it of prejudice, and leave it in a fit state to receive further revelation. It is a great step gained when mind in the body is prepared to receive with favor higher and further revelation. This is the proper effect of Davis' book, and I can assure all that no believer in the Bible as founded on revelation has ever been led out of

116

that belief by anything that Davis has written; no believer in the efficacy of prayer has ever ceased to believe in it or refrained from it, because he has declared it cannot move or affect the Deity.

It is true that prayer of itself does not move God, and it never can change His laws or His order. But it makes the sacrifice of the heart and of man's free-will to God, which by His law He requires of man as the means or preparation for the entrance of His spirits, high, holy, and pure into the corrupt and corrupting free-will established there. The work of purification goes on so long as there is a continuance of submission and sacrifice by prayer, or in any way, to God; as long as there remains an unswept corner in the heart, and until the man is so purified as to be a fit residence for God's spirit. When his heart has become the home of the spirit, the spirit leads and guides him into peace, that peace which passes all understanding and which can never be taken away by men or spirits, by devils or the world of evil. This it is that prayer does. And now I must tell you that you need not assume any particular position nor go to any particular place to do it. Be willing to join in others' prayers in the way most agreeable to them, if they pray with sincerity and for good. Be desirous, for yourself, to pray without ceasing, which you may do if every aspiration and thought of your mind be brought into subjection to God, and an entire willingness to submit wholly to His guidance and direction be established in your heart and mind. It is to this I earnestly call and sincerely urge you, O child of Earth who desires to be speedily a Son of God.

How, then, were the revelations of John or through John to be made useful to the church, when neither he nor they fully understand them? The words of his book were to be sealed up unto the time of the end. So were Daniel's. Neither of these revelations was understood by the prophets or by the hearers of the prophecy. Each knew, though, that the source was divine—the writing or words given by Divine Influx. Each submissively received and recorded the words given by the Holy Spirit to him, and though both asked for knowledge of the meaning of what they had written, neither of them received it. Daniel was assured that he should stand in his place in the last day, and John was told he should stand in his. They stand now side by side in the same circle of the same sphere. And the same union that exists in their prophecies now exists and will continue to exist in their spirits, which together have progressed from the fourth sphere, and together will progress to the highest circle of the highest sphere.

Were it necessary, I would confirm the truth of My revelation by miracles such as raising the dead, or healing the sick. But the time for these has not yet come. When the time comes it will be done, and through this holy medium first. But some will say, Why do you not give us a sign now? This book is a sign, and he who is wise will so regard it. Blessed are they who believe, but more blessed are they who believe, not because of outward signs, but by

117

internal evidence—who believe because the witnessing spirit within themselves declares to them the truth of revelations made to others for them. This will be the case with many; with all who desire primarily and principally the establishment of true notions of God and man and the future. Those who desire other things before this evidence will not receive it, for receiving it they would reject it and thereby commit the same sin the Jews did who rejected Jesus Christ, the Savior sent to them with power to work miracles and to teach as no man taught. This unpardonable sin I wish you to avoid, and I will not force it upon you by giving the internal declaration of your own spirit to you when you desire before all else the establishment of your own theory, your own church, your own doctrine, your own good temporally, or your sole good spiritually. The true worshipper worships the Father in spirit and in truth, and the Father will have such to worship Him. But he who blinds himself to the fact that God is as great now as ever, that He is omniscient and omnipresent as ever, that He is as willing to save men from sin as ever, that He is indeed unchanged as He is unchangeable, that He is now as ever willing to be His people's guide, by night and by day, and as ready to declare Himself openly or by outward sign as ever, can never experience or enjoy that peace which God bestows upon those who receive Him in the way of His coming. Too many ask Him to come in their way. Too many are exclaiming without authority, Lo! here is Christ, or, Lo! He is there! Whereas He is in you except ye be reprobate. And if you be reprobate, O child of Earth, turn, turn, to God; turn within yourselves, for be assured that God is not very far from you!

Walk humbly before God; do justice; love mercy; and seek by prayer to God, looking within yourself for an answer, what else He requires of you. Nothing else, you infer from the Psalmist's expression, and it is only that you so walk before Him, so humbly receive His counsel and direction in your heart or mind. For He will write His law upon your heart, and put it in your inward being. Flee, then, from the wrath to come: the time of the end draws near, when the heavens of men shall wither and roll up like a scroll and the sun and moon and stars of the imaginations of men shall fall to the earth in which they originated. Then shall appear the sign of the Son of Man, Jesus of Nazareth, coming in the clouds of great glory. His kingdom shall be an everlasting kingdom; His glory, honor, and dominion; power, authority, and praise shall have no end. For unto Him the Sonship is given, as Daniel declared it should be, and unto all men the Son is given, as it was declared He came to all.

This prophecy is now to be fulfilled. The day has dawned. The light of its glory is evidenced in the mechanical and physical advances, in the abundance of gold and silver, in the discoveries of sciences, and in the revelations of spirits. The establishment of Christ's kingdom in this nation of the people of the United States has commenced already. It will proceed, for who shall withstand

Him? The armies of heaven follow Him, and all the four and twenty elders (the Jewish and the Christian Church), the twelve tribes and the twelve apostles, and the four living creatures (the four kingdoms Daniel prophesied of), and all the saints of the Most High, from every nation, kindred, tongue and people, with one mind declare: Thou art worthy, worthy, worthy! Thou art worthy to open the seals of that book which reveals to man the future he must pass through to arrive at the glorious company of the Sons of God, who shout for joy at the announcement that now is come salvation and honor and power, praise and glory forevermore to the sons of Earth, to make them Sons of God.

What, then, shall I do to be saved! some of you will ask. Believe on the Lord Jesus Christ and thou shalt be saved, thou and thy house. Let us see what this primitive creed required and implied, for this is all the profession of faith that was necessary at that time to make a Christian, and admit that follower of Christ as a member of the great and universal church of God. And this same profession of faith is enough for the present time, if it was enough for that, for now there are fewer temptations to draw a man back from the good work. Then there was persecution by the Jews on one side, and by the heathern idolater on the other, the one declaring that he had the whole counsel of God; the other pointing to the wise, great, and good who had worshipped and sanctioned by their example the magnificent temples, the splendid ceremonies, the mighty oracles, and the everlasting gods whom their fathers in all former time, as they believed, had worshipped. And if a believer resisted both these, then came the philosopher, the gnostic, declaring that by reason man could find out God, for the prophets had never declared as much respecting Him as they could. When all these had exhausted their arts, then came the son of Earth, unwilling to admit any man to be better than himself, and declaring that Jesus was God, and that unless He were worshipped as very God Himself, the refusing soul must be condemned to eternal undying punishment in material and unquenchable fire. If not overcome by this, he must withstand yet another assault. He would be assailed by the Trinitarian who would assure him that there were three gods and one God, and he must worship all the gods equally, yet only one God, and should he fail to render this worship equally and justly, he could never receive the rewards of heaven, but must suffer the punishment of rejecting that Christ in whom he had professed his belief in the same manner as the jailer of Paul and Silas had.

But let us inquire exactly the meaning of this profession: I believe in the Lord Jesus Christ. Jesus Messiah is the true reading. Jesus was the Messiah whom the Jews looked for. He was also the undefined extraordinary manifestation that the eastern fire-worshippers and the western idolaters expected. He fulfilled all expectations, but not to the satisfaction of mankind. Each group expected Him to teach the doctrines they believed, and to maintain and establish their own respective modes of worship. He came in His own, or God's own, way, preached unheard-of doctrine, and established no church. He died without having left directions for the formation of a church or a confession of faith.

Still we need not infer from this that no church ought to be organized or that no creed should be established. But we do infer, and ought to be allowed to infer, that He did not consider them essential nor, indeed, of any considerable importance. The apostles found themselves without any outward guide in this matter when they began to gather the church, and they filled its offices as seemed at the time wise and expedient, by appointing discreet persons to them. Their superior claims to authority were of course uncontradicted at first, and in general always. Paul's claims were more disputed, as he had not been one of the twelve first chosen, though he was really the twelfth, taking the place of Judas, who fell by acting in his own will. For Jesus of Nazareth in his spiritual body appeared to Paul and called and chose him just as fully and even more extraordinarily than the other eleven. Paul, then, was an apostle, as he claimed to be, and the appointment of Matthias was an act of the eleven, done in their own wills and without the Divine Influx or command. When asked by the jailor, What shall I do to be saved? Paul answered: Believe on the Lord Jesus Christ and thou shall be saved, thou and thy house. He also told him what outward form he would require, baptism. And immediately he was baptized, he and all his household.

In these latter days men have not been so easily admitted, and allowed to bring their families unquestioned into the church. But are men outside churches any less Christian now than they were then? Oh, no; but you say the times now are different and experience has proved to the church that a longer probation or a fuller profession is necessary, or at least expedient. Well, then you admit Paul did not know so well as you how to conduct the church government, or else you claim that he did not act by inspiration and that therefore he could not be expected to choose a form that would not need amendment. Choose for yourselves on which horn of the dilemma you will hang. But you must hang on one or the other or admit that you yourselves are in error. This I shall take to be the case, for I would rather defend Paul than you. I will, then, suppose that some of you will agree with Me that Paul's creed was long enough, and then I will go back to its explanation.

Christ is put for Messiah, and you will find that the learned admit its substitution for Messiah, though it does not possess the same meaning. Yet in general I use it freely, for its true meaning, which is Spirit of God, is expressive of Jesus's distinguishing feature: the possession of God's spirit or guidance in a most remarkable and, on earth, unique manner. Belief in Jesus Messiah was the only requirement of faith to be professed in Paul's time; at least so he began his preaching and receiving of converts. Paul, though, did explain to the jailor who and what Jesus was, and showed in his persuasive way that the Messiah had been much prophesied of long before; that the time for Him to come had arrived and passed; that He had come in the very way, time, and place fixed upon in the prophecies; and that Jesus of Nazareth was He. Then he explained further to his attentive listener that He must necessarily have been crucified as He was, as I have already explained to you in this book. Then, as his listener, instead of asking to have the walls of the prison shaken or the gates thrown open again, persisted in being attentive to the explanation of the brief reply Paul had made to his one question, which, in fact, is to each son of Earth the great question—as he continued to be attentive, Paul willingly showed how it was that belief in him must necessarily include a belief in the reality of his miracles and of his authority to teach, and of the doctrines and precepts that he had taught or promulgated. Then he went on and showed him that if he believed these doctrines and precepts, he must practice them; that they were not merely matters of speculation or themes for congratulation; that they were true and praiseworthy, but that believers were required to do something more than to exult over the poor, ignorant, worldly-minded outward followers of burdensome faiths and sectarian religions; that believers had really a work to do: first, to sacrifice to God their own will; second, to obey God's direction whenever they received it; third, to desire to receive it continually in order that all they did might tend to God's glory, by being done entirely in His will; fourth, that with God directing them always they should ever be desirous to serve, and never to seek to lead the spirit within them, and that this course would make them joint heirs with Christ the Messiah in the Sonship to which He was destined; and at last, that he should fear nothing, but trust fully to God and to the direction of God's spirit within him for salvation from every earthly peril or spiritual depression, difficulty, or doubt.

When this discourse was ended, Paul left a man prepared for works and rewards, and a man who afterwards did do work and receive the reward of being chosen to suffer for Christ's, or the Messiah's, sake or name of belief. He died a martyr's death and received a martyr's reward, that of hearing in his heart the assurance from God through His high and holy spirits: Well done, good and faithful servant! Enter thou into the joy of thy Lord. Thou hast been faithful in small matters and I will give great ones into thy charge. And then he became an elevated and rapidly progressive spirit in the life to come. This man never had any other creed than: I believe in one God, and Jesus Messiah whom He hath sent. He was a Jew, and Paul did not ask him to profess what all Jews' believed with unfaltering firmness in that day, though in the days of their fathers it had been very different, and quite the contrary at times before the Babylonian captivity, as I have before briefly described.

This is the history of Paul's creed as then promulgated; from what source did Paul derive it? Was it his invention, or was he a divinely inspired medium through which God Himself spoke by His spirits, or was he only impressed with convictions of truths in spiritual matters? He was sometimes a holy medium, sometimes impressed, and sometimes he acted in his own will by his unaided intellect. But in this matter he had a more impressive convicton and direction than either of these. For when he travelled with letters authorizing him to torture and imprison believers in Jesus Messiah, he saw a light and heard a voice saying: Saul, Saul, why persecutest thou Me? It is hard for thee to struggle against thy convictions of duty. Thou art already convinced that I am Jesus whom thou persecutest, and that Jesus is Messiah. And immediately Paul was obedient to the heavenly vision. He did not ask for it to be repeated or for another test. He knew that it was pride before that had smothered his convictions of duty and of error, which had been growing upon him gradually after he had, in full faith that he was doing God service, commenced the persecution of believers in Jesus Messiah. His reputation was at stake. He was a young man of great prospects for advancement in the Jewish congregation, full of zeal for his church, and ambitious for himself of its honors. How hard a struggle it must have been before he saw the vision, when he found his mind wandering between his church, his nation, his teachers, his friends, his family, and his ambition on the one side, and the despised dogs of believers in Jesus Messiah on the other—believers in One who, though said by His followers to have worked miracles, had yet suffered Himself to be executed for blasphemy upon that awful and terrible tree called the cross. This punishment was to the Jews the most degrading and the most shocking to his feelings of the whole catalogue of criminal executions, so much so that every subject of it was accounted accursed forever. Can you imagine yourself in Paul's situation, and believe that you would have yielded to a single vision and then become a preacher of Jesus, whom you had so persecuted, without going back to Jerusalem to see what father or mother, sister or brother, teacher or friend, would have thought of your vision, and what they would advise you to do? Paul immediately, without consulting flesh and blood, began to preach Jesus Messiah and Him crucified. He did not strive to reconcile his former notions with his present knowledge. He did not care to know what the people of Jerusalem who had just sent him forth to execute vengeance upon true believers would say when they should hear that he, too, had become infected with the pestilental delusion! No! he had only the guide of his inward direction, the spirit of Jesus speaking in him to himself.

This guide every man may have if he will act as Paul did, and surely you will not venture to say that any man now can have such claims of right to hesitate as Paul had? You may think you can say he had a most extraordinary vision, and a further sign by being struck with blindness! But, O wicked and perverse generation, you are wise in your own conceit! You persuade yourselves that electricity, or magnetism, or odic force, or the will of the medium, or your own hallucination produces the outward signs I have allowed to be given in your day. Are you convinced by them, and led to turn inward to see what God has for you to do? Do you ask: What shall I do to be saved? Or: Lord, what wilt Thou that I do? No; you laugh to think how disbelievers must be confounded at this sound or that test. You chuckle at the idea that you are so far in advance of the world as to be thoroughly convinced that the manifestations are spiritual. You long, perhaps, to become a medium, or to have one in your family—-and for what? Why, that you can invite skeptics, terrify the fearful, shock the pious believer in the old theology, or shake the world by astounding revelations from the spirit world. Or perhaps more sordid views impel your desire to be or to possess a medium. Perhaps you would exhibit for money, or you would dig for gold under spiritual direction, or you would make some great scientific discovery or settle a controverted point in history or chronology, or geology or science, or art of some kind. But for none of these things do I work. My holy medium has been actuated by nearly all these motives, but I never gave him a sound except once when, with others, he paid his dollar for this purpose. And then under My knowledge I impressed him so that he prepared himself with written questions of which no test could be made. He received his answers and afterward believed no more than before.

So it has been with hundreds and thousands and tens of thousands of others who have witnessed the outward signs of manifestations. Yet they have their use, or they would not have been permitted to continue. They have their abuses, but so do all good things. And when they cease to be useful, they will cease to be. So in the early Christian church there were the outward signs of healing the sick and raising the dead, of speaking with tongues and the interpretation of them, of receiving the holy spirit of God into the heart or' mind of the believer by the laying on of hands. But all these signs ceased as Paul perceived they would, and now the church still keeps up the form of some of them whilst there is no resemblance of realization, and of others the very nature of which is a subject of conjecture, so entirely and so early did they cease to manifest.

So it will be with the outward signs of this new movement. They will do their work, which is all the good they can do, and then they will cease. There will then be found those ready to deny they ever existed, as there are now vast numbers denying that they do exist. And how shall the truth of revelation be established, then, if signs are withdrawn? In reply, I ask how shall it be established unless they are withdrawn? This very day My holy medium has been informed that the believers in this place do not think it interesting to hear revelations of God's will, but that if any outward signs or manifestations are to be exhibited they will all attend! O sons of Earth! O low and ignorant minds! Is the altar, or he that sanctifieth the altar, to be worshipped? Say, whether the temple or he that dwelleth in the temple is to be thought much of; whether Jerusalem is the place to worship, or the place where God is, which is the heart of man. *There* is God. And when you turn there for your sign you shall have the sign promised in olden time, the sign of the Son of Man coming in clouds of glory. As the lightning shineth from the east to the west, from the one part of the heaven even unto the other part, so shall the coming be. The instant you complete the sacrifice of the heart to God, it will be filled with His glory—the glory of the only begotten son of God. Then will you sing: Great and marvelous are Thy works, O Lord God Almighty! Just and true are Thy ways, Thou King of Saints! Wherewith shall I come before the Lord my God? Shall I render my first-born son, or shall I sacrifice the blood of rams or bulls or firstlings of the flock! The sacrifice of a broken and a contrite heart, O God, Thou requirest of me and of every man and of every spirit of man. But men cannot understand this till I establish the truth and certainty of this revelation. They have Moses and the prophets, and they will not believe them. The dead have been raised and they believe that, but will not believe Me. They have had the precepts of Jesus, and they will not act upon them. They have had the Comforter, the Spirit of God, but they would not be led by him into truth. They now have signs and writings and they will not turn to God, but follow after the outward, neglecting the inward. What remains? Shall the master of the feast say to his servants: Go out and compel them to come in that My tables may be filled! Shall He also turn out those who press forward to the table without the wedding garment, denoting that they had not been called or invited?

Yea, verily this will He do. He will not violate your free-will, but He will make one more effort to conquer your reason and prejudice; to seize your attention and nail your convictions. To lead you to living fountains where you will drink and thirst no more for the outward. To lead you to the pool and put you in whilst the waters within are troubled, and before any other man gets ahead of you, robbing them of their virtue and efficacy. My signs shall accompany My preacher. My outward evidence shall accompany My appeal to the inward monitor, and I will thus try to make you all holy mediums. Prepare, then, for greater things than these.

As for those unworthy servants who did not do their Lord's bidding, and who do not hereafter do it, they shall be seized and bound and cast into the outer darkness, where there shall be weeping and wailing and gnashing of teeth; where their worm dieth not, and where even the ungodly perish. They perish not by eternal death or annihilation but, as I have before explained, they perish as regards every good motive, every lofty purpose, and every desire for reconciliation and unity with God. Not forever, but to man's finite ideas it would seem forever, so long a period would it be. Repent, then, for the kingdom of heaven is at hand. Repent or fear condemnation to the left hand of God, to that position I have described, and in which so long as you remain your punishment will be everlasting.

The word everlasting does not express the true meaning I now have in view, nor the meaning it has in the original Greek of the gospels. Its true meaning rendered in English is: continuance to a long and indefinite period. Analogy of its use by other authors shows this. But do not think it it nothing, if it is not eternal. For from everlasting to everlasting—that is, from one long and indefinite period to another long and indefinite period—will this punishment and comparative suffering continue, until you do just what you can now do with less sacrifice, more facility, and as sure a salvation, which by this means may be immediately secured to you. Be wise today, for you know not what tomorrow may bring forth. Give your heart to God now for tomorrow conviction may be less strong, worldly desires more powerful, pride or friendship, perhaps pretended and perhaps sincerely felt, may remonstrate so effectually as to place you beyond the reach of God's mercy for this long and indefinite period. During all that time you must suffer the want of happiness— the deprivation of the realization of your desires. For the desires of man's free-will are ordained to be unsatisfactory when realized. All is vanity; vanity of vanities, said the preacher, and so every soul has found it, and so every soul will find it till time shall be no more.

Why then will ye die, O people of Earth? God's door of mercy is ever open, and He calls to you continually: Come, come unto Me, all ye heavy laden, and I will give you rest. For My yoke is easy and My burden light. Fear not, then, for when you have come to Me I am with you thereafter to the end of the world! Come, then, and be saved by God's infinite mercy in the only way you can be saved, and you will then be able to see that every attempt to reach heaven in any other way is only a robber- or thief-like attempt, and you will then know that no thief nor robber can break through into heaven to steal or enjoy the treasures in store there for God's willing servants.

O people of Earth, My heart yearns for you and I would consider the sacrifice of a crucifixion nothing if it would save you. Far more than that would any son of God be willing to suffer if by that means you could be saved. But after all that you would still be required to save yourselves by sacrificing your free-will. No man or son of man, or son of God can save his brother by giving a ransom for his soul. No, the only thing that can be done is to persuade you to save yourselves. Save yourselves, then, O unhappy, unwise, O ignorant, proud people! Save yourselves all ye who are not perfectly happy! Those who are perfectly happy—which includes the realization at the time of every hope or desire or wish or aspiration—those I say, who are thus happy are saved, and they are saved with an eternal salvation, and of their joy and pleasure, thanksgiving and glory, honor and praise, there shall be no end. Because they are sons of God, one with Him in power, will, and action. And they can never fall from that perfect unity and oneness they have reached by passing through all the circles of all the spheres from paradise to Sonship.

Well then, brethren and children, you have on the one hand ages of separation from God; on the other ineffable joy and unending happiness. Choose ye now whether you will serve Baal or God. On yourselves must rest the consequences which will continue for a long and indefinite period in all their vast and, by man, immeasurable consequences; and different as they are, not more different is the old idea of heaven and hell, except as it affected your notions of the benevolence, justice, and mercy of God.

What then remains? I have called and pleaded; I have persuaded and entreated; I have argued and pled in God's name, in My own name, and in the name of suffering humanity. I will plead and argue in My holy medium's name. I will ask you who have known him best, if he was ever the subject of religious excitement? If he was ever disposed to urge men to care for the future state? If he was fond of the assemblies of God's people, or those who professed to be God's people? Has he been active in benevolence or ardent in the cause of suffering men? Has he been devoted all his life to the good of others, or rather to himself and his own gratification? Has he led away the drunkard from his cups, or the fool from the house of destruction? Has he not rather taken care of himself, and left others to take care of themselves? Has he been an elegant writer—a pleasing or an active member of the social circle? Has he not, rather, been moody and quiet? Has he not been reserved and cautious in his deportment; tedious and tiresome in his wit? Yet out of this man, unpromising as he appeared, I have raised up one willing to do My work. Now, if I should sacrifice his business, reduce his family to beggary, and consign him to a dungeon for an example of patience, would you be edified? I believe he could bear it, though it would be a trial he would avoid. I believe he would come out of the furnace heated seven times hotter than it was wont to be, without the smell of fire upon his garments. But I should not promote your salvation by this course.

I choose rather to show you by him that business shall prosper, relatives be benefitted, children be increased in beauty and intelligence, and fellow men encouraged and impelled to follow his example. I shall use him freely and often, but I intend to take care that his temporalities do not suffer for your sins. That he will live so as to use without abusing My favors, I believe. If he does not, he alone must suffer condemnation. He has his free-will; whenever he chooses to exercise it, he can. Think you he will use it? Watch him and see. But if he does, only draw from that this inference, that whatever favors God may bestow upon His children in the body, they may be perverted to evil; and however a man may be made a partaker of His grace in this life, he may withdraw himself and return like the dog to his vomit, or be for a long and indefinite period a castaway. The last state of that man shall be worse than the first, for at first ignorance was an excuse, but at the last the sin was against knowledge, and unpardonable. Again, then, let Me persuade and entreat; let Me argue and reason, that you should choose for your own good what will bring you peace here, and happiness hereafter.

What will you give for these blessings? They cannot be bought with money or time. There is only one thing to be pawned or exchanged for them, and that is your free-will, commonly called in the Bible your heart. Make this a willing sacrifice on the altar of God's mercy, and you are saved. Saved as long as you make it and, if you choose to persevere, with an eternal salvation. O son of Earth, lay not aside this book till you resolve to seek God, and by prayer offer to Him your heart. No matter how much you have sinned. No matter how high you stand in the outward church. Go down into your deepest self and there prostrate yourself before God. Make the prayer I have delivered for you in the twenty-first chapter of this book. When you can make that with sincerity and faith in God, and with desires of acceptance with Him, you are saved. When you can desire to make it thus, you are almost saved. A little more effort will then be sufficient. One more effort, one more getting down deep, into your inmost deep, will save you. May God help you! is My prayer and that of the medium. But our prayers can do you no good. God is willing already, and desirous already to help you, and He will do everything but force your free-will to bring you into reconciliation with Him.

Let us Pray

O Thou eternal, incomprehensible, almighty and ever-loving Father and Friend! Oh, listen to the humble supplication of Thy deeply desiring servant, or if not Thy servant, O God, make me Thy servant. Grant, O most loving and kind and powerful Father and Friend, that I may have wisdom from Thee to see what way I should take, to feel what I ought to feel, to love what I ought to love. Be Thou, O most kind Parent, my helper, my savior, my intercessor, my redeemer, my friend! I know, O God, that Thou art all these, but yet, O most kind Parent, make me feel its surety more. Let me know the peace that the world cannot give or take away. Be thou, O Father, my helper in this world's affairs and my savior in spiritual matters. O God, I desire to serve Thee and to do Thy will. May it please Thee to help me to do it. Help me, O Father, to walk as Thou wouldst have me, and to pray acceptably to Thee. Help me, O God, to say at all times and under every dispensation, when troubles surround me and trials depress me, then O God, help me more and more till I can truly and sincerely and with perfect reliance on Thy goodness and mercy and loving kindness, all like Thyself infinite, to say then: O Lord God Almighty, not my will but Thine, O Heavenly Father, be done! Amen.

O God, hear us for Thy Son's sake, for Thy own sake, for our sake, and we will try more and more to become Thy servants and to be worthy of Thy kind regard. O God! Thy art good; make us good. Thou art loving; make us loving. Thou art happy; make us happy. Thou art ever merciful and forgiving; make us so, we pray Thee, that we may be like Thee and like Thy Son, the Lord Jesus Christ, to whom with Thee be all honor, glory, thanksgiving, and praise; power, will, and majesty, now and forever, world without end. Amen.

O God! Let Thy Holy Spirit be with us, and guide us and help us into salvation. Because, O God! Thy Son has declared to us Thy great mercy and loving kindness, we dare to ask these favors, O God! of Thine otherwise unapproachable majesty and infinite nature. For Thine is the kingdom and the power and the glory, now and forever, from everlasting to everlasting, and so on to infinity, incomprehensible and beyond our natures to conceive of. Amen.

Almighty God! Who dost from Thy throne behold all things in the earth below and the heavens above, look down in Thine infinite mercy upon Thy would-be humble servant. Grant unto me the desires of my heart so far as they are worthy and proper to be granted; and fill me with love for Thee, and help me to be kind and loving and affectionate and to do Thy will and to walk in Thy ways, in Thy peace and in quietness and obscurity if such be Thy will. And, O God! be very kind and loving to me and preserve me in the enjoyment of Thy counsel and guidance in all things, and keep me and help me to keep myself passive in Thy hands and in the hands of Thy spirits, so that I may work in entire submission to Thy will and walk always in Thy ways. And to Thee shall be the praise, honor, and glory, forever and forever. Amen.

This last prayer has been made previous to this time by My holy medium in his own will, and when you can make such a prayer in your own will, you will be a holy medium too. Be ye, therefore, ready and willing, for ye know not in what hour the Lord will come. Come He will. Have your lamps then trimmed and lighted for if you be gone for oil when He comes, you will be ranked among the foolish virgins or unwise men. O My People! hear My voice and listen to My call. I am the Son of God and God sends Me through this medium to speak to you to convince you and to lead you to light and life, and eternal salvation; to lead you to the mansions of indescribable bliss where joy, happiness, bliss unutterable and inconceivable by men, fills every son and daughter of man, every creature and son of God. Every soul that in the universe of universes and the whole concentration of universes of universes, in the whole great and illimitable creation of God, is enjoying this bliss or will enjoy it to the end of time, and in the beginning of the eternity that succeeds to the time when all are Sons of God. Then shall one universal shout of joy and salvation ring through the whole illimitable creation. That, My friends, will be the last trump. Then eternity will commence and it, of course, has no end. Neither will their bliss have any end. What God will then have in store for them no man, no spirit, no Son knows. God Himself does not know because He has not resolved to know. But in due time He will make known His law, that is, His will. Would you be the last one of all God's spirits to reach His throne, His Sonship? No, you would not, I know, if you could help it. You can help it if you will sacrifice your heart, your free-will Let us pray again the prayer I have transcribed for you from the

128

twenty-first chapter and its additions.

Let us now leave this part of our subject to return for a time to the Chronological Account of the Creation.

CHAPTER XXIX

THE DELUGE

The Relations of the Crust, or Surface, of the Earth

The time when the earth assumed the globular form was about the time Venus became a ring about the central body. The central body emitted light and heat as it now does and as it had done from its first organization as a globe. The moon was separated about the time Mercury was formed into a ring and at about the same time Venus became a globe. When Mercury became a globe, the moon also became one, and the solar system received its present appearance of development except that a moon has since been added to Jupiter in place of a ring, and Saturn's rings have been separated and the outer planets have had some similar changes of rings and moons. Now this may be regarded as mere invention of the medium by some, and it would not require much to produce it if the general law of development is admitted. I, however, give it for such as can be gratified by the truth respecting this subject, and I shall extend My observations somewhat further on the same general principles of geology and the formation of the crust of the earth and the production of the animal creation.

There was a great contraction of the body or matter of the earth after its first assumption of the globular form. But this was continued for so long a period that I will not undertake to write the myriads of years before the dry land appeared. The contraction was caused or accomplished rather by the change of gases to liquids and of liquids to solids. It is a great mistake to suppose that the arrangement of matter into gaseous, fluid, and solid has always existed. Comets are gaseous bodies and represent the original or early state of the sun or central body before concentration had commenced. Next fluids appeared; lastly dry land or solids.

The central portion of the planetary bodies is now gaseous and the theory of Symmes is not so far from the truth as generally supposed. It is, however, so far from the truth as to be unfounded by any hypothesis which will explain its phenomena. I shall give an hypothesis, but it will not confirm his main points, which were the openings through the crust of the earth. The openings could not exist because the shell would contract to close them, and if they existed, disruption of a ring could not take place because the internal resistance would not be sufficient to maintain the position of the crust outwardly.

Well then, the crust, being formed first of gas, secondly of fluid, and thirdly of solid matter, thickens and hardens, becoming more and more solid until its thickness is so great as to prevent a further contraction. Before this time the inequalities of surface are produced by crushings, by contractions forcing up ridges, or forcing inward depressions; on the outside, on the contrary, hollows represent inward ridges or uprisings, as men in the body term them. When the contraction can no longer proceed in this way the preparations for the removal of the crust commence, by a course of separation proceeding within the outer crust. The internal matter continues to contract, leaving the outer sustaining itself by its own strength. When the inner surface has become sufficiently reduced in size to furnish a strong shell, its contractions again begin to cast up ridges and make valleys. In the valleys the fluids rest; above the whole the gaseous particles that have escaped from within float or adhere to the more solid and fluid surface. Then the solid outer crust is liable to receive and soon does receive such further contraction as to rupture it into rings. These rings may conglomerate, or slide into each other and remain separate, for some will be smaller than others. The polar portions of the old surface or outer crust fall to the inner crust because they have no centrifugal force to maintain them at a distance. The more central zones or equatorial regions become more and more accelerated in speed by being relieved of the polar parts. Then, the acceleration of rotary motion proceeding, the crust—or ring, as it has now become—becomes more attenuated and enlarges its circumference until the motion that has resulted from all these proceedings is precisely such as will cause an equality of centrifugal and centripetal force at the distance to which the body had arrived when the equalization took place. Then the body or ring continues to revolve or rotate about the central body until an inequality of thickness becomes developed by the contraction of its matter, which still goes on, and the thicker portions drawing by more cohesive and contractive force finally make a rupture or parting of the ring. Then, as I have before related, the ring begins to double inwardly until it winds up, as it were, like a cord into a ball or globe. This globe is more solid. than that from which it separated. But it is not solid. Again the contraction and the disruption proceeds with the first body, and the satellite also proceeds in the same manner.

But the satellite is more nearly solid, and can scarcely again throw off a ring. There is however such an instance in one of Saturn's moons where a satellite has been formed from a moon or satellite, and it now exists in the form of a ring. It is the third moon from the central body that has this attendant, unique in this system. How then did the Ark save Noah when the moon separated from the earth? I will explain.

CHAPTER XXX

History of the Earth, and its Inhabitants, after the Deluge

The time for the disruption having arrived, and the Ark's having been prepared by the holy medium Noah in exact conformity to direction of the Divine Influx, the windows of heaven were opened. God's mercy was displayed to men of that generation by signs of warning for forty days and forty nights. Terrible to them were the convulsions of nature. The earth rocked violently and almost continuously. The waters rushed over the dry land; the mountains fell; the lakes dried up. The whole of the race of men perished then, except Noah and those whom he had received into the Ark. Noah was a young man, and his family consisted of only three sons and as many daughters—a small number for the long-lived and prolific antediluvians. The Ark had been navigated by Noah under God's direction to the polar regions where its safety was secured by the law of which I have informed you. These polar regions reached the under crust of the earth without any serious shock, but the inevitable effect was that the earth began to change its axis of rotation and thus the one pole became the central tableland of Asia, and the other the tableland of Mexico and Central America. This change, through gradual, was sudden enough to imbed in ice the carcases of tropical animals in the now frozen regions of Siberia, formerly the tropical or equatorial portion of the inner crust. But how came the tropical animals there if the outer crust passed off in those regions and became the moon? Only the outer solid part separated; the gaseous and fluid parts descended to the central body by the attraction of gravitation and with them the dead bodies of men and animals. But the dead bodies of men have not been found with the bodies of animals in Siberia! They have not, but they will be, and may be before long, for it is only lately that the bodies of the animals were found there.

Noah, now being safely landed upon the renewed or inner crust of the earth, left the Ark when he found that the commotions had subsided. But why did he not receive the counsel of God as to the time when he should leave? He did act by God's direction to the end of his life. And it was only because God does not choose to have a man cease to act for himself in reliance upon his own powers that He insists on his doing it. It is in this that a holy medium's submission is most fully shown—a willingness to serve God and obey His revealed will, and also a resolution to work for himself without requiring God to be his servant because he has rendered God obedience.

Noah found a country not materially different from the former surface of the earth, because it was in fact a part of it. But this old surface was found to be less fertile than the new, under the changed condition of the atmosphere. He, therefore, soon sought under God's direction a more fertile region which he found eastward in what is now called China, and on what was at that time the western coast of a continent reaching over the greater part of the Pacific Ocean, as its place is now called. But the rivers in China run eastwardly, and how could they do that if it were the western coast? The change of surface elevation that took place under the laws of contracting crusts submerged the great continent of the region corresponding to Asia and raised the space between that and the central tableland on which the Ark had rested. This made the water shed toward the opposite direction in China, and in Chinese history, fabulous though it is called by learned Europeans, is found a record of this extraordinary and unlooked-for change, and of the devastation and destruction of human life which it caused. Wonderful as it was, and unheard-of as it is to the present generation, these changes were not infrequent in the early ages after the disruption of the moon's ring. But now the hundreds of thousands of years have so established the crust that it no longer changes much or violently. Slow and gradual uprisings and depressions of surface take place, and have been observed and recorded.

The last great submersion was that of the continent or great island of Atlantis, which took place about ten thousand years ago. A distant tradition of this event has been preserved, both in the records of the eastern and of the western continent: one by the Phoenicians and Greeks; the other by the Mexicans and Peruvians, and also by the aborigines of Cuba, Hispaniola, and other West India islands. Sudden as it was, a few people escaped to tell the tale, and remains of its surface exist in Teneriffe, St. Helena, and a few other small islands. The great ancient Asian continent left the numerous islands of the Pacific and the continent or great island of New Holland. In the history of the surface of the earth many such changes, involving vast destruction of life and obliteration of ancient records and monuments, have occurred. Nations have perished like individuals. But these nations, people, and histories have no interest for you who have never heard of them. Let us return to China, the oldest, the primeval nation of the earth.

Noah lived 600 years after the flood, and at his death saw an empire or nation of many millions of happy descendants. His sons ruled provinces or divisions of the nation or race, and Shem succeeded him by Divine appointment. Shem lived for several hundred years after his father's death, and was above a thousand years old when he died. But why was no record preserved of his life, longer than Methusaleh's? Because the tradition was lost before writing was invented. Who succeeded Shem? The names of his sons given in the Bible have reference to nations and to colonies, and not to persons. But the Jews, and before them the Egyptians, delighted to trace their genealogies to the utmost extent and would not be satisfied without reaching back to Noah. They chose Shem's line of descent for themselves, and that was true enough, but Abraham, the son of the tribe or race of Terah, was an obscure man in his native Chaldea. He, however, by obeying God, became the founder of a great nation, and his chronology becomes nearly correct in the Bible. Then his sons and grandsons are traced with fair correctness to Moses and Joshua and so on down to Solomon and the captivity. Yet they are not traced with entire correctness, for these records were all lost at the time of the Babylonian captivity, and restored from the memories of the chief men of the nation by Ezra. The Egyptians had records besides other than genealogies. In that early day, the people could read the sacred writing as well as the priests, and long before they lost that ability the scheme for withholding from them the recorded knowledge of the past was concocted. Copies of these records of Egyptian tradition and history exist, as I have stated, but when found they will not tell much more than this.

Egypt was settled in early time, but inundations and physical revolutions destroyed its population so that its history as a nation commenced about twenty thousand years before Christ came, in Jesus of Nazareth. Many and various tribes at first formed the population derived from India and Assyria. At last Menes united them all under one government and left the nation strong and powerful. His successors attempted foreign conquests but, except in a few instances, with poor success. At home, though, they maintained independence and extended their empire in population and wealth. By marriages they also acquired kingdoms in Africa, which their superior civilization and careful and discreet policy enabled them to maintain for a long period. They were at last conquered by a race of shepherds or nomads whose original home, or at least their home for generations, was Scythia or, as it is now called, Siberia. These Yethos or, as more generally written Hychos, were a marauding colony like those Kimbri and Kelts and Germans that made such inroads into southern Europe in after ages, though they never have obtained such complete possession of the countries they invaded since history commenced its records. Before that, their ravages had been more powerful and destructive. The Pelasgians were Scythian tribes or emigrants, and the Teutonic nations also trace their origin to Siberia, faintly but correctly. The Yethos maintained their supremacy in Egypt for more than a hundred years—about as long as the Vandals retained northern Africa in the decline of the Roman Empire. The people of Egypt then rose as one man at a given signal and put to death every hated oppressor and made their memory accursed forever. The very name of shepherd was such an offence to Egyptians that Moses led his followers away from Egypt almost without opposition because they were, in general, shepherds.

But you say Moses was opposed by Pharoah, and at last followed with an army from which his followers were saved only by a miracle! No, My friend; Moses was opposed by court intrigues. His claims to the throne had been set aside most unjustly. The enemies of his claims feared he might wish to turn the strength of his followers against the nation and endeavor to obtain by force what he had been wrongly deprived of by intrigues. Moses therefore threatened and negotiated until he at last extorted from Pharoah a reluctant consent. But Pharoah felt the necessity of watching the movements of Moses and his followers, which he did with his whole army. And Pharoah lost that army by imprudently attempting to follow Moses and the Israelites, as they called themselves, in their march across the head of the Red sea. The account of this, as recorded, is somewhat distorted; however, its general features are true and by making some allowance for the exaggeration of rewriting them in Ezra's time we shall easily reconcile it with an intention to give the truth. I have already alluded sufficiently to the wars of the Canaanites and the wanderings in the wilderness. I will only state that the account of Eden's being eastward in the book of Genesis is a transcription from a Chinese record that had been translated to Egypt, and adopted as a literal one in their theology. In that the change for Eden's locality was made from westward to eastward, so as to suit the longitude of the place.

CHAPTER XXXI

LATER HISTORY

Origin and History of Commercial Nations

There is another history of a nation to be written in which you will feel an interest. That is the history of the Phoenicians. The Egyptians were a people similar to the Chinese in their institutions. They were separated into castes and each followed his father's trade of profession, and were quiet, unresisting, unenterprising, and indisposed to roam. They had no ships in early days and the fables of Grecian settlement from Egypt had no foundation in reality. Phoenicia was the great commercial nation of the olden time, or rather for the ten thousand years preceding the Christian era. Their power was broken by the Assyrians and their commerce ruined by the Greeks. Having commenced their settlements on the Levant by emigrating from Arabia and southern India, they extended their power over Spain, Italy, Sicily, and northern Africa, and in their earliest voyages reached the ancient continent or great island of Atlantis, where they had settlements or colonial trading posts. The inhabitants of Atlantis were a kindred people, but had left the original seat of the race or tribe at a much earlier period. From thence they proceeded by sea to the British Islands, to Denmark, Norway, and even to Iceland. From Iceland they found a way to reach America, though its productions in those northern regions had small value for them. The gold of Ophir was obtained from the interior of Africa, which is the richest in its production of all countries. They were supplied from the mouths of rivers along and below the Gulf of Guinea, since the commercial tastes of the African nations induced them to resort to the Phoenicians for the purpose of obtaining articles they sold.

But did not other nations share in this lucrative commerce? None did after the submersion of Atlantis until Solomon persuaded Hiram of Tyre to allow his ships to accompany the annual fleet that left the Phoenician ports and made a rendezvous at Sicily. Their voyages were performed under favorable circumstances, occupying three years, usually, in going, trading, and returning. But did not Solomon have his ships on the Red Sea? He did make a port at Ezion Geber but that was as a compensation to Hiram for the great privilege of being allowed to join the Phoenician fleet in its voyage for gold. The Phoenicians used Ezion Geber far more than did the Jews, and it was at that port that they did business and had commerce with India. Before that their supplies of Indian goods had been obtained from ports on the Persian gulf, which the Assyrians oppressed and from which the land transportation was much more difficult than from Ezion Geber.

The laws and language of Phoenicia have nearly perished; their history entirely so, to men in the body, except as it is connected during its later time with the Jews and Greeks. The Greeks themselves were the posterity of Phoenician colonies united with Pelasgian conquerors. These in turn were mixed with other tribes arriving from the Black Sea, or from what is now called Circassia and Georgia, a district which has long possessed the fairest and noblest physical specimens of man, and from which, too, the Saxons were derived.

The Saxons were not Germans nor Scandinavians, but Georgians. They left their original, or place of long residence, in the year of Alexander's invasion of Persia, and marched or travelled by slow stages and circuitous routes until they reached the Baltic at Riga. There they long maintained themselves, but because of new immigrations they were forced to fly from that country, and a small remnant of the bravest took possession of the peninsula of Jutland, the neighboring islands, and the almost impenetrable marshes of the neighboring continent. Here they arrived about one hundred and fifty years after the Christian era, and retained this region as their principal seat of power until they had by a long and persevering contest obtained possession of England and the greater part of Scotland.

The remainder of their history is well known to men in the body, but that they are to be the ruling and controlling nation in coming generations is only surmised by a few ardent imaginations who look upon British power as the manifestation of its development. They are not far wrong. But it is to *America* we look as the future seat of their power, for even England must fall before the combined power of the Dragon of Rome and the False Prophet of Europe. But the Woman that fled to the wilderness and was sustained by the two wings of a great eagle shall receive her progenitors and sustain the power of the Saxon nation in all its splendor until the fifth monarchy shall be merged in a universal brotherhood of all mankind under the government of Shiloh the Prince of Peace. But to return from this digression, let us briefly sketch the laws and government of the Phoenicians.

Their state or nation was composed of a considerable number of independent cities, under respective heads comparable to kings, but not possessing absolute power. They ruled through or with the assistance of a representative body appointed by the people at large. There was a federation, but its weakness cost them their existence as a nation when the Assyrian power became established and aggressive. One by one the small states were reduced until only Tyre remained. The colonies had never depended upon the mother cities for government and their most powerful one, Carthage, was only an ally. As is well known, they would have been glad to be left without the rivalry of Tyre, and they always had some excuse for refusing help in the time of her greatest need. Troy was one of the small Phoenician states and was destroyed by the Greeks, who even then began to prey upon the Phoenician commerce. The length of the siege and the cause given for the war in the Iliad are imaginary, but it was a contest that called forth the whole power and resources of the Greek states which had, like the Phoenicians and their ancestral branch, a confederacy, weak, it is true, but strong when all felt a common interest or desire to obtain a particular object. So Troy fell, a thousand years before Tyre.

The religion of the Phoenicians was a mixture of all that their commerce made them acquainted with. Its purity was maintained in their original land of residence, but they adopted whatever would make them more agreeable to people with whom they traded, and from them came such practices as human sacrifices and fire worship. They introduced fire worship into Asia after having derived it from Africa. Zoroaster was a Phoenician, whose works and preachings converted the Persians from truth to error respecting the origin of evil and the worship of fire, the sun, and idols. They still believed in God as the maker and preserver of all things, but their recognition of Him became less and less as time progressed and the priesthood declined in learning. At last only the outward form was left of this religion, and at the coming of Jesus of Nazareth as the Messiah, the knowledge of the one true God was confined to the single nation, insignificant and contemptible, of Jews.

Every nation that knew them, despised the Jews. Their own dissensions weakened them and they were the prey of every invader, whether he marched to or from Assyria or Egypt. Though living in a defensible country, they seldom resisted. Their spirit was broken by defeat and their power by dissension. Some adhered to an Egyptian party; some to an Assyrian. Some would trust to Phoenician alliance; others desired to call in the Greeks. The prophets urged them to trust in the God they professed to serve and to own for Kings of Kings, but their faith was too weak and their history, with a brief interval of the reign of David and the sway of Solomon, was a series of disasters following predictions of success had they submitted passively to their God, with whom their prophets or holy mediums always had communion through His holy spirits.

There was at last a cessation, under the Grecian Syrian monarchy, of the almost constant devastation they had experienced, and the successors of Alexander respected his grant of a nominal independence. The Romans in the beginning of their dominion were disposed to treat the Jews like other conquered nations, without any other rigor than what was necessary to secure their plunder under the name of taxes, but the Jews were so insolent and haughty and pugnacious, that their destruction as a nation became, to the Roman view, a necessity. From this impending event the Christians fled, and, warned by their holy mediums or prophets, secured safe refuge in various parts of the world. This event was so overruled as to be the means of aiding greatly the spread of Christianity, which was thus preached to every nation, tongue, and people of the western portion of the eastern continent. Paul was a prisoner in Rome for three years and there he made converts amongst the noblest of the Romans and their influence exerted itself to prevent a persecution of all Christians, as well as of Jews, which was at one time threatened by Titus and Vespasian.

CHAPTER XXXII

IMPENDING CHANGE

Changes of the Earth's Surface, Past, and Future

When the world was in its primeval form of a ring, the solid part was accompanied by fluid and gas, because the central body had retired by contraction to so great a distance from the outer crust that its attraction was less than that of the ring or earth itself. But for this the earth, like the moon, would have had no atmosphere or fluid.

The moon has, however, a gas surrounding it which suffices for the maintenance of life in organized beings. They are not like any in the earth and so I shall not weaken your faith by describing them, particularly as it is not properly within the limits of My title page.

The earth, then, having rolled up like a ball, at first retained sufficient tenacity and glutinosity to be moulded into spherical form by the laws of motion and centrifugal force. Its center was left hollow, so far as solids were concerned, and retained most of the fluids which had accompanied it. But as the contraction went on, the glutinous matter which was its solid portion soon began to separate itself into an outer and inner crust. It was, however, as yet too unconsolidated to maintain its outer crust far enough from the inner to form a ring that would separate into an orbit of its own. So its whole material fell back, as it were, upon the inner crust. It was by this process that the moon became large in proportion to the earth.

It was also this process which caused the great changes of surface I have alluded to in speaking of the submersion of continents, though it did not continue so long this time as before, nor so extensively. From this will result a smaller moon at the next separation which, as I stated before, will take place soon—that is, in a few thousand years. This second moon will be at first a ring and at last a globe, about half as far from the earth as the present moon. Before this disruption occurs we hope to be able to convert all men of the earth to a knowledge of their Creator and an understanding of His laws of being, action towards men, and salvation of men and spirits. But we do not know that we shall be able to do it, because man has, and will have, his free-will. We ask, then, that every believer of the truth shall work in submission to God, and passively, by His direction, extend this knowledge and secure this conversion. Outward signs of our presence will cease very soon, but we shall ever be ready and willing and desirous to work spiritually, internally, and by Divine Influx. The way to men's hearts is always open to this proceeding if they consent. If they are willing, we work; if they are passive, God rules. If they are passive, God rules in them and acts through them on other men. Who will rally to the help of the living God? Who will be on His side in the coming, the already commenced, contest between Him and man's free-will? Choose ye now whether you will serve God or the world; God or man; God or your own free-will.

CHAPTER XXXIII

ANTEDILUVIAN HISTORY

History of Antediluvian Life upon the Earth

When the earth assumed is globular form and became the residence of man, animals had existed upon it for hundreds of thousands of years. First, fish were the inhabitants, slightly above plants. Gradually the highest form of animal life rose in the scale of creation, by the development of the law that spoke the earth and the whole universe of universes into being. When quadrupeds existed there appeared man's type, the monkey tribe. But monkeys lacked the living soul; the spirit from Paradise did not enter their bodies. They were like the other animals and, like the monkeys of the present day, mere animals—mere sentiment existences. But they were not without some kind of intelligence, any more than are the higher animals of the present day. They had reasoning powers, though these were limited. They could form governments and establish laws. But their laws were simple, scarcely extending beyond the limits of personality. Property was recognized only by possession, and personal rights were nothing more than the right to roam unmolested. But there was a sentiment of justice instilled into their being which stood in the place of many laws, and the establishment of government is easy when beings are already under the control of justice. These animals were somewhat superior to the highest of the present relative tribes, or races, as men call them. They even assembled in large communities and erected huts in a kind of orderly arrangement.

Long before men appeared, these animals had subdued the earth's surface to orderly cultivation in large districts, and preserved its security and peace by the destruction of the animals that would, by their abundance, have injured the harvests or their domestic animals. For the various useful or domestic animals were trained into subjection by this race of superior animals which had thus prepared the earth for man's residence. Man is almost helpless without the aid of animals. He may be savage, but can scarcely be a civilized, a refined, or an intellectual man without having their services available for his wants and desires and whims. After the monkey or baboon race then existent had reached its highest development, man appeared by the operation of the same law of progress which had so carried matter forward as to make his presence a want or necessity to its perfection and beauty. Man made his appearance first in a few individuals of a lower form than at present. By lower, I mean more animal, sensual, gross. This primitive man was larger, stronger, and longer-lived, as I have before intimated. But the first creation or appearance of man was either by an act of matter, or else it was by an act of Deity or His Word. By His Word all things were made. So by it man was made.

Can you understand the process by which matter assumed form and being and sentiency as an organized body, prepared for the reception of an immortal emanation from God? I fear not. My holy medium found some parts of what I have before written too high for him, and wisely left it behind in reading the book, as above his comprehension. Not that he despaired of understanding it, but he resolved to take time to compare and weigh, resolve and combine, study and ponder, that he might understand it thoroughly. This you must do, O learned man, if you would fully appreciate and understand My next chapter.

CHAPTER XXXIV

DEVELOPMENT OF MAN

History of Man's Formation and Improvement, from the Beginning to the Present Time

Man, being prepared by the Word in the course of creation or development, was found in the bowels or matrix of a pure specimen of the highest order of animals that preceded him. He was then very similar in form, whatever he was in interiors, to the lower animal. But since his mother had been selected with particular regard for the event or circumstance, and since the mother of that mother had also been so selected, the possibility of improvement is evident enough to human reason. The father was selected as well as the mother, for by two consecutive proceedings on the part of the Word, the two mothers were induced to conceive an embryo without an animal congress. The result was a being highly developed, the admiration of its mother and of all the animal race.

How, you say, can such a result be possible without showing that Jesus was not the only begotten son of God? I will explain this by recalling to you that women in the present day do often so conceive an embryo. Virgins of unspotted, unsuspected, and real virtue and purity have born them. To be sure, they do not come to maturity, are seldom expelled, and are more often outside of the uterus when found. But the fact is well known that hydatids often occur in pure-minded women who have had no sexual congress. Then all that is wanting is to have sentiency impressed upon or imbued into the body thus formed, and it would progress to maturity. This the Word does, and the Word also disposes the constitution of the animal or woman to a state in which the hydatid must inevitably be found, and provides that it shall occur at such a time as will ensure its safe progress to the uterus. For it will always make such safe progress if it occurs at the period of the monthly return or catamenia. But at any other period than within forty-eight hours of that cessation it will not pass into the uterus. Then what is the reason that more do not occur in this way? First, the chances are as twelve to fifteen against one that it will not occur at that time; second, there is the further chance very much against it that it will pass off as a foreign body at or before the next return.

How, then, is this different from the conception of the virgin mother of Jesus of Nazareth? She was also operated on by the Word. For by the Word all things were made and therefore the only begotten Son of God must also have been made by the Word. But in this case the Word Himself took flesh; that is, He became the soul or sentient portion of the infant born to Mary. How, then, do we understand you when you say that the spirit or soul of Jesus was selected for that particularly prepared body because its desire in Paradise had been to do good? will be asked by many attentive readers—My holy medium included. It was by the Word that the spirit was selected, and the Word also joined itself to the spirit by an intimate parital union such as exists in paradise regularly and invariably with all spirits, but which is generally dissolved before leaving that state by the law of progress that leads the spirits to want to leave that place and situation of existence. It was, then, the Word and a spirit of man from paradise like other souls of men, except that its motive was to do good, that formed the interior of Jesus of Nazareth? It was. The Word took flesh and we beheld His glory as of the only begotten Son of God. Is not this plain now? You thought you understood it after the explanation in the first book, but now you see a higher meaning. There is a yet higher meaning which I cannot yet make you understand. So we will leave the subject and return to man's origin or commencement of existence on the earth.

When the body had been thus prepared for the first Adam or man's spirit from paradise, the adam or soul entered it in the usual way at its first inspiration. Animals do not cry out at the first breath; men do. What is it, then, but the soul's entrance that causes the manifestation of pain? Nothing else than that can account for it, though nurses and physicians have thought the reason to be pain the air gave the lungs. But if this were the reason, it would produce the same effect with all animals born in a similar manner.

There was then one, but I said there were several individuals at first. Some ten or more were selected for the first part of the process, which was carried on simultaneously in several cases. Their hydatids matured and were monkeys, to give them a name you are familiar with and which expresses the idea of their nature. They were monkeys superior to their fellows and chosen for their superiority to be the rulers of large bodies or communities.

From these ten or more, two were selected who bore bodies called or referred to under the names of Adam and Eve, in the traditions of that event. These were so decidedly superior to all other inhabitants of the earth that all the previous race then extant submitted willingly to their authority, and thus all the beasts of the field and fowls of the air, as it were, submitted to Adam by the submission of their masters the monkeys, or primary animal. Thus man appeared, or, as it may be called, thus was he created.

The man of the antediluvian world was a very different being from the present man, as I have before intimated. He was larger, stronger, and more sensual. He was also six-fingered, six-toed, and bull-necked, as the human neck resembling his is now termed. He had a tail, and it was the apparition of beings of antediluvian birth that caused the popular notion of the appearance of evil spirits with tails. He also had horns, short and straight, proceeding from his forehead. These grosser and more animal parts gradually lessened in development until near the deluge. Then again, Noah was produced by a hydatid from a selected mother, as was also his wife. This pair commenced a new era in the history and the form of men. As before, their superior appearance caused them to be promoted to sovereign authority over the great mass of a powerful nation, whose sway was almost universal and whose power and commerce was universally extended.

Here we close this branch of our history, saying merely that analogy will properly teach man that as he originated, the lower orders of being, descending even to vegetable and mineral, were also originated. As he originated, so may a higher developed body, and, he may infer, probably it will originate. The same analogy will lead him to infer that the various races on the earth have been found one after another, beginning with the lowest, and that there have been successive developments by hydatids from lowest of the negro, or still lower New Holland race, to the highest Circassian or Saxon type. Analogy also will teach him that a greater change of form will occur when a pair shall be got ready for the next crust of the earth, which will be when the second moon is disrupted, when the present or then-existent inhabitants must perish from their bodily existence by another confusion of the elements like the deluge.

But has not God set His bow in the heavens as a sign that the world shall never again be destroyed by water! He has. Though the confusion of elements will be similar, the outer crust of the earth is now so much farther removed and so much thinner, that the confusion will be less, and there will be large numbers of men left on the ring, which will continue to be inhabited by them. The polar regions as before will fall or be attracted to the central portion, and again the axial rotation will be changed. The newly formed race will people that body, and they and the present race in their new satellite will be subject to many great and destructive operations by the changes their respective habitations will necessarily undergo in assuming, the one a spherical, and the other an equalized, form and a solid crust.

Let us return to the history of the Spiritual Influx as manifested in the establishment of religion in the earth.

CHAPTER XXXV

SPIRITUAL DEVELOPMENT

Rewards, to Spiritually Minded Men, of their Progress

Man at first had no religious notions other than such as were common to the lower animals. But Noah was divinely inspired, and endeavored to awaken in them a desire for spiritual progress. He did not succeed in turning a single one of all the race to a surrender of his will to God. But when he taught his descendants, he warned them by the fate of their predecessors on the earth to be attentive to the Divine Influx, and for thousands of years they were. For many ten thousands of years they obeyed the warnings and submitted to the direction of God's spirits, transmitted through the various holy mediums who were trained for that purpose. They, too, were individually attentive to the Divine Influx or Word within themselves. Yet their disobedience and want of submission was so great that none of them made rapid advancement in the spirit world or existence. How is it that they were so obedient and yet so disobedient? They were obedient to the holy mediums but not to their own reception. It is the latter that effects the salvation of the soul. Then holy mediums ought certainly to make rapid progress in the next state! Not necessarily, because a medium may be used without his being passive. He may be passive to reception without being submissive in action. This is the state of My present holy medium. He receives passively but he reserves his action until I withdraw. Then he acts in his own will.

But do you not require this kind of action? Do you not say that mediums should act for themselves and not leave their whole efforts to God! No. I say mediums should desire to submit their actions to God's will, and that to be perfect they should have no will of their own. But this would raise them to the sixth sphere! Not quite yet, My reader. The will must be passive; then they receive correctly. Their actions must be in accordance with what they receive; then they are submissive in action. Because they are told they may do a thing once, they are not to suppose they may do it again or all the time. Because once they have resisted without evil results, they are not to suppose they may continue to be resisting. No, submission comprises not merely the surrender of the will, but a seeking for direction in order that it may be obeyed or followed. Then God will direct and the holy medium can act in the will of the fourth sphere, which is the highest to which man in the body can arrive.

What then do I lack? All these have I kept from my youth upwards! Sell that thou hast, give to the poor, and follow Me. This was the answer; this is the answer to all who think they have done anything. But does this mean that you are to work no longer? That you are to sacrifice your property, your business, your family, yourself, and to wander about as an object of charity? Oh, no. It means spiritually that you should have nothing: no will, no power, no action except as you are directed to have them; that you should part with everything that you suppose yourself to be the spiritual possessor of, so that you can offer your mind or soul as a pure blank tablet upon which God may then write what may best please Him; that you should ever be ready to dispense spiritual bounties to those who need them; and that you should do not only this but that you should do as Jesus did. He sacrificed Himself fully. He gave Himself, a ransom for many, for He sacrificed His time, His comforts and enjoyments of the home circle in order that He might preach and warn and persuade and threaten the sinner or the ignorant teacher or professor.

When shall you begin to prepare for this progress? If you do not begin now you cannot progress as fast as you may. "Time once past never returns." Let the past take care of itself. Let the dead bury their dead; do you press forward to life eternal. If you do not begin now, you may make no beginning in the body. You may do worse; you may retrograde. Then begin now, whilst you feel some inclination, whilst you can perceive, and I do see in your heart or mind some inclination to do so. Begin, and I will help you. Begin by making the prayer I gave you in the twenty-first chapter and progress to make it with the additions in the twenty-eighth chapter, and you will make progress; indeed, when by according with it you have made that prayer your own, you will already have made progress—great progress. Let it be for awhile your daily prayer, for until you have fully mastered every desire in you contrary to its spirit and meaning, you cannot have peace. When you have mastered all those contrary desires and laid yourself in submission of your free-will at the feet of God, you will have that peace which the world cannot give, neither can it take away.

Perhaps you have never had a taste of this peace. If you have not, you know not how great is the reward I offer you. It passes all understanding, and the reason of man can never comprehend it. It must be experienced to have any correct idea or knowledge of it. It is as far beyond contentment as contentment is beyond repining; as far beyond joy as joy is beyond sorrow. It is quiet in its manifestation but deep in its channel. It flows ever from the pure fountain of bliss which wells in the throne of God and proceeds continually from it to every part of the universal whole, a higher universe than I have yet spoken of. Before, I spoke of an association of associations of associations combined into a whole; now I speak of a combination of these last named combinations arranged into the Great Universal Whole. Does this comprise the whole of creation? No, finite reader, infinity cannot be described to you in language comprehensible by material minds, as all minds are to a greater or less degree whilst in the body.

Let us pause, then. God's bliss not only proceeds to every part of the universal whole but it proceeds to every part of the infinite creation continually. It is never in the least degree scanted or lessened. From everlasting to everlasting; that is, from one indefinite period, it proceeds unabated, without limit, without cause except God's will and mercy, without money and without price. It is this bliss which, entering into the soul of man when he has submitted himself to God, becomes within him that peace which the world can neither give nor take away.

Let us pray

O almighty, everlasting, and unappreciable God, may it please Thee to look with ineffable mercy upon this reader of Thy Revelation, so that he may understand and believe it; so that he may comprehend and receive it; so that he may feel and know the certainty of Thy Divine Word herein contained—Thy Word of Thy Power that took flesh eighteen hundred years ago and now desires to penetrate and pervade the bodies and souls of men in this transitory state of existence which they call life. O God! may it please Thee to aid by Thy power, sanctify by Thy grace, establish by Thy will, and confirm by Thy love and mercy, the good desires that sometimes arise in the heart of this reader. May it please Thee to help his every effort to control his passions, to overcome his base inclinations, his unworthy motives, his unwise resolves. O God! be merciful to him, a sinner. O God!! have pity on him, a low son of Earth who has aspirations at times, and hopes always to reach forward to something better without knowing how to progress or what to desire. Prepare him, O God, for advancement into the life to come, and for the union and communion of Thy holy spirits who desire, O God, to be his helpers and to serve him, as willing servants of all whom it pleases Thee to raise to the high and holy calling wherewith are called all the spirits in Thy Paradise and in every stage of their existence. Help us, O God, to do Thy will and perform Thy pleasure, and be our Mighty God, our Everlasting Counsellor, our Prince of Peace and not only, O God, to us, but to this reader of our revelation of Thy will. Save us and be our redeemer, O God, and help this reader with Thy sure power so that he too shall be speedily redeemed from the law of sin and death. Be his comforter, O God, even as Thou hast been our comforter, and be our helper to help him. Amen.

CHAPTER XXXVI

THE MOON

Changes of the Moon's Surface

The causes that hold the moon in an orbitual revolution at a rate precisely equal to its rotary one are interesting and instructive. Their explanation will also remove an objection or argument against the revelation I have made known of the formation of this body from a ring. It is easy to suppose that bodies having a rapid axial revolution might become round after winding up into a ball or spherical body. But how could the moon get this spherical shape and have an axial revolution precisely equal to its orbitual revolution?

At first the moon's rotary or axial revolution was quite rapid, produced in the way I have described. Then it ceased to revolve in consequence of a flattened pole. This became so flat as to be thinner than would be self-supporting. It collapsed and fell to the inner crust. The inner crust, again having no counterbalancing attraction to sustain its equilibrium, also met by attraction the opposite side of the outer crust. So the moon became really a shell open at one part of its periphery, containing a ball resting on the inner part of the shell opposite to the opening. It then presented this heavy side, where the two crusts touched or joined each other, to the earth, and by the earth's attraction it is ever maintained in that presentation. The opening, which is about eighty degrees across, is consequently ever invisible to the earth's inhabitants, though it is seen from other planetary bodies and from the sun. It is also so small as not to interfere with the presentation of a globular shadow during eclipses. This form of the moon is an anomaly in this solar system. But other systems have similar cases though they are comparatively rare.

CHAPTER XXXVII

Nature of Heat, and Condition and Climate of the Sun and Other Bodies

The cause of the supposed increase of heat towards the center of the earth is the concentration or solidification of matter, which is continually going on. By this the latent heat of gases, liquids, and softer solids is set free. This heat then reaches the surface of the earth's crust by degrees, by transmission through the solid matter. When it reaches the surface it is dissipated again into the gases and atmosphere, which retain and multiply and guard it.

But the atmosphere does not grow warmer; at least it has not within the memory or historical records of man, but rather the contrary! The caloric, or heat, which is a definite substance as much as is a gas, extends itself in an extremely rarified form in the upper or outer regions of the atmosphere, and would in time become luminous like the sun if it were not returned to the earth by the sun's rays, which thus obtain their heat. There is now no more heat brought to the earth's surface than formerly, because formerly that derived from the interior was much greater, since the changes from aeriform to liquid, and liquid to solid, proceeded then with great rapidity and nearer the surface. The changes are now more distant, and are also fewer there.

But the reservoir of heat in the atmosphere has increased and the sun's rays are more fervent than ever were experienced since before the deluge. The luminous appearance of all the stars is obtained from this source. The faint luminosity of the moon and of the other planets, as may be observed to exist when portions are visible to us unilluminated by the sun, is caused by this collection or reservoir of caloric in the higher or outer region of atmosphere surrounding each.

The spots on the sun are caused by depressions of its calorific stratum which themselves result from an attraction by its internal crust of solid matter, which at times draws into itself a vast portion of the outer crust, and into this chasm the atmosphere rushes. For the consequence of the solidification or concentration of the interior matter is the formation of a vacuum between the two crusts, and until the outer shell is strong enough to sustain its own gravity and form, it is liable to these collapses.

Now, with a brief sketch of the climate of the sun, I will close My explanation of the solar system. The sun receives no heat from other bodies as the planets do from it. But it possesses great internal heat because the process of solidification or concentration proceeds rapidly and on a large scale. Its surface, therefore, is warmed by its internal heat. Its atmosphere is also highly rarified and warmed by the same cause. Its light is derived from its own luminous atmosphere and it is only through the occasional openings or spots, as men call them, in its luminous atmosphere that its inhabitants can look out upon the glories of the great expanse. Their knowledge of it, therefore, is very limited. But the beings existing upon its surface are of a high order, because they are the result of successive formations like Adam and Noah, taking place after each successive departure of its attendant planets. In all other respects of its scenery and inhabitants it resembles the earth and other planets.

Comets are fragments of atmosphere arising at the times of disruption of planets or planetary rings from the sun. When they approach the sun they become luminous by the reflection of its rays from their denser portions. But this denser portion becomes elongated by the powerful attraction of the sun, which brings its more solid portion into an accelerated progress as it reaches nearer and nearer the focus of its orbit. None of these bodies extends far beyond the outermost planet's orbit, though some reach so far as to be lost to the sun's attraction and fall into the atmosphere of some other body of the solar system.

The Aurora Borealis is caused by a movement in the stratum of the atmosphere, which is highly calorific, and the movement of its particles makes the calorific stratum luminous, thus forming a faint representation of the manner in which the sun is heated and lighted from its own luminous atmospheric stratum.

Now a word upon aerolites or falling bodies which occasionally reach the earth and are often seen in their progress through the luminous stratum of atmosphere, where their rapid motion produces such a disturbance as makes visible their course but not their mass. These foreign bodies are the fragments of planets and of the sun, set free at the time of various disruptions of the rings of those bodies, and since then revolving in erratic courses about the sun or the earth. At first they are gaseous, then fluid, and finally solid. They are, in fact, comets solidified and, like most comets, small. Very few of the comets would weigh twenty tons if placed upon the earth.

CHAPTER XXXVIII

PHYSICAL PROGRESSION, CONTINUED

History of the Future of Anglo-Saxondom, and of the New Jerusalem

I might call this Part III, but I refrain, as it would seem so formidable as to size whilst it will be brief. I shall briefly sketch the future progress of mechanical or physical discovery or art. *But not by such particulars as will enable men to make the improvements referred to in any other way than they have been made previously. That is, by patient thought and aided by Divine Influx in

*1851—Ed.

their own endeavors to benefit mankind. Small success is ever the result of sordid motives of action in these departments.

Though railroads seem now to be fast arriving at the highest possible speed of travelling, yet ships will be built to excel in speed the swiftest railroad train now or hereafter to be established or operated. The Atlantic between New York and Liverpool will yet be crossed in twenty-four hours by power acting upon its waves. Balloons will be produced that will navigate the air with considerable success. But their results in voyages must always be irregular and they will bear to the rapid ships the same relation that sailing vessels do to steamships. A few occasional extraordinary voyages may almost equal the more perfect form or manifestation of power of movement. But the great average will be far behind.

Shall land travel then be stationary? Oh, no. Railroad trains shall yet reach a speed of one hundred and twenty miles an hour for loaded trains.

When will these things be and what shall be the signs of their coming?

The signs are evident from the past progress of men. Look back fifty years and see what has been done. Look forward fifty years and imagine an accelerated progress. For acceleration is the inevitable result of progress unless some other principle interferes to counteract it. But you do not know whether that other principle may not interfere now or soon in this matter! Well, let that pass. I tell you what will be. You may judge hereafter how worthy I am of belief. And if you are wise you will conclude that this entire book is truth and nothing but truth.

The time for the greatest of these improvements will be after the downfall of British power, which must and will fall before the last great successful effort of the dragon of prophecy, the seven-headed and ten-horned monster, the last phase of the fourth or Roman monarchy! Yes, the mighty power, the vast empire that the genius of Anglo-Saxons and the favor of providence has so rapidly established, and now so wisely sustains, must be resolved into another form.

Let us recur to the Book of the Revelation of John the Divine. There we find the beast, properly rendered the living creature or the seven-headed monster, which is there put for Daniel's fourth kingdom, will combine with the false prophet. The dragon will give his power to the false prophet and they will place a mark upon men so that no man shall buy or sell unless he have the mark. That is, they will restrict men from preaching any other religion or doctrine than they please to have preached. No man shall buy or sell any other spiritual matter or thing than they have marked out for him. That is the meaning of the passage. For the prophecy relates mostly to internal or spiritual matters. This is rapidly becoming the case in reformed church government in Europe, as well as in Roman Catholic church government. They are beginning to combine to tolerate no other.

The British Government forms the only European exception to this state of progress, and this will the more incite the combination of the dragon and false prophet. They will persecute the Woman. Britannia is the Woman. Her child is America or, more particularly, the United States of America. Her child is upheld or protected by its national emblem: the two wings of a great eagle. But the Woman was not destroyed, for the earth helped her and drank up the flood which was cast out of the mouth of the dragon. The earth is the flood-drinker which is the absorbent of all that the dragon casts out after the Woman.

The earth is the continent of America. It will receive and absorb all the armies which the European nations shall send forth from their shores. It can absorb them without injury—indeed with benefit to itself. It will thereby be rendered more prolific. What, will Britannia be in America, that America shall absorb the waters or floods of men which shall be sent forth to fully overwhelm and completely destroy the Woman of Britannia? Yes, there will be found the refuge of Britannia's nobles, royalty, and riches. There will be found every true Englishman, every high-minded Anglo-Saxon, whether England, Scotland, or Ireland is the land of his birth. There will all seek refuge when invasion shall have conquered and power overthrown that liberty of conscience, that security of personal rights, that guarantee of property and of liberty of speech and action, which is the boast of the native Englishman, the glory of the British Constitution, the first of the Anglo-Saxon laws, institutions, and character. Will this be in our day? Yes. The day is near at hand when in an hour all shall be destroyed. The modern Babylon shall become the prey of the spoiler. That city, never yet conquered, shall fall to rise no more. It shall become the residence of every unclean thing which the foulness of Europe can pour forth. In it shall no more be found the peaceful pursuits of industry. It shall decline and be heard of no more. And all the spectators standing afar off upon the shores of America shall say, Alas! alas! that great city, for in one hour is all her glory destroyed. And the shipmaster and those who go down to the sea in ships shall weep and mourn for no man will buy their merchandise any more. Yes, freights will be dull. Ships will rot in the ports, for commerce will be destroyed by the fury of the war and the ships of Britannia shall seek refuge in America's ports.

The colonies of Britain will gladly coalesce with the United States when the British Isles shall be ruled by the seven-headed monster. One mind and one thought, one government and one nation shall then comprise the Anglo-Saxon race. The mind and thought that pervades it shall be resistance to tyranny and the destruction of tyrants.

Then will commence the real struggle between the past and the future, the fourth and the fifth monarchy. Then will all the powers of earth and hell be arrayed against heaven and God's spirits. But the armies of Jesus shall follow Him. His sword will bear the inscription of the WORD OF GOD. And can you doubt as to who will be victorious? But if the earth be America, will not that be on the victorious side? Only when America by her inhabitants shall have submitted to be led by Him. But He goes forth conquering and to conquer. He has already mounted His courser. He is riding now His white horse. He is King of Kings and Lord of Lords and in Him is salvation and power and glory. Submit then, reader! to Him. Give Him your heart now. For the great day of battle is at hand and the blood shall flow so that it shall be up to the horses' bridles.

The earth here stands not for America, but for the power of man. Men under their own guidance. But the armies of heaven will be composed of such as are led and guided by the Lord Jesus Christ or His servants. And such guidance and leadership is the same as that of God, as I have before shown. Death and hell shall be taken captive and Satan shall be bound a thousand years, after which he must be loosed a little season. Death alluded to here is death of the soul or separation of the soul from God, not its separation from the body. Hell is the punishment received for sin which is, as I have shown, the want of happiness, the existence of unsatisfied desires, the realization of man's hope which never satisfies him or makes him happier.

And Satan is the accuser of his brethren, which is also the outward desire, the free-will, of man. This leads him to glorify himself at the expense of his consideration of his brethren and to accuse them in conversation or thought of evil desires, bad motives, and unworthy actions of which there is no other proof than the desire in his own heart to do the things so charged upon the brother man. He will be bound a long time for the Day of the Lord. For a thousand years are as one day, saith the Lord. So declares the Psalmist, and so this was intended to be understood. That so long as the Day of the Lord continued to exist in a man, so long Satan, the accuser of his brethren, would remain bound, and when that day ceased by the man's leaving his state of submission to God, then Satan or the accuser would be loosed for a little season. He would then go about as a roaring lion seeking whom he might devour, and the last state of that man would be worse than the first. He would gather together the opposition to God from every place in which it could be found, and in the valley of Megiddo, or of slaughter, he would be overthrown and the camp of the saints of the most high God would be established in safety after the death of the body. Then Satan should be finally bound and placed in the bottomless pit and a seal put upon him that he should deceive the nations no more. Now I have explained this in the past tense, for it has taken place with men continually for a longer time than since John wrote, but it is also true in a future tense, for such will continue to be the course and experience of men in the body.

At the last will descend the New Jerusalem, arrayed like a bride for the arms of her husband. It will not be an outward city but an inward residence for the saints of God in the heart of man. When man yields his free-will in submission to God's will, he will find this city coming down from heaven. It will be to him as beautiful as it is described by My servant and holy medium John. But it will also be the purified and sanctified residence of Myself. For I will be the Comforter to him who submits to God and becomes passive to My holy influence. To him will I be King of Kings, and to him will I be Lord of Lords, and to him will I be Kings of Saints, and to him will I lead the armies of Heaven with the Word of God upon My sword.

But is there not to be any other sense found in this revelation or vision! Yes, there is also an outward sense, for in all that I delivered to John there is an outward and an inward sense. The outward sense has been seen and declared by Protestant commentators as far as the prophecies have been fulfilled. The last is now near fulfillment. The fifth monarchy of Daniel, the holy city of John, is about to be established on the earth as an outward form. The United States already exists as the fifth kingdom. The holy city is proclaimed in you by this book. When I shall have proclaimed it still further, I shall make you willing to have it come outwardly.

The signs of its coming will be a general belief in My revelation. I will establish them by signs and miracles in My own time which is near at hand. I will raise up servants or holy mediums in all parts of this kingdom who shall declare its truth, who shall be willing to sacrifice their fortunes, reputations, lives, and families for it and for their faith. Verily I say, they shall have their reward. Well done, good and faithful servant! shall be their great and exceeding reward. But not a hair of their heads shall be harmed. No smell of fire shall be on their garments. I say unto you that he who shall give up father or mother, wife or child, lands or houses, ambitious hopes or political consideration, shall receive a thousandfold in this bodily life or state to come; in the spirit world, life everlasting; life eternal in due time. Fear not, I am with you to the end of the world. On the Peter, or Rock of Faith in Me as the Christ, the Son of the Living God, I will build My church and neither the gates of hell nor man's opposition shall prevail against it. Be ye also ready, for I am coming soon. Be ye also ready, for ye know not the day nor the hour when I shall come. Be ye also ready, for as soon as you are ready, I will come. I will enter your heart when you submit to My will. And My will is God's will.

Let Me, then, once again entreat that you lay aside every prejudice of education or tradition, every worldly excuse of want of time or opportunity, every desire of self-gratification like love of ease or power, or of consideration amongst bodies of men, every form of church censure, every reliance on worldly judgment—that you resolve to go down into Jordan, the lowest valley of your country or heart, and be baptized with the Holy Spirit and with fire. This is the baptism I called My followers to eighteen hundred years ago. And this baptism by fire is a baptism of God's love, that as consuming fire will purify your wicked heart of every impure desire, every unworthy motive, every unholy aspiration, every desire to do your own will, and implant in it the ashes of joy for mourning, and the oil of joy for consolation.

Let Me entreat you to submit whilst you have the free choice Accept My invitation now, whilst you can refuse. Do not, O hardened heart, refuse to admit Me because you have the power of reason and can argue after you are convinced. Do not refuse Me because you would show your stronger mind, your really rebellious disposition. Submit to Me as a little child submits to its father's teaching. Receive My authority as parental. Be ye as little children, for of such is the Kingdom of Heaven, and unless ye become as such, passive, obedient, loving, and reverent, ye can in no wise enter into the Kingdom of Heaven, though that Kingdom of Heaven is within you except that ye refuse to have it there.

New let Me appeal to you once again by every consideration of your own and others' good, by every desire you possess for true happiness, to turn once more to the prayer of the twenty-first chapter and strive with all your power to enter into its spirit, and in reading it to make it your own. It is only your free-will I ask you to surrender. And I ask you not to give that to man, who might make a bad use of it, but to God—to His holy spirits who will let you work in their will, which will be a great deal better. God is wiser, happier, better, and lovelier than you, and if you act in His will, you must be brought to such resemblances to Him and His nature, and your manifestations must come to be so much like His as to make you declare with joy: I give thanks, O Most High God, Father Almighty, that Thou hast been pleased to make known these things to babes and sucklings, in men's opinion, and to withhold them from all who will not cease to be strong men. Now, My dear reader, let Me again ask you to turn to the twenty-first chapter and make the prayer there your own. You will so find that God is good, and that in Him is no shadow of turning. Read it as yours, and say Amen in your heart as if you had composed and offered it by your own intellect. Amen.

CHAPTER XXXIX

THE TIME OF THE END

*Present History of Anglo-Saxondom and the New Jerusalem;
Present Call on All Men*

When I left the theme of America's future, I said I would portray some of the features of the future greatness of her extent and power. Let us, then, once more return to the consideration of Daniel's two visions and his interpretation of the vision of Nebuchadnezzar, and the part of John the Divine's Book of Revelation which refers to the fifth monarchy of Daniel.

First I will recall to your memory that the fifth kingdom was to have no end, and that the fourth kingdom was to exist until the commencement of the fifth. The Roman or fourth kingdom has continued by a constant succession of princes under the names of consuls, emperors, exarchs, and popes, and has been distinguished always as the Holy Roman Empire since Christianity was the religion of the state. Was not, then, this empire the universal reign of Christ when His worship was extended over all of it? By no means. Where do we find the city of peace, the New Jerusalem which was to come down from heaven? Where do we find the great gathering of armies alluded to as to be in the latter time when the dragon and his angels fought and prevailed not? Nowhere in the history of the past. Let us see when Daniel declared the time should be that the fifth kingdom should commence. Unto twelve hundred and sixty days or years of men would the time be, after the daily sacrifice should be taken away and the abomination that maketh desolate set up. These are the times of reckoning. From the time first mentioned and from the time last mentioned we may derive the exact time when the existence of the fifth kingdom shall commence. From the time when the power of the Pope of Rome was fully established as an abomination that has since desolated Christendom, to the declaration of the independence of the United States of America is 1260 years. And what was the daily sacrifice that was then taken away? It was the sacrifice of the heart which was then no longer required, but indulgences and pardons for sins were granted from that time by popes, bishops and priests. The Greek branch of the Christian church, too, went astray at the same time. They, too, declared the head of it to be infallible and endowed with power to forgive the sins of his fellow-men. This was not unguardedly claimed, as by the Roman Church, but still the claim was made and established.

But then Daniel has referred to another time: the twelve hundred and ninety days or years. Blessed are they who continue to wait for that time. Then the last period given is the thirteen hundred and thirty-five years. At this time should the end begin. And this time has expired. The year 1851, so called, of the Christian Era fulfilled and completed the prophecy. But the armies have not yet appeared under the leadership of the dragon and the Lamb! The New Jerusalem has not yet descended like a bride adorned for her husband! But the time has come when these will occur and have occurred individually. I, however, admit there is also an outward signification which must equally be true. The armies are assembled. They have had one great battle in Europe during the year 1848. They will have another presently. The last great battle shall be in the coming time, but very soon. Then the time has not yet arrived when the kingdoms of this world shall be the kingdoms of the Lord Jesus Christ! Not outwardly. Spiritually, His kingdom is established in some minds. But it is near at hand now with many.

Where, then, shall we look for the outward New Jerusalem! In America. It came down from heaven in 1776. In the succeeding thirty years it acquired strength enough to declare war against the dragon, then represented in its temporality by Bonaparte, Emperor of France and of most of Europe, but certainly Master of Rome. But did the United States declare war at that time against the Emperor of France and Italy! Yes, in effect they did when they threatened war if their demands were not complied with. But a peace had just been concluded and a territory acquired by the United States from France! It was wrested from the dragon by fear of its loss to the Anglo-Saxon mother country, and by the demands of the government of America. Its cession and acquirement, though peaceful outwardly, were none the less an outward triumph. Again let Me remind you that the last of the times set forth expired in 1851. In that year liberty, extinguished in Europe, fled to America. In that year the last remains of religious toleration began to be extinguished in Europe, whilst even England was driven to further resistance to the spiritual and temporal assumptions of the power of the dragon.

But, then, how were they so blessed who waited and came to that year? Because in that year My revelations commenced through My servant Hammond. I caused lower spirits to deliver to him *Light from the Spirit World*. Did this produce great consequences? It awakened some; it confirmed others; it led to the establishment of My holy medium in passiveness. He as a consequence became qualified for his high office, that of being passive in My hands and delivering to the world or inhabitants of Earth what I choose to reveal. He is improved by his reception of this book and has resolved to serve Me only hereafter as I may direct.

I shall use him more. But not merely in writing. I shall use him to declare verbally and orally My revelations. When called upon, he shall go forth with power to perform miracles and to make outward signs even as I may direct him to reveal their coming or intended performances. He shall have power to raise the dead in sin to a knowledge of God, and to reconcile or heal all who are sick at heart, lame in spiritualities, from hostility or opposition to Divine Influence. He shall be also a worker of outward signs such as healing the sick and raising the apparently dead. But when shall these signs appear? Whenever he shall declare them as at hand. I will speak to him at the time they shall be done and he shall obey Me in making known their intended performance. But shall he not fail to succeed at times? Yes, he is not so entirely submissive to My will as he will be and as he should be to be free from rebellious desires and unwilling performances.

Let us pray

O God! Almighty Helper, and Everlasting Father, may it please Thee to make Thy servant L. M. Arnold a patient, submissive holy medium of Thy communications to mankind, so that he may be passive in Thy will in the hands or will of Thy Holy Spirits. May it please Thee, O God, to accept him with all his imperfections, with all his shortcomings, and to pardon him for all the manifold sins which a long period of worldly-mindedness and mingling with the world as a part of it have impelled him to, and his own free-will has helped him to perform. But, O God, may it please Thee now to let him atone for them by being Thy servant in this life in the body and Thy son in the life to come, in the spirit. And may it please Thee to manifest through him Thy power and wisdom, so long as he shall clearly and fully give to Thee the praise, honor, and glory of all his works as of right it belongs to Thee both now and forever. Amen.

The holy medium accepts as his own the prayer which I have made for him to the Father. Will it be granted? All power is given unto Me both here and in Heaven. Why, then, need I pray to the Father? Because the Father's will is that all His sons or spirits of every degree shall have all power through Him when they submissively ask Him for it. And because I am His son, possessed of this power, extensive as I have previously shown it to be, I am in possession of it as knowing how to use it, as having My will in such perfect submission to His that I always act in His will and never in My own. But is not the prayer in Your own will? Not at all. It was God's will that I should pray to Him and it is pleasing to Him not only as a manifestation of My submission, but because it is a pleasure to Him to grant the desires and petitions of His servants and sons.

I have not written the explanation of the prophecies as I desired to. My medium was not in a perfect state of passiveness, though he tried to be. This I shall have to leave until a future time. I will only say that the Time of the End has commenced. The fifth kingdom is established on a firm foundation which will withstand all assaults. Let earnest seekers find the truth by looking to their own internals. There I will enlighten them. Let them read the prophecies and compare them with each other and with the history of the past, and I will help them to understand. The Lamb with seven horns and seven eyes in each horn is He who is now advanced to the seventh circle of the seventh sphere, and He is worthy to open the Book of Seven Seals. He has unfolded it or broken its successive seals. Its successive trumpets, seven for each seal, have sounded. The last trump has sounded and the kingdoms of this world have become His. To Him be glory, honor, praise, now and forever and ever, world without end.

Let us pray

O God, who art Worthy to have all honor, praise, thanksgiving, and glory! be thou the Enlightener of those who seek knowledge. Let knowledge be increased, O God, as thou didst cause to be declared to Thy servant Daniel it should be at this time. Be Thou, O God, the fulfiller of the desires of Thy servants, and lead and help them to desire such knowledge of Thy hidden things as may be profitable to them and to their fellow-men, and to Thee shall be eternally honor and glory, thanksgiving, power and dominion. Amen.

Be merciful, O God, to those who do not believe this revelation. Let Thy power not destroy them by the destruction of their wills. But let Thy power so manifest itself as to overpower and master their reason. Let them be satisfied, O God, that this book could only have come from Thee and that Thy servant, the holy medium, had no other part in it than to receive what I, Thy son, formerly called Jesus of Nazareth, now the Son of God, of Thy Love gave. May it please Thee so to show forth Thy power through the other holy mediums of Thy spirit that the eyes of all believers in them may be turned to these truths, and that they may thereby be led to sacrifice to Thee their own wills, and hereafter to act in Thine. Let us all unite, O Lord, to establish the kingdom of Thy power, the reign of Thy saints. And to Thee they shall ever give praise, honor, thanksgiving, and glory without end.

Almighty and most loving Father and Friend, be Thou very gracious to me, Thy humble and unworthy servant, or would-be servant. Make known to me Thy will and help me, O God, to do it, for I am desirous to serve Thee in Thy own way and as Thou mayst direct and guide me. O God! help me for I am weak. Give me Thy strength and help me by Thy wisdom, for to Thee shall be the glory, honor, and praise forever and ever. Amen.

Let us pray

Be pleased, O Most Kind and Benevolent Father, to grant the above humble petition of Thy servant the holy medium L. M. Arnold, made as Thou knowest it was by his intellect after writing in Thy will in this book, and after having been confounded by the revelation he had received and written. Be his Helper and his Guide, and lead him into perfect submission to Thee, the only sure steadfast Supporter, the only true and perfect Counsellor and Guide, the ever-sure and ever-perfect Lover and Bestower of gifts to those who ask them in submission to Thy will. Be the Helper and Friend, O God, of all the mediums Thy lower spirits have educated, and as they submit to Thy will and cease to act in their own or other men's wills, may it please Thee to raise them to Thy right hand and establish them as Thy holy servants.

Let us pray

Almighty and most loving Father and Friend! I, Thy unworthy servant, most humbly beseech of Thee that it may be pleasing to Thee to lead me to full submission to Thee and to Thy Holy Spirits, and may it please Thee to support me in every time of trial, relieve my every doubt, and console me in every affliction. Amen.

This last prayer has been made by My holy medium's intellect and is written as an example for others who may have to pass through some of the scenes or times or experiences of trial by which he has suffered and been purified. For God works by various means upon spirits in the body. His most loving dispensations are sometimes the hardest to bear. But all things work together for good, and to him who is fully persuaded of this truth, sorrow has lost its sting and the grave its victory. For what is sin but sorrow, and what is sorrow but joy, when the soul recognizes the hand of God in its punishment! What is death but the entrance into life, and what is that life but an eternal progression towards the perfection and love of God! The High Holy, Ever-Loving, All-Powerful Creator, Preserver, Savior, and God Almighty, Eternal, Incomprehensible, Omniscient, Omnipresent, All-Pervading, Infinite!

Amen.

PART III

The Spiritual State of Man, from
Death of the Body to
Knowledge of God by which
All Men are Saved

INTRODUCTION

This book is the highest production of spirits given to men in the present age. It is the conclusion of its series, and forms, with the first and second parts by the same author and through the same medium, a whole of history most interesting and instructive to mankind. But it is only a History of the Origin of All Things. The history completed would be too voluminous for the present state of man's belief. The faith even of those who call themselves believers is faint and weak. Those who ought to be its most strenuous supporters are often the stumbling-blocks which prevent the approach of others to the Great Fountain of Good.

I shall however continue to make known from time to time further revelations to mankind. And though the abundance of books now being published upon these subjects, given through spirits, may seem to some to preclude the extensive circulation of any one kind, even this series, I would assure them that these books must sell because I shall have the testimony of every holy medium given in their favor. All who write or speak or receive raps or sounds or movements of any kind from spirits shall be assured by spirits of their truth and of the propriety of earnest and profound attention being bestowed upon them. Every sincere inquirer, too, shall receive such an answer through any such holy medium as will satisfy him that he ought to be earnestly engaged in their study.

Be, then, diligent, faithful, earnest seekers after Truth and you will be established upon its Rock, and on this Rock the Eternal Church of God shall forever rest. Resting upon this Rock, no man's work shall be overthrown. It shall endure until the end of time and throughout the Great Day of Eternity. May it be your lot to stand in your place on the last day, praising God for His mercy, glorifying Him for His wisdom, and rejoicing in His love. Your place will be a happy one, and no man shall be left any longer unhappy than until the time when he can sing the new song:

> Great and marvelous are Thy works,
> Lord God Almighty;
> Just and true are Thy ways,
> Thou King of Saints.

> Amen.

PREFACE

It is left to man's own will to decide whether he will or will not receive the Good Tidings of Great Joy thus thrice proclaimed to the earth's inhabitants, in the three parts of this book. Reader, that you may be willing to serve God I have prayed that you may desire, with every power and aspiration you can summon to assist you, to receive the Truth. It is My hope and wish and prayer, but all depends in the first instance on yourself. You alone can do nothing but resist God's influence. You can resolve to serve yourself, to maintain your old faith, to refuse to consider and weigh the announcements and arguments here and elsewhere made to you, by which you could arrive at Truth and a knowledge of your duty if you would only desire it, and ask God to help you. You need not ask Him to help you to believe what I have written or to arrive at any certain conclusion. What you should do is to ask Him to help you to perceive and know and practice and believe the Truth. Pray for right direction, not for support in your present or any other particular course, except that you may be enabled to serve God by doing His will here as He would have it done on earth, which is as He would and does have it done in Heaven.

This series is complete; taken and read connectedly it will lead every sincere inquirer to the knowledge and love of God, and of His Son the Lord and Savior Jesus Christ, and of the Comforter or Spirit of Truth, which last will remain in you and abide with you so long as you submit to His teachings and desire with good desires and pure motives to serve God. It will also lead you to know God and His Son whom He has sent, for by the Word all things are made known and by the Word all good resolutions are established and helped. The Word of God is the great Comforter, the Prince of Peace, and through the Word from God proceeds to the mind or soul of every servant of God that peace which passes all understanding of him who hath not experienced it, and which nothing earthly can disturb. May it be yours hereafter in this life. So shall it be yours in the life to come!

Reader, farewell. I shall have more for your assistance and instruction, but if you can not receive this with faith, that which is to come will not benefit you but will only increase your sin and add to your future unhappiness. Farewell. May you believe and have faith. Amen.

CHAPTER XL

The Call to Men to Obey and Serve God in this Life without Delay

There is a proceeding from God now progressing in the earth which comes from God through His Son Jesus Christ, one with all the spirits of the seventh circle of the seventh sphere. From Him it descends through various spirits or circles till it is manifested in the outward form of men. This proceeding is called by men the Rapping Delusion. It commenced its progressive movement in western New York. It has extended itself over nearly all the northern states of the union, and will continue to proceed until it will be manifested in every part of the United States. No county will be without its sign, no town without its holy medium. It will be spread by outward manifestations, all of which will be of the same general character though various in details. Some mediums will receive raps; some will write automatically; others will be speaking mediums. Again, there will be those who will receive mentally as this holy medium does. And the more passive and unresistant, the fewer doubts, the greater faith the medium shall possess, the greater his power or his display of spiritual manifestation will be. The highest holy medium now used by spirits is the one I shall use on this occasion. He is not so good or perfect a one as I desire, but until I find a man or woman—for we do not make sex any distinction—more willing to be used, more submissive and patient when used, more faithful and obedient at all times, I shall continue to use him in preference to inferior ones. But whenever a better one in the respects above named appears or by training becomes so, I shall leave this one to be used by inferior spirits acting in My will.

The next manifestation of this proceeding from God the Almighty Father through Me, the Lord and Saviour of men of Earth, shall be the preaching of the word or gospel or glad tidings of great joy to all men, commencing in the United States and extending to every nation, tongue, and people or community in the whole earth.

This will commence immediately wherever by the outward manifestations first described the minds of the people have been prepared for the glory of this manifestation or revelation. Let all men, then, who desire to see the Millennium, or Great Day of the Lord, the Kingdom of Jesus Christ established in its glorious outward manifestation upon the earth, prepare for it by submission to God, by desires fervent and ardent for the coming of the Lord Jesus Christ in that spiritual form in which He must appear preparatory to His coming in clouds of glory and assuming to the eyes of outward men a bodily form similar to their own, but refulgent with light and manifesting in its appearance the glory of an immortal Son of God. For all these things must come to pass shortly. Verily I say that this generation shall not pass away till He, that is, I, will appear in My glorious appearing and heavenly effulgence. But this expression is much like that which I used in the body when I told My disciples that the signs of My coming should appear before their generation passed away! The signs did appear, My coming was expected, but outwardly I did not appear. The signs were given as an earnest that I would appear and men were thereby incited to reformation and perseverance in good.

The signs were promised and the signs were given. That is acknowledged by all. But the coming was not promised then to be outward but inward, and that coming took place. I entered the hearts of such as were willing to receive Me and have continued to do so until the present time. In this work I have not been alone. The Spirit of God, that was the manifestation of My mission to men when I was in the body, continued to aid and instruct Me after I disappeared from the wondering gaze of My disciples.

This spirit, as I stated in My second section, was a production through Saturn. He had been similarly chosen and similarly aided by another spirit from an earlier developed planet called by men the ninth great planet, or Le Verrier's discovery. For there are in all twelve planets now existing that revolve around the sun of the earth's system. The outermost will not be discovered by any instruments men now possess, but the others may be and will be very soon.

From this brief sketch let Me return to the subject which I desire now to unfold to men. It is a subject so comprehensive in its character, so glorious in its nature, so Godlike in its manifestation, that I may well pause and hesitate to try your faith and this holy medium's faith by its announcement. He desires at this time My help to make him receive it with faith and if you will also ask, O reader, you shall receive help. I will write for you a proper prayer which, if you can join in, or make your own heart breathe a fervent and perfectly acquiescent Amen, will secure you from doubt and bring you to a knowledge of the great things of God.

Let us pray

O our Father and Friend! O most kind and affectionate Creator and Parent! be pleased to give unto me Thy holy love and Thy ever-flowing kindness. Grant, O most kind and benevolent Father and Friend, to me, Thy humble and unworthy son or servant or would-be servant, that aid which shall secure me from the perils of doubt, from evil and unwise counsels and persuasions of friends or relatives here, and the influence of any extraneous matter upon me. O God! Thou canst aid me, and without help I can accomplish nothing. I, Thine unworthy servant, have

163

desired with great desire to know the truth. I ask not, O God, to be established in my present belief or in any man's belief, nor in the belief of any combination or association of men, but only, O God! that I may know the truth and serve Thee as may be most acceptable to Thee, so that I may be deemed worthy to be Thy servant, and so that I may at last, when works have justified it, hear from Thee: Well done, good and faithful servant, thou hast served Me well in small matters; thou shall now serve Me in greater and more arduous ones; so that I may be deemed worthy to receive Thy revelations as truth and Thy desires as commands. Be pleased, O most charitable Friend, to aid me in such ways as seem to Thee best and to help me to know Thee, the one true and only God, and Thy Son Jesus Christ whom Thou hast sent to me and to every other child of earth, to invite them and me to the great feast of the marriage supper of the Lamb of God that takes away the sins of the world of earth; and invest me, O God! with the wedding garment of Thy love existing in my heart for Thee and for my fellow-men.

Then, having made the prayer, say Amen heartily if you can. If you cannot, expect no benefit from the perusal of this book. He who cannot allow the Father to help him cannot have My help. But he who has the Father has Me, for I am one with Him and He is one with Me. Not that we are one being, but that we are two beings, He Infinite, I finite; He Incomprehensible, I comprehensible to men in the body. He great beyond and above all and every thing; I inferior only to Him, but equal with an innumerable company of other sons of God, as I have explained in the second part.

How, then, am I one with God? I am one with God and am to you the same God. Because, first, I have no will of My own; I only seek to do My Father's will and I do His will. In order to do it I have His power. He is the Director of all and the Controller and Sustainer of each of His sons. And not of His High and Holy Sons only but of every part and parcel of the great whole of His illimitable creation. Where then are you, the reader? Are you on His side, on My side, or are you acting in your own will, resolved on trying the Infallible Truth by your fallible reason? Are you resolved to contend with Me or God as long as you can find in any corner of your heart one rebellious thought, one unsatisfied desire, one proneness to destruction? For what is separation from God and rebellion against His revelation but destruction? What is life but to know the Father and Him whom He hath sent? Oh, man! with what perversity you resist, with what perseverance you oppose, and with what destructive power you maintain rebellion against God. One says the manifestations are undignified; another says they are incomprehensible. One says they are the operations of known agents, such as electricity or magnetism or odic force; another says they are spiritual but evil. One says they will lead to something one of these days and then he will deem it soon enough to trouble himself about them; another says they are delusions that will pass away and leave no trace: that they have been, and therefore he need not inquire into them.

Oh, people of Earth! awake from your supineness, from your indifference to the spiritual, from your absorption in the material. Arouse yourselves, ye professors of God's ministry. Try yourselves, ye who think ye stand on the church's platform and believe it to be the Rock of Ages. Try yourselves, ye who delve in toil of gathering outward treasure, who add farm to farm, house to house, money to money. Who ask continually, What is the price of stocks? but seldom, How is my soul now? And yet what will it profit a man if he gain the whole world and lose his own soul? O unwise people of Earth, God will persuade, entreat, reason and argue with you, but He will not contend with unwillingness. He will not be thrust aside into a corner. He will have your heart if you will give it, but He will not take anything else.

What then will you do? Will you give Him the only thing that He will take, or will you offer Him money for a pew, money for a preacher, money for a church, money for a missionary society, money for a Bible society? The last you can give with pleasure, but the former is a sacrifice. The latter is popular and will return to you in the esteem of men, the praises of the newspapers, or the honor of the church. The former is obscure. Men will not know of it; perhaps God will accept it, but its reward may be all laid up in heaven for you. You will perhaps get nothing back for it while you are in the body. You will, to be sure, have peace which nothing else can give, but men will think no more of you for that. Your notes can not be paid by your contentment. Your losses by unfortunate ventures in trade will not be returned to you by that inward spiritual reward of having from God, through Me, a declaration: Well done, good and faithful servant; thou hast given Me thy heart, now give Me thy money. No! but I will tell you again that he who gives to Me or to God will not find an ungrateful recipient. God will not long be your debtor even from a worldly point of view. The treasure you lay up in heaven will return you a better interest on it regularly paid here, but you will find the principal and the interest all added to upon your arrival in the spirit world. For there are no failures when God takes property to keep. Then, since it is profitable in a temporal and a spiritual view, and since the chance for reward is so great, the loss certainly nothing, will you not resolve to make God your choice and discard the world? Sacrifice what men call pleasure to duty? Sacrifice men's opinion for God's favor? Will you not be despised by men in order that you may be God's servant and that you may indeed all the sooner be His son!

Will any man say: You assure me I shall be eventually His son, whatever I do here! and therefore I will go on sinning or whatever you choose to call it but at any rate doing my own will! Yes, some will blindly and unreasonably say that, but how short is life in the body? How long is eternity! Let us place the one beside the other and see their disproportion.

It is a long while since the first settlers from Europe came to this continent. It is much longer since England received its first Norman king. The Saxon conquest of the Britons was long before that, and the Britons had long been independent after they had ceased to be a Roman province. To go back to the times of Caesar is a long stretch indeed, but to go back to the foundation of Rome by Romulus is reaching into the darkness of antiquity. Yet beyond all that the Bible record reaches to Abraham with nearly a correct chronology. And this, compared with the chronology I gave you in the second part of more than a million years to the deluge, is but a 250th part of it. And what is a million or a hundred millions of years to eternity? It is less than a grain of sand compared to the whole solar system, which is a million times greater in volume than the earth on which you stand and defy God!

O Thou eternal and incomprehensible God, forgive the reader; he knows not what he does. Oh, grant to him more knowledge, more help, more manifestations! O God! do all but take away his free-will; that I know is inviolate for on that depends his eternal and present existence. Grant, O My Father! that he may be shown by Thy power and wisdom the folly of his dreams of enjoyment in sensual pleasure when the spiritual delights of reconciliation with Thee are so overwhelmingly great and so far beyond what the heart of man has conceived of.

Reader, can you seriously declare that it is better to enjoy your life in the body in your own way than to serve God, even if this life ended with the body and the spirit, like the body, then returned to the earth? Can you really believe this world of materiality can satisfy your desires? Have you ever known a man who trusted in it for happiness to secure his object, to obtain his last wish? Can you still go on in the path broad and straight, leading to destruction of every spiritual aspiration which God implants in man to point to the life to come? Can you resolve that nothing from God shall enter your heart here in order that it may be given up to the things of time, in the assurance that some time before eternity ends you will be brought into subjection to God and be raised to His right hand as His high and holy son? No! I am sure there will be in your heart a condemnation of such resolutions. Then perhaps you will next say: God is so merciful and I am not so bad but that I will hope that I may get along pretty fast in the next life! I see a great many worse than I am! I do not violate any law! I do a great deal of good! I try to be benevolent; everybody calls me good!

Now this is really the state of a large portion of professing Christendom. And yet this state is little better than the worst. It is that lukewarm state which John the Divine declared existed with some in his time and which I declared I would spew out of My mouth. For I would that such were hot or cold. I would that they were better, for they can not be worse. To be unpretendingly cold or lovingly warm toward God is better than to be self-satisfiedly lukewarm. The first sort may improve easily; but the last will remain in the second sphere a long time because they will there, as they now do here, feel that they are a pretty good kind of people, that they did no harm; they were pleased with themselves, and their associates were equally satisfied with them. How, then, will they answer the higher spirits who will then call on them to submit to God? They will say, what does God want? Have I not kept the law? Did I not follow the inclinations He gave me? Did I commit any crime? Was I not a good neighbor, a kind friend, a loving husband, an affectionate parent, a dutiful son? Did I not belong to the church militant, as my servant the preacher or minister or clergyman called it? And did I not cheerfully pay my full portion toward supporting and advancing the creed and the doctrines of that church, and was not that what God wanted me to do and what the Bible enjoined me to do? Such will be your plea in the spirit world, for such is your plea here. As the tree falls so it lies. There is no repentance beyond the grave; there, all is atonement. Yes, there you must atone for the deeds done in the body. And how shall the atonement be made? By suffering the deprivation of happiness. By having every desire fulfilled and finding every one to be dust and ashes. By finding all to be vanity—vanity of vanities!

When you shall have recovered from the rage of disappointment time after time experienced, when you shall have suffered every pang that remorse can inflict, and mourned and wept over your misspent time, then you will try with longing heart to submit to God. Then the task will be more difficult than now, for then the time is longer, the work greater, the establishment of good desires more difficult, because the temptation to do evil is no longer present.

It is this which you can with difficulty understand: that when you are no longer tempted you can not so easily progress. When you can not recede at all you can with difficulty advance. But I will try to make you understand this also.

The Effect the Present Life must have on the Life to Come

There is in the mind of man an idea of self, a consciousness of AM, a realization of individuality which I have shown in the second part to be a gift of God, indeed a sphere of seven gifts. This quality, or essence, or gift, declares to him his identity when he has passed through what men delight to term the Dark Valley of the Shadow of Death. The spirit or soul does not for an instant doubt its being what it was. But it does not instantly perceive in all cases that it is now endowed with higher powers and that in leaving the body it left the fetters which had chained it to relations with matter, now forever dissolved. Thus it has sometimes happened that spirits have returned to their former residences and associates, and manifested their continued existence and presence in various ways, but each always blindly and unknowing, to itself, that it was now a spirit in form and material. For spirit is matter, though it is matter more refined than bodies can feel or in any way appreciate. Spirit is invisible, but spirits can make themselves visible by assuming such an agglomeration of the moisture ever present in the atmosphere as makes manifest to bodies of earth their form and figure in such guise or dress as they may will to appear in. Such apparitions or ghosts have often been seen. But they have generally been, heretofore, such chance visitants of the earth as I have described.

These have unknowingly to themselves used laws and made manifestations belonging to their new position. The continued appearance or repeated apparitions which men have believed to be made, were only the results of a different process. This process is impressing upon the mind of the supposed seer, or hearer of the ghost or apparition, a psychological or mental idea that it does really experience by its senses the actual view or sound, or even feeling. The same psychological process has often been shown in the body in public experiments lately.

But there is yet a third way in which spirits in the second or higher spheres may act upon men in the body. That is by having the use of such agents as electricity, magnetism, *od*, and of a still higher or more refined quality or substance called by spirits *adamic force,* or Adam. This last is the material of the spiritual body which invests the true soul or essence of man, which is the proper Divinity within him, the part of God separated to a separate existence by God from Himself in the beginning and then placed in paradise, as declared in the first and second parts and as will be more fully declared and explained hereafter. It is this last adamic force which enables them to control matter by means of the other inferior, but to men extremely subtile, forces. The end of their being is to control these forces with intelligence and wisdom of God and, by controlling these, to control all matter in the universe or in the great whole of creation.

The spirit, having reached the second sphere by dissolution of its connection with body, becomes a spirit of the first circle of the second sphere. There it remains an instant or a long time, as its life in the body has made it attract or repel the spirits already in the circle. If it does not immediately leave this circle, it must commence its reconciliation with God there. As I stated in the second part, there are spirits still there who were among the earliest born of men upon the earth. Adam, or the first man, has progressed much beyond this sphere, but many antediluvian descendants of him are found in this and all other circles of this sphere.

168

The first part of the process of reconciliation is effected in the first circle and may be known or described as the Reconciliation of Will. Will is the highest form of man's power and the highest attribute of God. But yet this must first be subjected to God because, inasmuch as man's whole nature and self is under the dominion of His will, he can not be separated from himself in any way so as to be free from the power of his own will until that will has submitted to God. But then the submission of the will in this circle does not include its entire and perfect submission. It submits in part and so in part the whole of the lower faculties or qualities of the man become subjected to God. The part here submitted or subjected to God by man's free surrender of it is the power of will to do wrong in thought, of the very first intention to do wrong to other men. Revenge is its name among men. War is its highest form on the earth. Death is its great punishment—death to God, for it separates men as far as possible from God.

Hate is the next circle's work to remove. The third circle is Reconciliation of Love to Men. The fourth circle is Reconciliation of Love to God. The fifth circle is Reconciliation of Power to God. This means reconciliation to God of the power to do wrong to men, of the power to harm or disturb others in or out of the body. The exercise of this power has caused the belief in a being called the Devil, or Beelzebub, the name Beelzebub being a corruption of the name of Baal, or the Sun as worshipped by the Canaanites, its termination being significant of power to harm. Baal that wrongs or injures, is the signification of the term. The other terms, Devil and Satan, I explained in the second part. Having also declared there that the popular idea of his material form arose from the appearances of antediluvian men, I will further state only that this power of appearance continued longer in them than in others because their grossness and sensuality prevented them from realizing their changed condition sooner. Even up to this time some of them ignorantly believe themselves to be in the body and try to make men believe them real actors on the stage of outward things.

The belief in a material devil or Beelzebub, has, then, some foundation other than invention or imagination. It has just so much foundation as I have described. And almost every great popular belief, though scouted by science or theology, has about as much. Where, then, are we to look for truth, if neither science nor theology can establish or declare it? Look to *revelation*. That is always sure. But how shall we know what is and what is not revelation? Men will very seldom declare any thing revelation that is not. They are too fond of glory and honor. To declare revelation makes a man humble. He must deny to be his what men are willing to declare to be his own work. He must say that for it he deserves no honor, no reputation. That he is no more than a humble instrument, no more to God than the pen is to man. When a man writes a thrilling story or a learned treatise, do you attribute the work or the honor to the pen with which he inscribed it, or do you regard with veneration or love or admiration the mind that composed it? The latter, of course. Then would you expect the man who has thus produced the excellent work called the second of this series, containing so many new and startling hypotheses, if not truths—such profound deductions from known facts, if not revelations—such eloquent preaching and entreaty, if not divine essays—such cogent argument and convincing well-arranged logic, if not the intuitive responses of man to the touch of God's finger upon his spiritual essence—I say, would you expect a man who had produced such a book, written in less than three weeks, without previous preparation, without neglecting his daily business duties, written entirely before breakfast or after tea, the one early and the other late, would you, I say, expect such a man, having written such a book, to declare that he did not do it? That he was incapable of it and that he could not even now write it if deprived of a copy; that he could not even copy it in double the time he occupied in writing it? I say would you believe it possible that he should deny publicly and totally what men would thrust upon him, the honor and reputation of having taken a place in the annals of literature without a precedent?

No; the temptation is always the other way. Men gather from the labors of others and claim themselves to have accomplished the work. Men steal the literary work of others and say, "Look how much I have done!" and "How quickly I did it!" When men refuse them credit for the labors of others, they still the more pertinaciously insist that they deserve all the honor, all the glory, all the reputation, all the reward that such work of right should receive, and that they can produce plenty more as good.

But My holy medium claims nothing of all this. He declares on all occasions that he was only My instrument and that any man of ordinary intelligence and attainments could have done the same. And you won't believe it! You never before suspected him of such smartness, but now you begin to try to make yourselves believe you had seen some evidence of it: that he had read a great deal; that he had thought much; that he had had a varied experience; and that, after all, he must have written it or picked it up from his reading and agglomerated the results of the long devotion he had maintained for knowledge. He had then produced a remarkable book but one unworthy of reverence as truth or revelation. But you who know him best know that his memory is defective and his command of ideas for conversational or essay purposes very limited. You know he could not have written it, though you do not like to admit that one you know to be so much like men in general should be so selected for such a work. Some say "We do not want any more revelation; the Bible is sufficient for us." But such forget that the Bible declares *there shall be more,* that

God would pour out His spirit on the people and that old men should dream dreams and young men see visions, and that all should prophesy. And you know also that there must be a beginning; that God has always chosen the humble among men to do His work and that He can raise whom He will and put down whom He will. All this you know and admit. But then you cannot believe it should happen that God would select a man you knew, a man with whom you had associated so often and had treated so familiarly as a common man. It seems that it should rather be some one from a distant country with imposing mien, strange apparel, and elevated above the common wants and failings of common men. But a moment's reflection must tell you that every man is familiarly known to some circle, and you may also call to mind that when I was in the body performing My work among men, to which I had been similarly called but with greater advantages of nature and preparation, that I was regarded by those who knew Me familiarly as a common man and they exclaimed then, "Is not this the carpenter's son and have we not his brothers and sisters and cousins and friends with us!" And those brothers and sisters, those cousins and friends—with one exception, that of a cousin—believed not that I was inspired but that I was deluded or, in their form of expression, that I had a devil.

My holy medium is similarly situated. He, too, treads the winepress alone. No word of encouragement, no manifestation of sympathy cheers or incites him. He has a cousin who believes not especially in him but in the general truth and divine origin of the outward manifestations. But though thus left to rely on Me only for support and consolation, he has not found any want, nor shall he experience any. Every consolation, every cheering promise, every vivifying hope is furnished him. For he who gives up father or mother, wife or child, friend or acquaintance for My sake or to do My will or work, shall receive a thousand-fold here in this outward bodily life. The peace and contentment that I give passes all understanding and can no more be explained to the understanding of those who have not experienced it than can the joys of heaven or the bliss prepared by God for those that love and serve Him. And this is because it is a part of heaven: a part of that same bliss or reward prepared by God for the faithful servants of His will and the sons of His love.

What then remains? Shall I portray the future of My holy medium and take away from him the pleasure of enjoying the succession of events which make up the experience of this life and the joys of the next state of existence? Shall I give him power to make all men bow down before him and worship him when they ought to worship God? Shall I leave him to neglect and poverty as a mark of My power to sustain a man under every dispensation and suffering? Shall I cause him to be offered up as a martyr to the faith he teaches by influence and command? None of these things will I do. He shall live as he has lived and die like other men. But he shall be helped and aided in his temporalities as in his spiritualities, and so long as he gives to Me the glory and honor of My help he shall have it. This he is now willing to do and I believe will continue willing to do. But do I not know? No spirit, not even God Himself, knows what any man in the body will do, because man possesses free-will. But inasmuch as God and His high spirits know what has been, can see and know what is, and can perceive all the causes now existing which must influence the future, a judgment nearly sure may be formed as to what any man will do throughout his life. And as God and His high spirits further have power and will to affect man's reason and urge his passions or propensities or incite his aspirations, they can secure such outward results as it may please God to resolve to have occur. So far we go and so far God goes, not because of the impossibility for Him to go further, but because His will is that He will not and that we shall not go further.

After this long digression, necessary to your understanding and appreciation of this part of My book, I will proceed to inform you of the course by which spirits become reconciled to God after leaving the body.

The next circle is the sixth. This is the circle of Love of Good Works as a manifestation of reconciliation with God. For the spirit or soul, having come to a knowledge of God's love, is now desirous of offering a greater return than its own love. It then is taught that only by benefitting others can it serve God: that He needs no help but that He graciously pleases to permit and approve of the efforts of His created beings to help each other, to benefit one another by their love, their labor, and their time. They take pleasure in thus serving God, and God is pleased with their offering this part of themselves to Him. So their pleasure and happiness is vastly increased by having progressed into works of love and love of good works.

Then the last circle of this sphere of reconciliation is the seventh circle, the circle of Good Resolution. You thought that acts or good works were higher service to God than these! In some respects they are so. But, viewed as a whole, good resolutions embrace Action. The soul executes in the spirit world its resolves. It wills and performs or executes. Thus good resolutions comprise good acts, and good acts comprise works of good. For, as I explained in the second part, the higher circle always comprehends and has the power of the lower. Thus the highest of all circles is the combination of the whole complete knowledge of all God's creation. And not only Knowledge but Power, and not only Power but Love, and not only Love but Action, and not only Action but Will, and not only Will but the Power of Will, or of forming Resolutions. But did I not teach you differently in the second part? Did I not say that the highest spirits could not know more than the Intention of God? I did. And that is the Power of

Will, for they enter into God's unity so as to will with Him the execution of His intentions.

The sphere above the second or sphere of Reconciliation, is that of Memory or Remembrance. In this sphere is first brought back to the consciousness of men or of their souls the memory of the paradisical state. But this is unfolded or developed in them by degrees. In the first circle the memory of good works done in the body is developed. This is a great satisfaction to those who have done many, but it is clouded by a knowledge or memory of the motive or motives, for generally men act under the influence of several.

Motives thus apparent sometimes take away all the pleasure of remembrance. But you may think if a man does good it is of no consequence what his motive is. So it is to the recipient, but to the doer it is of great consequence, for he reaps as he sows. If he sows for men's approval he may get it and that is his sole reward. If he acts for God's glory he receives a return from God, proportionate not to the good performed but to the intention to do good. So you can see again how much surer is a man's reward who works for God than if he worked for men. For men always give the glory in proportion to success; God gives it in proportion to intended accomplishment. In the one case the man is rewarded according to his desire if he succeeds. In the other, whether or not he succeeds, his reward is the same. Here is another great inducement to choose Me or God (which is the same) for your guide and counselor and king. For not only is My yoke easy and My burden light, but My reward is sure and steadfast and never passes away. He that drinketh of the water that I shall give him shall never thirst more. You see, here is a new meaning to this text. True, you understood before that I was to give you some spiritual consolation that should quench your desire for it, but you did not know that it was to last to eternity. To the end of time would do very well for some, but to eternity never-ending is a glorious reward and one worthy of its Giver.

The second circle of this third sphere is Memory of the Good Intentions you or the soul may have formed while in the body. This is ever a great satisfaction to those who have them to refer to. Some have few though all have some, as I showed in the second part in declaring that all were saved or reconciled to God at least in part during their bodily existence. But some have much of this joyful delight and it forms one of the purest sources of enjoyment in all the future life. Here you have a further illustration of God's sure rewards. He rewards not as men do, for work done, but for work intended to be done. Do any say, Let us form intentions which will reach to Heaven! they shall be scattered and dispersed in confusion as was symbolized in the tradition of the building of Babel. Remember, God knows the heart of man; He sees into every latent desire; and He can resolve every mixed motive into its component parts. More than this, the memory of good intentions embraces a memory of motives for them, and no intention seems good to a soul unless the motive for its formation was wholly, or at least partly, good.

The next circle is that of Memory of Good Desires. A desire is very different from an intention, because the intention can be performed only by an act, while the desire may be fulfilled by an intention; and the desire to do good may even exist without an intention ever being formed to execute the desire. How, then, can the desire benefit a man when he does not even will its accomplishment? He does not will its accomplishment but he desires it. This God takes as it was meant. If the desire was good and the intention not easily consequent upon it, God accepts the desire for the intention and the desire for the proper consequent of the intention; that is, for the *work*. Thus you have a further evidence that God's ways are not as man's way, for man would scarcely bestow a second thought upon one who desired to help him but failed even to resolve to do so.

The fourth circle is that of Memory of Events or Opportunities. Memory of opportunities usefully employed, events properly turned to account, is a great pleasure and joy which is clouded indeed by the greater number of opportunities unimproved and events for usefulness unaccepted. But still every man has some good to remember of himself in this respect and he gladly leaves the neglected events and opportunities to oblivion. And can the man, then, leave what he pleases behind him? Will he not be obliged to remember the whole class? May he forget when the province of his state of progress is memory? He can. But not until all has passed in review before him in the fifth circle. Here he judges himself. It is thus that God's books are kept in every man's heart. Thus He judges men with righteous judgment, for the man's interior or soul, disencumbered of the body and acted upon by the efforts of the other spirits to show it the goodness and glory and justice of God as well as His mercy and His benevolence, becomes qualified to see its errors and its right actions, resolutions, desires, or neglect to form desires or intentions or resolutions.

So the past goes before the man in memory. So he judges himself and repents in dust and ashes for his misdeeds or neglects. Does he indeed repent beyond the grave! Not repent in the true sense of the word; he bewails or regrets his performances or non-performances. But he has not the power to repent. Repentance is regret for misdeeds but it is accompanied by resolutions for reformation, with resolutions to resist temptations, which is the true repentance and the action which constitutes its distinction from regret. Then let us not say the soul repents in dust and ashes! I will not take back what is written. It was My sentiment and expression, though My holy medium has been alarmed, thinking his inattention had caused the error. But the true repentance is in the body. In the spirit-world there can be only regret and wishes that

174

it had been different. Repentance is an act of will; will is an act of freedom; and free-will is left only partially to spirits in the third sphere. They are not at liberty to go back to evil and they must, if they move at all, advance. They repent then in effect, though not in deed. They desire to undo what wrongs they did, to do good they neglected to do. They regret their inability to do it but they also perceive in the next circle the compensation experienced through God's mercy and justice to be sufficient for their punishment and for the correction of all their doings or neglect.

This next circle is the Memory of the Whole Past Life including its first circle in paradise. By viewing as a whole their actions, thoughts, desires, intentions, and resolutions; by seeing the combinations of circumstances and the course of events which influenced and controlled them, they see that they have suffered and atoned for the evils they did; and now they may look forward to the highest happiness as their sure reward for reconciliation with God.

But what then remains for the seventh circle of the sphere of Memory? You will think with My holy medium that I or he has made a mistake and told the experience or progress of the seventh circle for the sixth, and that I ought to or did intend to put in another advance before the sweeping one I gave to the sixth circle.

But God's power, His invention and wisdom and knowledge, is great indeed. It is infinite, which is as much farther beyond great as great is beyond little, or nothing. The seventh circle of the third sphere receives the memory of others into its own, the Memory of Memories. This you may think as scarcely a proper expression, that it is more properly an acquirement of knowledge than a return of it. But this is not so. Memory of Memories is the proper and true expression of it, for as the spirit passed through its experience sphere and its reconciliation sphere it obtained by association with others in its respective circles an impression of their mentals or interiors, which now assumes to it the form of memory of their experience or, properly, memories. For the memory of each associate having been imprinted on all other souls in the same manner that the course of one soul's own experience had been imprinted on it, a similar process restores to its consciousness the memory of all that had been done by every other member of the circle, and also the memory of all that had been done by every spirit that was associated with it in each of the other circles of the sphere of Memory while it was passing through them.

It does not stop here, but each spirit with which it was associated in the sphere of Reconciliation, and also in the sphere of Experience, had so imparted its memory to every other spirit that everything previous to such association in the mind of the soul or spirit of each member of each circle of each sphere becomes possessed by each other spirit associated with them at every period of their sojourn or continuance in the respective circles. I present this matter with such variation and particularity and with such apparent repetition in order that I may cause you if possible to appreciate the greatness of this gift and the greatness of the Being who has so abundantly provided for the employment and enjoyment of all the vast and infinite number of His children, or emanations of His substance. And now, when I unfold to you that this process of receiving memories of others goes on in every higher circle under the simple law I have before stated—that the higher circle possesses always the whole power or will or knowledge or experience or memory of the lower circles, you may be able faintly to realize a conception of the vast resource that exists in Memory to give happiness or employment to the advancing spirit or soul of man.

CHAPTER XLI

THE OBJECT TO BE ACCOMPLISHED BY THE SPIRITUAL MANIFESTATIONS NOW BEING MADE

Prophecy and its Manifestations

The first part of this chapter I shall devote to an explanation of the term *fifth monarchy* and the prophecies which foretell it; the last part to the revelation of the future course of events, political and religious, which are necessary to the fulfillment of those prophecies.

The first part comprises the past; the second, the future. The past is not understood, though it has been much written of. The future, though prophesied of, can not be conjectured but must be revealed if known to men in advance of its occurrence. Prophecy and revelation are different, as I have already showed in this book. Prophecy is not understood by those who declare it in all cases, but revelation is plain and perspicuous and comprehensible by men— at least by intelligent and earnestly inquiring men. Daniel always asked an explanation or revealment of the meaning of his prophecies or visions. So did John the Divine. Isaiah did not. But none of them obtained it with such clearness and understanding as to enable them to comprehend it, or as to enable others who have studied them and who have been aided by the light of events occurring since the prophecy to comprehend them fully. Parts have been well guessed at and terms have been well explained by some, yet the guesses and the explanations are alike rejected by others equally learned and as qualified to judge or form an opinion. It all illustrates the truth that man by reason can not find out God nor anything that God does not choose to have him know.

What, then, is the declaration of the fifth kingdom or state? For kingdom is used to denote nation or government. Not because republics were then unknown but because the kingdom was to be under the rule of Shiloh, the Prince of Peace. It is to be the

Kingdom of our Lord Jesus Christ, as John the Divine expresses it. I am to be its king in virtue of My title, King of Kings and Lord of Lords. I am also to be its king because the inhabitants will elect to have Me so and will voluntarily submit to My rule. I am also its king because God has given Me all power both in heaven and earth. I rule with absolute authority the spirits of those who once lived on earth and I am permitted in the Will of God to proceed to establish My rule on earth as it is established in heaven.

For a long time and, at various intervals, with intensity, men have expected Me to appear in bodily form and assume the government by force, by overpowering the resistance of unbelieving or wicked men, by marshalling the armies of heaven as an innumerable company of spirits again restored to bodies and marching with carnal or outward weapons against the powerful array of the various earthly potentates who should be unwilling to submit to My rule. Something like this will occur, as will be seen in the sequel to this chapter.

But the appearance now to be made is spiritual and in the internal or soul of man. It is the same procedure that long has been maintained to be in existence by Quakers and which I know is now believed in various forms and types by other professing and non-professing Christians. Hide it and cover it up as they will and as their creeds do, they yet depend on it for their ministry and for their individual guidance. Yet as they generally accompany the belief by rules for its coming, or manifestation, they do not often get it and more seldom get it so unmixed with their own willful additions as to be reliable. From this reliance on their rules for its coming they can not receive the rapping manifestation as divine in its origin or as spiritual in its existence.

Four monarchies or kingdoms are plainly mentioned or described by Daniel so that there is nearly a general consent to designate the Roman state or empire as the fourth. But the fifth or last one was represented by a stone that broke in pieces and dashed into fragments all the others, and that should endure to the end. His kingdom shall be an everlasting kingdom, Daniel says in another place. I have explained to you the meaning of everlasting to be an indefinite period. This period shall, however, be to the end of this condition of the earth. It shall last to the disruption, or great catastrophe, described in the second part. That will occur suddenly and every one should be prepared in all future time until it does occur. The last shall be first then and the first last. And so they are now and will be till then. (But this you understand if you have read the second part; if you have not, read it now and afterward read this again). The last shall be first then, though in another sense, for the first with men shall then be first with God. But the last form or development of man will then be first in honor amongst men because of its superiority, as Noah and Adam were made supreme rulers of the inferior or previous races.

The rulers of men now are desirous of maintaining splendid palaces, armies of servants, and splendid luxury in everything relating to themselves or their families or their immediate attendants. In this nation of the United States a more simple republican form prevails, but the same desires prompt the corruption of all the leading politicians and hurry the downfall of public virtue. The simplicity inherited from the formal Puritan and austere Quaker, from the poor immigrant contending with the wilderness and savage beasts and more savage men, has nearly departed. Unless the kingdom falls under the government of Him to whom it rightfully belongs, its progress must be downward until it falls into disunion, dissension, and destruction. It must follow the course that all free communities before have pursued from liberty to anarchy, from anarchy to despotism, from despotism to destruction by exhausted nature and oppressed people, falling at last a prey to a barbarous or, more properly, an uncorrupted nation or handful of people who may resuscitate its energy or reanimate its people for a brief period. But nature and man together fall exhausted at last. So it has been; so it will be till the end of time.

What, then, shall save the nation that I have declared to be established already as the fifth monarchy which should last till the end of time, to the end of the present time, to the disruption of the present surface of the earth! It must be saved by submission to the government of My holy mediums. It must disown every man who does not own Me for his Prince. It must let no one administer its laws or legislate for it that does not acknowledge Me in heartfelt submission to be King. Do you begin to think that after all My medium is not so disinterested as he would have you believe? That he is seeking more than the honor of writing a book, that he is seeking the control of a nation? Verily, I say he is as much surprised as you who read. He is incapable of such a magnificent scheme and were he to form it how should he get help to carry it out? Other holy mediums would not unite with him unless I desired it. And if I desire it, why should he not be ruler as well as any other man? Is he not as honest and as capable as some who have filled the highest office? If not, then he can not write this book. If so, he may with propriety aspire to the office. If not, I must help him to write this book; and if I help him to do this could I not direct him how to perform with credit to himself and profit to the people his official duties? If I do not help him to write this book, how shall he be helped to obtain the office? For he can not influence the other holy mediums or spirits by his own intellectual effusions. So if he himself writes he will not obtain that reward; and if I write for him he will not seek that reward, for he will no longer be My holy medium when he asks pay for his services. But, you ask, does he not make money by his books? He has not yet and if he should, he will hold it as My steward to be disbursed by My order. He may be unfaithful! Then I shall know it and I will take care that he harms no one but himself by such unfaithfulness. No, My captious reader, do not so easily take alarm at the bold expressions I must use to awaken your attention and fasten upon your conviction that I am one who speaks with authority and teaches as man never taught.

I call, then, on every man to investigate, to weigh carefully the evidence already obtainable, to collect more and more, to cease not to investigate while the manifestations are made unless he becomes assured and convinced beyond wavering that there is a procedure from God especially at this time and for a particular purpose; that the purpose is the establishment of a kingdom or nation or people which is already designated to be the United States of America as the kingdom that was hewn out or established without hands; and that it was declared that I should take the government of it and now I am ready to do it. Now I call on you, O reader, to submit to Me. I ask you to give your heart to God; to promise from henceforth to try to serve Him faithfully; to walk humbly and submissively in whatever path I, His Son, shall designate. Do you ask to be ruled by God and not by Me, as I know some of you do? Then I will accept your offering exactly as if made to Me if you make it to God. Pray to God; I will answer for Him. Submit to God; I will rule you for Him. You need never mention My name unless it is enough for you to serve God in His own way and in the way He marks out for you. His way is to rule through His Spirit, or Son, or Sent—all synonymous. But you admit that God enters you and impresses you by His spirit, and yet you would not like to have that Spirit (though it declares itself to be of God and from God and one with God) also declare itself to have once inhabited a body on the earth. You don't want to have any other than God playing with your internals, you say in a half-joking, half-earnest way. Well, to you I am as God, because when you ask for God I present Myself; when you pray to God, I answer your prayer; when you open your heart to let God's spirit enter, I Myself enter. When you shut out Me you shut out Him that sent Me. For I am one with the Father and I ask you to be one with Me even as I am one with the Father. But you do not understand this Trinity! Yet this trinity is explainable, though the Orthodox Trinity, as it is called, is inexplicable. That I am one with the Father is fully explained in the second part and to that I refer you. That you can be one with Me I have also shown there, but I will again briefly say that you must become one with Me by the sacrifice of your free-will. You must act in My will and try to bear the same relation to Me that I bear to the Father. I will help you, and if we both try our best be assured that we will not and can not fail.

The next point to which I will direct your attention is that of your submission to the directions of My declared or revealed will through holy mediums, the name that will hereafter be used to denote such as the Jews called prophets. But yet the Jews seldom obeyed their counsels or followed their directions. The history of the Jews is a history of rebellions and punishments. The terrible and fatal one which followed and was consequent upon their rejection of Me was the last except the constant, ever-enduring one of dispersion and disruption as a nation. True, this occurred in part before Titus sacked Jerusalem and burned the pride of the nation with fire. But it was then a consequence of their sins and rebellions, as was the final dispersion the consequence of their crowning offense.

The last great call I make upon you now and in this relation is to hear this holy medium. For I will establish his truth by the mouth of many witnesses. If any doubt and sincerely desire to know whether to believe the whole of these writings, or a part, or none, let him go to a rapping or a writing medium and he shall have an answer. If his inquiry is sincere and made with a resolution to be governed by it he shall have an answer. If not so made, he shall get none. By this hereafter shall you know that a medium is operating in unison with Me. Every spirit that confesseth not that I have come in the flesh manifest by My writing this three-fold book, which, however, is not all I shall write thus; I say, every spirit that confesseth not that this is of God is not of Me, is not so far advanced in the sphere of reconciliation or else is so much controlled by the medium as to be unworthy of regard. This test was given by John the Divine in his General Epistle to the churches then in being. But he knew it would continue to be true even after the signs or manifestations of that day should have ceased.

Prophecy and its Future

The second part or sequel of this chapter is of the future. Relative to this is the explanation of the prophecies of Daniel and John the Divine so far as they are unknown to men in the body in the present day.

The first prophecy was derived from the vision of Nebuchadnezzar, the second was when Daniel stood or dreamed that he stood by the banks of the river Ulai or Peace. The third vision was by the Hidekel, or Tiver, river, and the fourth was in the banquet hall of Belshazzar. But you say I have not named them in the order given by Daniel and that he gives the time very particularly. Blame by holy medium, not Me. He perverted the stream of My revelation by his faulty recollection of the second, and as it was of no consequence I let it pass in his own way as a punishment to him for his want of passiveness. The last vision was that of John the Divine when he was in the spirit on a day appointed by Me in the isle of Patoms, or Peace.

Daniel declares the fifth kingdom in the figure of a stone rejected by the builders, John as a city coming down from heaven. But the stone was hewn without hands; that is, without outward hands or labor. The government of the first settlements of America was of this kind. It proceeded from Britain without the aid of Britain. It made its own laws and established itself in a wilderness. The colonies grew when the parent country neglected them. They were oppressed when it undertook their care. They established their independence by the aid of prayer and thanksgiving to God for victory. They formed their constitution by its aid and by the earnest supplications of devout, sincerely pious men. The last shall be first hereafter and the first last now. The efforts of pious—sincerely pious—men shall always avail much and hereafter shall avail more, because more confidence will be reposed in their efforts by the nation when it shall have placed itself under My government.

But the true reliance must be upon the holy medium who declares My will by revelation. The Hebrew always admitted his obligation to God as his supreme ruler, and expected His will to be made known through His selected mediums. Many were educated to be mediums, and a proper training does much to fit a man for such work. But the true training must at last be completed by God's spirits ere the man is a proper or truthful holy medium. The kind of preparation required is only what will be most useful to every man and what will fit him to be a judge of the truth of the revelations made through or declared by the holy mediums, because it will fit him for the reception of Me, or of lower but elevated spirits, into his heart who will impress upon him a judgment or conviction of the truth and genuineness of all revelation that is divine in its origin and unperverted in its transmission. This power ought to reside in all men and will reside in them if they will submit to God. The training is to render them patient under affliction or burdensome commands, to render them carefully attentive to their interiors, to make them free from the pollutions of sensuality, to educate their intellect to a comprehension of nature and of God's created beings, to make them in a word willing and passive and submissive servants of God. Then He will use them as He deems best, some to honor and some to dishonor. For He claims the same right and authority over these works of His spirits that the potter does over his clay when he forms it as he pleases or as he has occasion for it.

The New Jerusalem is represented to have been a perfect city abounding in good works and filled with pious men who continually praised God. They were the same 144,000 that stood on Mount Zion with golden harps and sang the praises of the Lamb of God. They were a perfect number, the result of the multiplication of twelve, the great number of the Jewish polity, and twelve the complete number of the Christian Church; the result of these multiplied by God's great mercy a thousand-fold. By so much will the numbers of the New Jerusalem church exceed those of the former true members of the old forms of worship. The worship of God by submissively doing His will, which is that men should live in harmony and delight in helping one another, must be general in the United States or the fifth monarchy can not exist till the end of time. But the prophecy was sure and so is the interpretation of it; therefore it will continue and men will agree to come under My rule and government. But a beginning must be made and I ask you, O reader, to be My subject. I ask you to give. Me your allegiance, to be willing to serve and obey Me, to be willing to learn of Me, to be willing to take up your cross and follow Me. Then when you find the burden heavy, appeal to Me for help. I will make it light. When you thirst for knowledge I will lead you to fountains that will satisfy you. When you become dispirited I will encourage and re-animate you. When you are afflicted I will console you. When you want friends or companions I will be with you and yield to you a friendship that will never be jarred or disturbed in any way by Me. I will be not only your King but I will be your comforter, your Prince of Peace, your everlasting Counselor, your ever-present Guide and ever-ready Helper. You shall enjoy My society always and be My faithful and obedient servant and submissive son and joint heir with Me to God's sonship. What then shall separate you from the love of God? Shall height or depth, earth or sea, love of man or love of the world keep you from the enjoyment which is so richly provided by the everlasting Father through His Son, the Shiloh or Prince of Peace?

But I must proceed with more specific details of the present and of the future. The last effort of the Dragon or seven-headed monster, the last phase of the Roman Empire, will be to overthrow the liberties of the United States. An alliance miscalled holy will be formed by a union of all the Ten Kings that shall unite to give their power to the Beast for one hour for this occasion. They will fail. They will return to their own dissensions, they will fight amongst themselves, they will turn upon the Whore that they worshipped and rend her flesh with fire and sword. The Beast shall come to his end. The False Prophet shall go out to deceive the nations no more. Gog and Magog shall perish also, for these stand for those who are outside the fourth kingdom—who have never owned its authority, who have worshipped idols of flesh and who have not repented of their sins but believed they could have absolution from men.

It is now time to expect the combination against Britain to be formed. It will overthrow the government and scatter the inhabitants, as I have described, in the second part. It is one of the signs of this that all the great powers of the European continent have such cordial interchange of courtesies. France is the only exception to this harmony. But France will make her peace and secure her good fellowship with the others by joining in the conspiracy against liberty of conscience, of personal rights, and of popular representation. All will unite to form a so-called holy alliance. Blessed by the Pope and consecrated by pretended Protestant ministers, the armies will land on England's shores and the peace and security that has so long left the Englishman's hearth inviolate will be at one fell swoop demolished. There are no fortresses that can sustain a siege, no disciplined armies of sufficient strength to meet the invading force. There will not be. In short, the enemy will find the nation unprepared and the country undefended. Then the people will rush to their colonies and to their fellow Anglo-Saxons of America or the United States. There will be no opposition to the union of all the North American British possessions with the United States but what proceeds from the same dragon power. They will strive to prevent it. They will threaten war. They will make war with the United States but they shall not prevail. Here is no capital to be secured as the heart of the system. Here is a vast surface which can not be marched over or occupied before armies can be disciplined, and the very proclamation of welcome and discharge from service of deserters will thin the ranks of the floods of men that will be poured upon the land of freedom. This will help the Woman which has fled to the wilderness for they will be absorbed into the general population of America and contribute to the resources of the government of the kingdom of the saints, or spiritual believers, or servants of the Most High.

There is a consistent account. But is it a real prophecy—a true declaration of the future? It is not. I have before declared that no outward sign shall be given through this holy medium. I have given it because so many have desired an interpretation of the prophecies respecting the fifth monarchy or kingdom. But why give it untruly, says the reader and the medium. Because I would show you how easy it is to give a startling and interesting declaration respecting the future. And is this the only object? By no means. I also wish to prepare you for spiritual revelation or explanation of the same prophecies.

The saints of the Most High are those who believe in God's power and love, who have submitted themselves to His government. They are they who have washed away their sins by repentance and who now, taking no merit to themselves, give to God praise, honor, and glory forever for all His mercies and loving kindness, for the peace which He gives which passes all understanding and which is to them the New Jerusalem; which is to them the Head of the Corner of the temple they have erected to God in their hearts, a stone hewn without hands, shaped by God Himself by the power of His will manifested through Me, the Lord and Savior of men, the Lord and Savior Jesus Christ. These redeemed from amongst men shall sing to God a new song which is nevertheless an old one. Great and marvelous are Thy works, Lord God Almighty. Just and true are Thy ways, Thou King of saints. This song is old because it is that which has been sung by every spirit that has arrived at a knowledge of God. It is new because it is never sung by the redeemed of God until they are redeemed from outward views and impure desires. My medium can not sing this song yet. He is not redeemed from sin. But he is passive in My hands and submits to My directions and influence. How, then, is he not free from sin if he thus submits? Because he has not repented and atoned for the many shortcomings and misdeeds of his former life. Because, though at times he acts entirely in My will, he also at times does not wait for My direction but proceeds in his own will. How, then, shall you succeed any better, if as well, inasmuch as I say he is My best holy medium? To be a good medium requires a peculiar constitution and character, and resolution and desires to be submissive and passive and to endeavor to walk in the paths of virtue and peace. But cannot this good constitution succeed better than another inferior one? With God all men are equal. He judges with righteous judgment and He tempers the wind to the shorn lamb.

The first part of My book has been declared publicly to be incompatible with the truth of the Bible. One or the other, some say, must be wrong and of course they will adhere rather to the old friend than leave it for an untried one. And I boldly say there is not a single sentence that has a signification contrary to the Bible in that first section. I say further that it is not only in entire and cordial agreement with the Bible, but that its tendency is to cause men to value the Bible more because I have in this book interpreted several of its most difficult or apparently irreconcilable parts. So it is evident that men will not sacrifice their established easy chairs of faith until I arouse them by some other method.

CHAPTER XLII

REASONS FOR BELIEVING THIS MEDIUM TRUTHFUL

The last chapter will no doubt be thought very unsatisfactory to most readers because I led them to expect revelation respecting the future glory of their beloved country, and because I seemed several times to be on the verge of great disclosures. But, O reader who thus objects, are you prepared to believe what I could thus disclose? Is it not worse than a vain curiosity that impels you to desire Me to make such disclosures? Do you not desire them for the purpose of doubting, combating, overthrowing them? In order that you may urge against them the opinions of this commentator or that father of the church, of this society or of that preacher? If you are one of the few who have desired to have the knowledge in order that you might glory in the power, love, and wisdom of God, then read the Book of the Revelation of John the Divine and I will help you to understand at least all that can aid your motive to satisfaction!

But if you are one of those more numerous spiritual believers who are hearers of the Law but not the doers thereof, I shall commend you to be quiet. You are only adding to your responsibility by an increase of knowledge; hereafter when in the third sphere there will pass before you in review the knowledge you possessed, the good you might have accomplished with it, the opportunities wasted, the talents buried in a napkin or left to waste in unprofitable pursuits. You, too, are actuated first by a vain curiosity. Some of you, too, will try by your reason to measure God's revelations. And having formed a Procrustean bed from your former prejudices and chance reading of the works of lower spirits through other mediums, you will have Me and My books tried by that and stretched here and lopped off there till I shall be disfigured by error and destroyed by man's wisdom.

What, then, shall I do to reach you? Ye seek a sign, as did My followers and hearers 1800 years ago. I refused one then; I refuse one now through this holy medium. Not that he is unwilling to give them or have them given through him, but that I reserve him for internal manifestations solely, as I have declared before. Yet I also declared he should raise the dead, and heal the sick or lame or blind. What will I do to overcome the opposition that the Dragon and False Prophet will combine to make to My spiritual kingdom? I will cause other holy mediums to bear testimony as I said. Ask those who communicate the knowledge of lower spirits who and what I am and what regard should be paid to these My books. Ask them. By their answer judge Me. Ask not with a desire to hear them answer in a particular way but with an earnest desire to receive a truthful answer and with a resolution to be governed by it, and you will not ask in vain. You will indeed bless the hour in which you became so passive. That will indeed be a first step in submission. It is in such a state that prayer ought to be made. It is in such a state that the Bible or any other book that claims to be revelation ought to be read.

Now let us proceed to examine the book of Daniel's prophecies. He declared to Nebuchadnezzar the meaning of his dream, the dream of the image with the golden head. He also declared that the stone hewn without hands should destroy the fourth kingdom. So it will. Truth will overcome error. The remains of the fourth kingdom, its last phase, is the hierarchy of empire, whether Roman Catholic, Protestant or Greek Christianity. They are all founded on the sandy and earthy foundation, and when the stone of truth falls upon them it will crush them to powder. It will grind them to the earth from which they originated and leave not a trace of their appearance or form in the whole image of God, which is man. The earth shall receive their fragments and they will no more arise to disturb men in the body, and they never have disturbed men in the spirit who were beyond the second sphere.

This morning I permitted My holy medium to be seriously disturbed in his faith in Me and his own truthful course. He felt all the responsibility, all the weight of obloquy or ridicule or reproach with which men might and should treat every pretender to revelation such as he claims, or permits to be claimed, through him. He felt also, though perhaps in a less degree, the responsibility he must bear in the judgment of Heaven, if he suffered himself to be led away by a vain imagination or deceived by his own desires influencing his writings. Yet he went to work like a faithful follower who has lost his guide or leader. He started to work for the great and all-wise Creator and Former of himself, determined that humility and submission and a willingness to do anything, to retract or to proceed, to destroy or to publish, should leave upon that Creator—who certainly knows all things and certainly has promised that He will hear prayer and grant the requests of His servants—I say, he so proceeded as to leave upon God the burden he found so great as to weigh down his spirit. God through Me reassured him. And how?—by merely telling him again who I am and asserting it over and over? By no means, on the contrary I set his memory and his reasoning powers to work. I showed him that he certainly could not write as he had without some foreign influence. I reminded him that he had perseveringly sought Divine guidance by prayer and submission and appealed to his intuitive perceptions of Deity to assure him that if God does not rule men they must act only in their own wills or in that of him to whom they subject themselves; that he had never given his will to any one but God and Me; that if he could not rely on Us he could not be accountable for mistakes or deceptions imposed on him by other superior beings that must either be made in defiance of Us or by Our consent; that there was but one God and that He either cared or did not care for men. If He cared for one, He did for all, and that if, as orthodoxy declares, men can be saved only by the help of Jesus Christ, he had been submitting to Him and had by every faculty of his mind tried to do My will because I had convinced his reason and assured his affection and veneration that I was in unity with God and that in serving Me he was serving God. Then I let him pray some of the prayers previously written, including those for such as would be holy mediums, such as are mediums, and such as would be servants of God. He also prayed by his own intellect. Then I set him to read what had been written in this part to the twentieth page and I dismissed him cured, reassured, resolved to do My will without hesitation or doubt or wavering. He came out of the furnace, which was greatly heated, purified somewhat, rather more humble, and stronger in faith than ever. Amen.

So may it be with you, O reader! For I related this to you not as his boast but against his will. I gave it to you not for vain curiosity but that you might go and do likewise. I gave it to you not as a piece of news or as an anecdote, but as an evidence that My holy medium has not been disregardful of his great responsibility to you and to Me and to God. It is evidence that he has not thoughtlessly proceeded in a conjectured mission without searching his heart to its bottom, without regarding the consequences of his acts, without having endeavored to secure the evidence of God's authority and the truthfulness and reliability of the communications of which he is and has been the holy medium. Further I will say that this is not the first or second or third time that he has gone down into the lowest valley of his heart and there sought for the conviction of truth and the assurance of his duty. Willingly would he have been excused from publication and been satisfied to have walked in humble quietness to his grave. But I called him and he obeyed. I appointed him and he accepted. I enlarged his knowledge, his responsibilities, and his duties and he manfully strives to keep the faith, to fight the good fight and be willing at any time to finish his course on earth with joy and thanksgiving for the mercies of God and the peace of the Comforter.

CHAPTER XLIII

THE EFFECT UPON SOCIETY OF A GENERAL BELIEF IN THE DOCTRINES NOW PROMULGATED

The present subject of My discourse is the vast consequences depending upon the reception by men of My teachings and the joy and peace and happiness that will fill the earth when My will is done in it as in heaven. Some ask, Where is heaven? This is a question which can not be answered in a word, a chapter, or a book. It involves so many spheres of matter, so many preconceived opinions must be shown untrue, so many long-cherished theories must be demolished, so many false prophets must be overthrown, that I must attend to more pressing wants first. Be satisfied that heaven *is*. That it is a place as well as a state of mind. It exists wherever God is pleased to establish it, and whenever He chooses to enlarge or to circumscribe its boundaries He does so, and we give thanks always to Him, for all His works praise Him whenever they know the hand or power that made them.

Let us see what would be the state of things on earth if men received My teachings and submitted to God. Hate, envy, desire of revenge, cruelty to men or animals, and every passion or emotion which incites men to injure themselves by discord, or others by their acts or non-action, would cease to exist. Then men would gladly serve each other. Scarcely any man would need help because no one would oppress or attempt to injure another. War would cease and all preparation for war would cease too. Kings and magistrates in general would have little or nothing to do, because where all were governed by the law of kindness, by a willingness to suffer rather than to cause suffering or injury or make resistance, there would be no appeals to law or to strength or force to establish any man either in his rights or supposed rights or even wrongs, which are now too often supported by law and appeals to man.

Let no man say this is impossible, that Quakers have tried it, that it is Utopian or Socialist or that it is Fourierism. It may be Utopian but we may certainly place perfection before ourselves as an aim. If we fall short of its accomplishment we shall at least have reached further than if we had attempted less. If it be socialist doctrine, then real socialism must be the practice of the precepts of Jesus. If that be all they desire to accomplish by procuring an acceptance amongst men, by all means join their ranks. But if they desire to overthrow the primitive institutions of society which are founded upon the revelations of God to Adam, to Noah, and to various holy mediums since; if they war with society. instead of reforming or purifying it; if they establish an association to strengthen each other, then they do not rely on that only sure and steadfast support that can sustain a man or an enterprise or an association, that can not only sustain it but bless it with that internal peace, that consciousness of union and communion with God through His spirit and the soul, which is, after all, the only un-alloyed happiness which men ever can have or ever will have in the body experience. So, too, if it be Fourierism to practice these precepts, very well. Be Fourierites so far. But if Fourierism further include a reliance on man's wisdom and his perfect power to form perfect rules and lead a perfect life by following his own or other men's inventions, then be not a Fourierite. The only wisdom that is perfect is God's. The only laws that are perfect are God's. The only happiness that is perfect is that which flows from Him and which, proceeding ever in unremitting streams of rays, per-vades and blesses every spirit from highest to lowest that will allow the light of its love to enter his heart, otherwise in dark-ness.

This proceeding includes God's love and approval. God can give even a greatly sinful man a consciousness that he has performed a good action even when that action is a solitary and isolated one. How much more, then, can a man feel conscious of a succession of uninterrupted good acts inspired by a desire to serve God and benefit others! How much more since God has declared long ago through Me: To him that hath shall be given and to him that hath not much even that which he hath shall be taken away! Then seek first the kingdom of God, for the kingdom of Heaven is within you and so long as you are in the body you can enter no other heaven than that spiritual and eternal one of dwelling with God and agreeing with Him, serving Him and enjoying His blessed and holy communion. This is the true Lord's Supper which I declared should be enjoyed by him and Me, when he opened his heart to Me. The outward supper is its type. But the type profits nothing. The letter killeth; it is the spirit which giveth life. The outward supper was to last only until I should come to My disciples inwardly.

This occurred on the day of Pentecost when My presence was first manifested to the senses of men through the holy medium Peter. After that others, apostles, disciples, evangelists, believers, and seekers experienced in their own hearts that I supped with them. The outward form became a snare which led to revelry and to lasciviousness, as Paul declares in his epistles. But the church, acting in the wisdom of the body, reformed and modified the feast to a mere sign or slight manifestation of eating and drinking. Now some have perceived that the outward was only a type and, having reached to a knowledge of the antitype, they are and have been willing to trust to that and enjoy that. I do not, though, condemn the outward institution, though it was not and is not conducted in My will. For as I stated above, it was intended to last only until I came to those who were enjoined to practice it then. But that it is symbolic and gratifying to sincere believers and practicers of it I know; and it is to them innocent or useful in leading them to search their hearts and endeavor by all means and earnest prayer particularly to purify themselves and cast out every sinful desire or unworthy motive, and to make themselves fit receptacles for the Spirit and Son of God.

But to others who are more outward in their views it is injurious, inasmuch as they are led to take the form for the substance and accept the type as a fulfillment of the antitype. This not only perverts the institution but it destroys the health of the mind of the man who so receives it. I say then with Paul, Let him that eateth, and him that drinketh, and him that eateth or drinketh not, all do it all to the glory of God and the praise of His holy name. So shall they live and enjoy My communion which is only in the heart or spirit or soul of the man who desires to have Me there, and who prepares the table for Me by submitting himself to God and yielding passively and perfectly his own will entirely to Mine or to God's.

CHAPTER XLIV

WHAT DID JESUS OF NAZARETH, THROUGH THE AID OF GOD'S SPIRIT, TEACH?

The next subject I have to elucidate is the Origin of Evil, a subject which has puzzled the intellects of theologians in all ages and has in fact caused the origin of idol worship. But how, you say, can God permit such a difficulty to occur, such ignorance to exist, when the consequences are so momentous as the destruction of true worship? It is not always true worship that people arrive at when they are persuaded to leave idol worship. What God most requires from man is the sacrifice of his heart to service of others. The sacrifice should be made to God, but if mistakenly made to an idol, personage, or to a being supposedly superior, God accepts it as if made to Himself. So idol worship is not the worst infidelity that can befall men. The worst is a worship of *self*. When men worship themselves, their own plans or purposes or wills, establishments or churches or creeds, then God is neglected indeed. Then the sacrifice of the heart is not made and man serves himself, not God; a scheme of his own or other men's, not an imaginary superior intelligence, but a form or idol that is as senseless and as powerless to bear him to heaven or confer upon him peace as is a block or mass of wood or stone or metal.

He who looks beyond himself and his own or other men's schemes, or forms, to the spirit that they represent, to the Being whose honor and service they may have been instituted to represent, is accepted as a worshipper no matter what church owns or disowns him. He acts according to the best he knows. He has sincere desires to do good, to serve somebody, feeling that he owes service. To whom much is given, from him much is required. So the reverse is true: To whom little is given, from him little is required. But a man must not consequently expect that he may shut his eyes to the light he knows is shining for him. He must not refuse to hear, to read, to see, when truth and opportunity for acquisition of the much is offered him. He must thankfully receive all that God gives. He must earnestly strive to get knowledge that will profit his soul. He must seek, so that he shall find. He must knock and have the door of God's mercy opened to him. So shall he have peace. So shall he serve God. But he who refuses to investigate, who desires to stand still and to keep back others who would otherwise press forward, sins against God and His holy laws.

His laws are laws of progress. The precepts of Jesus were not directed to men in the form of laws or rules of conduct, applied merely to the existing circumstances by which men should be required to perform a certain routine of action or worship, or of form called worship, but they were living principles intended to govern and direct men how to form rules of conduct for all circumstances and all states and every age. They were to be the exponents of progressive religion. They were the perfect principles to which human effort should desire to attain and to which humanity's effort should be made to conform. But they were not to be refined or roughened into other shapes and fixed as barriers to advancement. They were not to be the foundation of unchangeable creeds or unprogressive churches. Their followers were to seek continually to be the willing servants of the God from whom they came and the obedient subjects of the King through whom they were given. Only the next world can realize their perfection, and the apparent unfaithfulness or impracticability of obeying some of them is only outward. The internal or mental obedience is compatible with every form of government or state of society.

The oath that Christians or the followers of Jesus should not take is not the judicial oath or pledge to speak the truth as a witness between conflicting interests, but it is that unnecessary or light and thoughtless form of calling God or some part of His creation to witness the truth of a meaningless or unimportant or vindictive or malignant expression. It is this which, like telling an untruth as a witness, is taking the name of God in vain. No vain or light appeals should be made to the Deity, though He is ever present as a witness to every assertion or testimony of men in the body and He sees when the error or lie is voluntary or involuntary. He will punish the one; the other needs no pardon. Ask not pardon till you make restitution. Ask not correction till knowledge is obtained. Be merciful as you would have mercy. Do unto others not merely as you would have others do unto you if your situations were reversed, but do unto others as much more than that as you think will in the present instance be beneficial or just, useful or consoling, softening to hardened sin or leading to love or faith or hope, or to the honor or glory or praise of God, or the advancement in any proper way of His cause.

My holy medium has been instructed as much as you can be by My revelations in this book. He had received much from Me previous to his writing for others. The first was for his training, the last is for the training of all men. The last, though, is higher, newer, more filled with heavenly intelligence than the former. I did not allow him to show the former except in a few special parts and cases and for special purposes. But he has passed that state of training and I now rebuke him more seldom but not the less severely. I can now speak of him freely to others without exciting his nervous sensibilities, at first very easily deranged, but now very firmly fixed and enured to exposure. For this is the result of training and of submission of desires and of will, of passiveness and willingness to be passively used. He does not flinch but I shall not try him too severely, for though he might be benefitted the great cause might be retarded by his rebellion or restlessness.

Let us, then, proceed to the subject of bearing arms. Shall a man comply with the law of the government under which he is protected in life, in security from personal or pecuniary injury, and whose institutions require, or are deemed by the ruling minds to require, resistance by force to intruders or to domestic assault? In brief, is war ever justifiable to nations or is partaking in it justifiable to individuals?

War is a great evil. War calls forth the exercise of the lowest passions, and those engaged in it are often actuated by the basest motives. Its injuries are incalculable and its destruction of public and private virtue lamentable and almost irrecoverable to the generation in which it occurs. But wars have been, like other evils, permitted by God to exist, and have been, like other evils, overruled to good results. God makes everything in creation good and He brings harmony out of the GREAT WHOLE, as I have declared. God has even directed His servants to make war and to exterminate nations. So the Old Testament says and so it was. A doctrine or belief so abhorrent to many minds in this day of refinement and high notions of God as benevolent and merciful and loving to all His children, has been a stumbling-block to many and a puzzle to more. But all this is reconcilable with truth and with God's justice and mercy and, in a word, with all His attributes. I will explain. I will first take the case of the wars of the Israelites under Joshua, where it is plainly declared that God ordered the fight and watched the slaughter with satisfaction and sanctioned the extermination of an enemy unoffending to the Israelites and innocent of injury to them.

God beheld that the Canaanites were sunk in debased and debasing religion. It was carrying them continually to lower and more sensual states, to deeper and deeper depths of destruction in which they would suffer complete separation from Him and their last good motives would have expired in the darkness and despair of the lowest condition of virtue and goodness. For virtue and goodness are never wholly extinguished in man, because, as I have said before, man's soul is an emanation from God. It is the breath of life that was breathed into Adam. What, then, would enlightened mercy lead an all-powerful king to do for these suffering subjects in a state of rebellion? He would perhaps first offer pardon and amnesty for the past and endeavor to persuade and entreat them to return to their allegiance. God did this. While the Israelites were wandering in the wilderness and being purified from the grossness they brought out of Egypt, God was sending His holy mediums over the Promised Land and calling on men to repent of their sins and come out of their selfish sensual courses. No entreaty or argument or sign sufficed to produce a favorable change. God's mercy was then shown in their removal from the earth to the spirit-world. There they could not become worse and there they must at last be happy. There they will sooner be happy than if left longer to pursue their downward destructive course.

But could He not have removed them by pestilence instead of war, which I have admitted to be such an evil in itself? He could. He did. He also completed the work by the hands of other men because those other men were to be led to trust to Him for deliverance and for victory. Those other men were also gross and sensual. But God desired to raise them to be His servants or subjects. Nominally they became so but really they were selfish and fearful. Fear is not in God's servants; perfect love casteth out fear. After all the great deliverances, after the extraordinary victories and the apparently miraculous aid they experienced, the Israelites were still faithless to God and disobedient to His requirements. What good, then, was derived from this plan of God's to settle them in Canaan? Was not the war after all useless? And could not God tell without trying the experiment what would be the effect of His operations?

These are all proper questions and when asked in a desire for reconciling the difficulties before named will not be left unanswered. God did know what the result would be. He knew that out of all would result good. That notwithstanding their manifold transgressions and their numerous relapses into idolatry, there would be a continued progress in faith in Him. That they should at last know and worship Him as the one true and only God and that they would thus prepare the way for the introduction of a purer and brighter faith which might be preached to them in their improved condition, and in them produce fruit and through them bring other nations to the same or a better knowledge of the One True God. This war then, directed and controlled by God, was for a good purpose and effected a present and a future good. The Poet says—

All partial evil, universal good.

The evil was, in fact, only apparently so. He that was wicked before the war might be wicked after it; he that was righteous continued to be so. He who gave God the glory was strengthened; he who lived for self, and glorified self, was cast down—perhaps slain—perhaps mortified—perhaps left unreformed. Worse he could not be as regards himself, though he might affect others more by his wickedness. But God continued to care for those others and to make all they experienced conduce to their elevation and instruction. At last they were established in the land promised their forefathers. God's covenant with Abraham was kept. Not because Abraham was by any means a perfect or a moral man, but because he obeyed God and sacrificed his own will, God accounted him righteous. This is, too, the key to the favor with which other prominent personages in Jewish history are said to have been regarded. They were men not perfect but in advance of their associates; not so good as they should have been, but comparatively good. They fell often into sin and yielded easily to temptations, but they struggled heroically to recover from their lost condition and again sought to be reconciled to God by confession, repentance, and good works.

But some wars are made without God's sanction and in the will of men. Shall a servant of God obey the order of his earthly ruler, of the law of his country, and take arms, enter the army and fight in the unauthorized, probably unjust, certainly unwise, war? He may, first because the powers that be are ordained of God. This is hard for some to concede, that wicked tyrants and bloody miscreants who have so often ruled nations should be of God's ordination. Nevertheless, God sanctions their elevation. He can create and He can destroy, as declares one of the oldest hymns in existence amongst men, and as declares the Psalmist. But He does not interfere, perhaps, in order to elevate this evil man to power. He permits men to choose their own rulers or allows the wicked to execute His resolves so that they may be the means by which He brings trials and temptations upon His subjects—so that, though their externals may suffer, their internals are benefitted. Man's free-will is ever left to him. Man chooses, God sanctions the choice, not perhaps by approval, but by strengthening the man in his performance of his own evil resolutions, so that the man of evil resolutions may experience their vanity—may see that his own desires when realized do not bring happiness—so that he may be led to reject the earth and its hopes and wishes and turn to God in submission to His will. But you do not see such changes taking place amongst tyrants or despotic rulers! No, but beyond this life you cannot see. God and His Holy Spirits can. It is to the WHOLE that a being must turn his attention who would form a just and perfect judgment.

Shall a man then, sincerely desirous of serving God, of sacrificing to Him or for Him his will, his life, his worldly substance, his every earthly possession, shall he bear arms at the command of this tyrant who holds his position as the head of his nation by force, by armies overaweing the peaceful inhabitants? Shall he so bearing arms obey the commands of his officers and seek to destroy the enemy of the tyrant who calls on him to fight for his own enslavement? No. The precepts of Jesus do not require this. When a man sees so clearly the causes and the effects, or thinks sincerely that he so sees them, he should act on his convictions of right. He should seek the guidance of God as to what is his duty, and he should be willing to suffer unto death or torture the consequences of his obedience to God rather than to man. In this case God indeed ordained the power that exists but He also directs His servants in what they shall do. He may see fit to destroy the power of the tyrant or He may let him proceed in his triumphant career as a scourge of God. If so, it will be for good ends and God, who sees to the ends, sees the harmony that will result from the discords, turmoil, and contentions—the suffering, the misery of the separate parts. But the glory, the honor, and the praise of all the good that comes from the evil, of all the harmony that results from the discord, belongs of right to Him who produces the good and the harmony, while all the punishment and unhappiness consequent on the individual's own acts fall upon him who is evil or wicked.

But all cases are not so clear as this. Most wars result from misunderstandings rather than from intentional injustice. So the cases will shade from self-defense to aggression. What is the Christian's duty? To bear arms in general, or to refuse ever to do so, or to wait until he is himself personally aggrieved? *Let every case be decided by the Higher Law.* This is the pith of the precepts of Jesus. Man should endeavor to do right. He is placed here for his good, and the experience which the proper exercise of his judgment gives him of right and wrong, of good and evil, is the very thing for which he left paradise. But in many cases the law requires a man to arm himself and march against his nation's antagonist, and the man may not understand the causes which lead to the rupture sufficiently to judge of them! What then shall the man do? Fight or suffer? Refuse to obey the law, or obey and kill his opponent who had never injured him?

I say again, let the man seek God's direction. Let him seek it with submission and willingness to be governed by it, and he will not be left in uncertainty. If the injury is against himself, only he should forgive. If it is against the law, he should in general support the law as the power ordained by God to exist. If against his nation, let him judge from all the light and knowledge he has or can obtain as to what the rules of justice and good national policy founded on justice and right require, and act according to his convictions of right and justice. Do sad consequences result from this determination—does he kill his arrayed opponent—or does the law condemn him to imprisonment or suffering of any kind for not obeying it? Then suffer with cheerfulness or resignation. If the sad consequence falls on others then let him find, as he will find if he has a good motive, consolation and justification within himself. If all war and fighting had been interdicted, Jesus would not have told the soldiers to be content with their wages. If all arms-bearing had been wrong, He would not have allowed His disciples to carry arms. For Peter informed Me there were two swords amongst them when I told them to provide themselves with arms. I knew they had long been carried about with Me by My followers. But for My kingdom, My servants do not fight either with arms or contention—by words or any deeds. If My kingdom were of this world, My servants would fight. When I am King of the fifth monarchy then will My servants fight. But then they will only withstand assaults. I will tell them how to overcome without bloodshed.

But some will say I do not settle this point, that I do not lay down a general rule by which each case may be squared. That is so. I did not intend to give any more settled guide than the principles of right, truth, justice, mercy, true benevolence, and uninterrupted desires for God's glory, honor, and praise, would give. Let these influence you in all cases and at all times and you will find the crooked paths made straight and the true vine fruitful. You too will bear fruit abundantly because the true vine will be in you. The Father, who is the husbandman, will prune and train it and cause it to be protected and cherished, and its fruits to ripen into eternity, undecaying and indestructible. From that fruit will be derived the wine I will drink anew with My believers and doers in the Kingdom of Heaven, or, in other words, My kingdom. My kingdom is established in Heaven—pray ye that it may be established on Earth.

Let us pray

O our Father who art in Heaven, hallowed be Thy name. Thy Kingdom come, Thy will be done on Earth as it is in Heaven. Give us this day our daily bread, and forgive us our trespasses, as we forgive those who trespass against us; and lead us not into temptation but deliver us from evil. For Thine is the Kingdom and the Power and the Glory forever and ever. Amen.

Now this prayer was the only one I taught My disciples when I was in the body. I have given through this holy medium many others, but after all, does not this prayer include them all? Let any make this prayer his own and he will include the substance of all others that he could with propriety make. But then should no other be made? Others should be made because, as I have shown, prayer is the expression of our desires and of our submission to God, of our wants and of our passiveness, and though God knows all we want and what we feel before we express it, yet we may by our own expression of it better realize to ourselves what we are and what we feel and compare these with what we should be and what we should feel. It may also benefit others as well as ourselves, and should be made at proper times by us for others' good and in various ways for that purpose.

Then prayer is not to be confined to what we want for ourselves! We should of ourselves desire the good of all men and if we love our neighbor as ourselves we shall pray for our neighbor as we do for ourselves and regard his wants as our own. But then who is my neighbor? I can not pray for each and every man I have ever known, nor even for each one who has ever done me a kindness! Spiritually speaking, he is your neighbor who is connected with you in some act; carnally speaking, he who lives near you. In society it is he who is your friend, your associate, your companion. In families it is the one who is wife or husband, father or child, uncle, aunt or cousin. In fact, any one with whom you have relations of business, friendship, love, or association. But if I pray for all these as for myself and for each separately as for myself, I shall not have time to do anything else! Pray without ceasing, was the command of Paul by My will. But can a man do that and live as others do by his labor? He can, for prayer needs not words to be prayer though, as I have explained, words are often useful. But the heartfelt prayer and thanksgiving which is felt without expression into words or even ideas is equally acceptable and understandable with God as the most eloquent and refined form of words, when they also come from the heart. And yet it may often be a benefit to repeat a form of words that you can not fully join with, so as to make them your own. Because if you are convinced you ought to be able to make them your own and desire to, the repetition of them will help you to adopt them, and cause God to grant you the double prayer; that is, the desire expressed by yourself to be able to adopt the form and the desires expressed by the form of words. True prayer is from the heart, and when a man has earnest desires, his heart at least will feel them, and those desires do reach in their aspirations and in their effect to the Throne of God.

The next subject I will treat of is one which has been the subject of much dissension amongst Protestants and has at times shaken the Roman Church. And that is, what compensation, if any, should a man receive who devotes himself to preaching the Word of God to his fellow-men.

The Word of God is preached by men, then, to other men? It is, when they speak *by the aid of the Word,* and this every preacher should earnestly strive to do. But should any one speak without it? They should not, because without its aid they declare only their own cogitations or imaginations and do not preach the Word of God. For though the Word of God is extensively found in the Bible, an exposition of the Word without the Word's aid is like applying a candle to sunshine in order to illuminate its shadows. Human reason is but as a candle so applied to the glorious Light of the Word. If even the darkness can not always comprehend or perceive the Light of the Word, how can the light of reason perceive or comprehend what it does not at all equal, though it is of the same nature? For reason is a gift of God as well as is the Word. But reason is contaminated or clouded by its union with free-will and with other perverted gifts of God to man. Reason, therefore, is full of fallacies. But God's Word is a part of Himself unmixed with error, undefiled by passion, uninfluenced by any other will but His own.

The Word of God, then, is infallible because it is equal to God in all but that it is only a part of Him. The man who preaches in reality the Word of God will also be *directed* by the Word of God if he will listen in the cool and quiet of the heart's twilight. For God teaches His people Himself when they are willing and desirous to be taught. As in the case of bearing arms, each case of payment for preaching must stand on its own merits and be governed by no other rule than that men reap as they sow. He who preaches for hire or divines for money shall have his reward according to his desire. So if the desire be only to serve God and to do it entirely in the will of God, the reward shall be the thanks of God and of His Son saying: Well done, good and faithful servant; thou hast been faithful thus far, I will try thee more. Not always by increased temporalities will appear the additional trust from God, but often by increased cares, more confining duties, more overwhelming responsibilities. So God's rewards are rather in the nature of trials? They are very generally so. For this life in the body is for the express purpose of trial and of proving to men what they are and of providing them with a knowledge of good and evil. It was for this that they left paradise to return no more—but to progress in a life to come, ever nearer and nearer to God's own perfect nature without ever arriving at it; and to be ever able to compare the happiness or immortal bliss of eternity with the frail and finite life in the body, when happiness itself was so like misery, was so nearly despair.

CHAPTER XLV

THE PREACHING AND LIFE OF PAUL

The most important subject I have left for this book is the History of the Church in the Apostolic Times, a most interesting subject to all sects and orders of Christians and one that has engaged the learning and talents of some of the most worthy servants of God in every century of the existence of Christianity. Unfortunately, learning was at a low ebb during the early age and after the simple and beautiful and concise account written by Luke of the Acts of the Apostle Paul, there was no history attempted until the materials were, if not lost, so confused with errors and false traditions that despair of ever recovering the truth might well have prevailed. Attempts were made, however, but by men unqualified to sift out the false from the true, and the results of their labors were soon disregarded and lost. It was better so, for the truth was so adulterated that the account would have been only a stumbling-block to pious inquirers and true servants of God.

This history I shall briefly write, for My holy medium has this day renewed his submission and proved his passiveness. I believe, therefore, that he will write without apprehension or doubt what I deliver to him. Few can conjecture the trial a man's faith receives when he is made a holy medium of disclosure of novel doctrines, theories, or arguments. His nature is earthy and it struggles against the bond of the spirit. He desires to believe, but his reason fights or resists. He lives in faith and keeps in faith by a recurrence to first principles like these:—first, God rules all the creation or He is regardless of it. If regardless, whence its order? If He rules it, He watches the movements of all His creatures. If He ever made any revelation to men that has been recorded for us now, He made it through Jesus Christ; for if what Jesus Christ declared was not true revelation we may rest assured we can never know when we get revelation. It was attested by so many outward manifestations, and was so much in advance of all other teaching then or since given to men, that we must believe we can never have more convincing proofs of Divine Origin in any revelation than accompanied that. Jesus Christ says that not a sparrow falls to the ground without God's notice, and that the very hairs of our head are numbered. Then it is evident that if God has declared anything to mankind He has declared that He extends His care over them.

Second, if God does watch over man, His highest creation—at least the highest evident to man—then He, being good and benevolent and merciful, would not let a man sincerely desirous of serving Him in submission to His will and from praiseworthy desires to render Him glory and honor—I say, He would not let such a one be deceived by invisible beings assuming wrongfully and falsely to be other than they are and using the confidence inspired in man by their superiority to him to lead him away from duty or happiness. It would be inconsistent with all we believe of God to suppose that He would allow a man to be ruined in reputation, business or social relations, because he yielded to a superior intelligence which proclaimed itself Divine in its origin and as acting in the will of God, its creator as well as man's.

Such an argument does not need the many supports it can call from the Bible. The promises of God are Yea and Amen for ever. He has in all times, if He has spoken to men at all, promised them that He would hear their prayers and that He would take care of His servants. But some may say, as My holy medium says, that the motives of men are so inscrutable even to themselves that no one can feel entire certainty that he is God's servant and that he is impelled to action or to any partial or particular kind of action solely by sincere desires to serve God! There is always danger that other motives may mix with or cloud the purity of the true or proper motive and that he may be found, after all, to be acting in his own will instead of being singly desirous of doing God's will. But God has declared that he that seeks shall find, that to him that knocks shall be opened. And though the man may fail to be perfect even as his Father in Heaven is perfect, yet God will be a helper to those who call upon Him and no man shall ever see the righteous forsaken or his seed begging bread. And who are the righteous if they are not those who desire with all their knowledge to serve God and to obey every intimation they believe comes to them in His will and power. Who but these shall be found to have washed away their sins, their short-comings, their impurities, their peccadilloes, in the blood of the Lamb? That is to say, they have made the same sacrifice that Jesus Christ the Lamb of God made when He gave Himself to God and continued to serve and obey Him even to a bloody end.

But then, if Jesus was crucified, did he not suffer for His obedience to God? He came into the world to save sinners; He came as I stated in the first part to do good. And it was His mission to do what He did. To set an example to all future time, that everything should willingly be sacrificed to God for His glory and honor, and that man should follow the obedience which He gave to the Word of God within Him, though it should lead to torture and death. But though men are thus required to be willing to be led by God, and to this extent, God does not call on every servant to suffer without cause. His rod will guide and His staff support those who serve Him, whether they are called to make temporal sacrifices or left to enjoy temporal blessings. All creation is God's. All that His servants have is His. Will He not delight in conferring upon those who thus hold their all for His glory and honor the abundance He has to bestow? the rich gifts that it costs Him nothing to lavish upon them?

But are not all these gifts spiritual and must not the devoted Christian expect to meet with trying dispensations and deep afflictions! By no means necessarily. God's servants or would-be servants are often purified by trials, are often deprived of their idols, are often reminded that the love of God is unfailing and that it alone passes not away. But all these are bestowed upon the man as kindnesses and will be perceived by him to be so, either at the time or afterward. Yet this does not make it evident that temporal affliction is not a fate properly expected for a servant of God. Then try it by reason, and see if God will make any rule that might discourage men from becoming His servants. See if He will not draw them by all means to Him, even by blessing them abundantly. Not that a man should serve God or pretend to himself that he serves God for his reward or expectation, but that he should feel such confidence in God's protecting care and all-powerful benevolence and foresight as to believe that he may trust in Him for his daily bread. See the lilies of the valley, how they grow; they toil not, neither do they spin, and yet Solomon in all his glory was not arrayed like one of these. And does God so care for the grass or weed of the fields, that is at all times liable to destruction by man, and shall he not care for you, O ye of little faith? Be, then, no longer fearful or unbelieving, but know that God lives and reigns and rules with justice, kindness, and affection toward all men but most certainly toward those who live, move, and have their being in Him in the hope and desire that they may attain to a sonship by being first admitted to be His servants in this life and by seeking to do His will by sacrificing their own desires, wishes, and will to His.

The history to which I have resolved to devote the remainder of this book is grave in its character and lifeless in its style. It will be pronounced by many dull and stupid, because it will be a simple narration of the prominent features of the course of reasoning which led the Church of the Apostles to dissensions and of those which caused the rapid decline and almost the obliteration of all true service to God in succeeding times.

The Apostles were, in general, led and guided by Me. You think perhaps that it is strange that they were not always so guided. But that they were not is evident from the account in Luke, where it is related that Paul and his companions could not proceed harmoniously together, and so they separated. It is evident by the account of Peter's conversion to Paul's views respecting the duty of obedience to the Mosaic laws regarding outward observances, such as circumcision and refraining from certain kinds of animal food. If Peter had always been acting under immediate inspiration a vision would not have been required to change his practice or his course or his opinion. So, too, we read that in the first council held at Jerusalem, when Paul's authority and teaching came up for man's approval, there was great difference in the views of those who were then considered the heads of the church; and that the resolution finally adopted was a compromise which suited neither the requirements of the Gentiles nor the prejudices of the Jews.

I do not say that it was not the best thing possible then to do, but it is a proof that if it was the best, it was because the heads of the council did not receive and obey My direction, but used their reason and abided in their former traditions. Otherwise there would have been entire unanimity and the resolution would have been different. For I must always act in submission to the will of the holy medium, or in God's will, or not at all. I can never force a man to be a medium of mental reception; he must seek to be by submission; then I accept him. But I accept him in his state, which is necessarily imperfect, and endeavor to reform and educate him so that he shall be more useful and reliable; but I do this by counsel and advice rather than by command or force. For man has perfect free-will, and he is never deprived of the least portion of it while in the body. So when he receives passively and submits perfectly to the direction or counsel I give him, he may be entirely led and guided by Me in every action of his life whether that action relates to spiritual or material things. That the Apostles, or some of them, had not arrived at this state of perfect submission to and entire reliance upon Divine guidance is evident from the facts referred to above, and though they were the light of their own age, the present may open clearer views of God's purposes and requirements to man without detracting from the reverence with which all revelation made through them or others should justly be regarded.

Paul was, as I have stated, the Twelfth Apostle and filled the place from which Judas fell by temptation. The preachings of Paul were attended by extraordinary results even in that time when all the Apostles preached with great power and success. His first address was always to his countrymen, then already scattered through every Grecian city of Asia Minor and Egypt. He first endeavored to convince them that the Messiah promised to their nation and so long looked for had visited them and had been the rejected Jesus of Nazareth; that it was their duty to leave the rest of the nation to act for themselves, but for them to embrace the cause of truth and believe in the Lord Jesus Christ. It was for them to believe that He was the Messiah and that His precepts were above the laws and traditions by which they had heretofore been governed; then the delivery of these precepts had been accompanied by many miracles of which there were abundant witnesses, one of whom was himself though only in a secondary manner; that God must be served not by pilgrimages to Jerusalem but by submission to His will, which He would manifest to them by His Spirit if they would but turn to Him and seek His aid and counsel and direction within themselves; and that he who did this should live and never die, but should never be separated from God the Father, and should jointly be heir with Jesus Christ to immortal life and perfect bliss.

But though this simple preaching was often accompanied by outward signs of God's power, he more often failed than succeeded in converting the Jews. Then he turned to the Greeks, who were their fellow citizens. Here his arguments were of a different kind, for he was all things to all men by addressing himself to the capacity and knowledge of the men he had to deal with. He would endeavor here to show the heathen idolater that God was one great overruling Creator; that all other objects of worship were powerless and unworthy of regard. He endeavored to show the Greeks that it was proper that every man should render obedience to that powerful Being who had bestowed upon them life and capacities for enjoyment, and that if they desired to serve God they should, by addressing their prayers to Him, seek Him and they should find, when they sought for it, in their own hearts an evidence that they were heard and that God would so assure them of His invisible presence and ever-watchful care and love. This discourse might or might not contain an allusion to the life, death, or resurrection of Jesus Christ, but he never failed to teach to every believer of the first preliminary discourse that the beginning of this movement had been the appearance of a long promised Messiah, or Sent of God, at Jerusalem; that He had been rejected by His own nation and friends but that nevertheless He did perform wonders and miracles and that He was eminently the Son of God, and that His Birth, Life, and Death were a succession of miracles all showing that He was indeed what He professed to be, the Son of God. Paul told them that Jesus had declared that He would be the Comforter of His followers and that wherever two or three were gathered together in His name there He would be with them; and that the manifestation of His presence would often be accompanied by an outward sign.

So Paul preached and on this simple platform laid the foundations for the church of Christ. At Antioch, the capital or head city of a very large territory, where heathen worship was supported by the strongest pecuniary motives and where the whole population felt particular pride in the existence of their great temple and believed that their city was really favored by the residence of a god who had been a man, the converts of Paul first received the distinctive name of Christians. Before that they had had no appellation that classed them as a body. Previously they were merely called Jews, or Gentiles who believed in Jesus the Messiah—who believed in Paul's preaching, or in some other teacher's preaching. But names are nothing; it is the sacrifice of man's will that is required of him and God requires all men to seek Him now as Paul then preached they should.

Paul speaks of his having a thorn in the flesh that continually reminded him of his own littleness and prevented him from appropriating to himself any of the glory, honor, or praise which men delighted to bestow upon him. He gave God the glory as being the director and guide of all his actions through the Lord Jesus Christ, who acted in unity with God and in accordance with what He had declared to be His intention when in the body. He was the inward revealer of God's Will, and Paul desired to be passive in His hands or direction because he believed that all power is from the Father, and that the Son received His power from God in common with every other manifestation of existence. Paul believed that even Jesus Christ was nothing except what it had pleased God to have Him be, and that the converts who received his teachings and obeyed submissively the Voice of God in their hearts, given through the Spirit of Jesus Messiah, should be one with Christ so as to be joint heir with Him to all the gifts and blessings which God in His love and wisdom had conferred upon Him, who had been a man without guile and had gone about doing good and obeying God's command until at last He suffered the extremity of horrid death upon the cross.

CHAPTER XLVI

THE ORIGIN OF THE CHRISTIAN HIERARCHY

The last chapter treated of the manner of Paul's preaching, but how and where he preached I have yet to relate. In all the principal cities which he visited he was admitted to the synagogues as a Jew, which he claimed as his right. But as the doctrines he preached were often unaccepted by his hearers he was sometimes driven forth tumultuously, sometimes arrested as a disturber of the public worship sanctioned by law and custom, sometimes allowed to depart peaceably but for the future refused admittance. Then he felt himself released from further appeals to his brethren as the already favored people of God and, laying aside his character of Jew, he became the Roman citizen. He preached in the market-places, in the theaters, in the hippodromes, or wherever he could find people collected. But when any considerable number believed he formed them into an association and appointed one of them deacon or elder, an office that made him the supervisor of their outward profession and practice and the administrator of the revenues of the church or association. But the revenues were trifling, consisting merely of the offerings made weekly, as Paul directed in his Epistle to the Corinthians. The Church was then established, but where was its preacher to be found when Paul left, as he usually did very soon? Paul would also select a man for that office and him he would call presbyter, or deacon of spiritual matters. The same office was called by others *episcopos,* translated bishop. These titles now so numerously conferred and so eagerly sought were not then places of ease. They put a man up as a mark for persecution and for all the venom of bitterness which enraged men so lavish upon those who are converted from their ranks.

The episcopos or presbyter was, however, yet to be endowed with his ministerial power. That he could receive only from God through Christ, His Sent Spirit. This was done ceremoniously by the laying on of hands; that is, the Apostle placed his hands upon the head of him so selected and prayed God to give unto the man His Spirit abundantly so that the man and the church should glorify God the Father and Jesus Christ, the Son of God, and the Spirit that guided and inspired the man was believed to be of God and, like God, true and faithful. Faithful to what? Faithful to the wants and holy desires of the church and addressing to it words of counsel, advice, and direction. But why did Paul further interfere by giving directions personally or by letter? Because the episcopos was not always faithful to the guidance of the Spirit and there was no way to insure that he would be so, for as I have before told you, the free-will of man is never restrained in the body by any other power. If it were restrained, man would then immediately cease to be a free agent and would sink into a mere automaton. Yet this free-will you would have us surrender to You or to God! Because you surrender it and also retain it. So rich is God that He accepts man's offering and leaves it in his own possession, and what is more, He leaves it as fully the man's as ever it was. If the man is a faithful servant, he is continually sacrificing to God the only thing God wants him to offer. God always accepts it with pleasure and the man has it ever to offer. So the sacrifice must be continual and is continual when it is made as it should be, and thus the incense of good works arises continually from a good man to God. But when the sacrifice ceases, the man ceases to serve God. He then serves himself. God does not reward a man for serving himself, but for serving Him. His spirit in the man must either withdraw or act imperfectly when the perfect sacrifice is not made.

Thus it was that differences and dissensions arose in the church. The presbyters or preachers or episcoposes, becoming imperfect mediums of the truth because they imperfectly sacrificed their own wills, differed from each other as well as from the truth. Then councils were held and the majority assumed to decide forever what God's Truth and Will was and should be. So was built up the hierarchy. But the hierarchy is a false prophet. It is the False Prophet referred to by John the Divine. It is that False Prophet which must be destroyed before Jesus Messiah can reign as King of Kings and Lord of Lords. But did I not give another interpretation to it in the second part? I did. But I also informed you in Part I that throughout all the Bible there are passages of double signification; the one outward, the other spiritual, both to be understood by the help of the Spirit and neither to be arrived at with certainty except under its guidance. And have not many commentators sought this guidance, and if they sought why did they not find? They did seek, but not with a perfect sacrifice of will. They desired to establish their faith and that of their associates, and they often obtained their desires.

So it will ever be, as I have explained in this section already, that the incomplete sacrifice is accepted as it is made, and God grants the desires of His servants or of His would-be servants even when they strive to obtain untruth. Then how can a man be sure that he is receiving the truth if he is so liable to be deceived by his own imperfections? By the assurance of the Spirit of God freely given him and, without his own desire or request, its declaring to him when the surrender is perfect and complete or that he requires still more effort to obtain a full sacrifice even for a time. Thus Paul was obliged to interfere and sometimes to depose an episcopos and establish another. So too the second one chosen might prove unfaithful, for the church was made up of illiterate and newly received members whose passions and traditions assailed them with frequent success.

But how did Paul lead so uncomfortable a life if he had such authority and could so easily silence error! The result of a deposition was not always to silence the former preacher; often the man's imagination or revengeful feelings would induce him to disturb and distract the church. For its members were not all obedient to their outward teachers any more than spiritual believers in the present day are obedient to the Divine Influx or the declaration of their duties made by holy mediums, so that confusion takes the place of order and an attempt is made to be governed by reason in the place of God's Will or Wisdom.

Let us once more return to the early formed church. We read in Paul's letters or epistles that there were various signs in the church. There were some who spoke with tongues and Paul was one of these. Some had the gift of being able to interpret the speaker-with-tongues. Some could heal the sick or lame, outwardly. Others were obliged to be content with the gift of prophecy. We read that these gifts were not united in one holy medium but that in general each one possessing any such power possessed but one kind, or manifestation, of God's action. But what were these gifts so abundant then, so rare now? The gift of tongues was speaking in a language unknown to the medium, or speaking so as to be heard by different hearers in different words or distinct languages. The first was by far the most common, yet the latter was its first great manifestation and was the first miracle or manifestation of the Spirit given by the Apostles after My ascension. This was through Peter. The gift of healing was also performed by many beside the Apostles and it was done in the same manner as an episcopos was appointed; that is, by laying on of hands and prayer to God to manifest His power and make known His mercy by healing the disease or restoring sight or removing lameness. The gift of prophecy was not a power to foretell events but to discern spirits and to interpret dreams.

The spirits of the prophets should be manifested by truth and agreement with others of the same powers. So if one's interpretation did not appear to another consistent, he gave a different one. But the resolution of Paul and the church was to let the other prophets present declare when a prophecy was correctly given. The proof would be the response of the Spirit in the internals of the other prophets. But then if this kind of judging led astray the episcoposes, why were the prophets kept in the truth by it? Because in the one case the assembled prophets gave their testimony from the intimation of the Spirit, and in the other were led by reason which alone can not guide a man to truth. Nothing can guide a man to truth but God's Love manifested through his spir' causes truth to be declared and bears testimony to it ' of the humblest members of the church or heare for even hearers who hear a declaration of the firs' have encountered can find within themselves the truth declared respecting what they have heard, if they will submit to God with earnest desires not for their own satisfaction but for the truth and its general establishment.

The last matter that I shall notice in this book is the fact that Paul had a thorn in the flesh given to him that he should not forget his dependence on God and glorify himself. This thorn in the flesh, you no doubt understand, was some affliction appertaining to the body. It was a lameness caused by rheumatism, at times very painful but always annoying except when engaged in God's service. Paul was not then always engaged in serving God! He was not always victorious over his own will but sometimes allowed desires of his own to precede or obscure the gift of the Spirit. But this thorn in the flesh was given to Paul so that, when he lost sight of purity and grieved the Spirit by neglect, the Spirit ceased to maintain that equilibrium of vital invisible odic force or essence by which every part of the body is kept free from disease and of which the derangement is so quickly declared by pain. This was the thorn in the flesh so briefly alluded to by Paul, for which he was thankful and for which he had reason to thank God.

And now having led you to see how Paul established churches and how the churches became the generators and supporters of error, let us thank God that He has been pleased to manifest Himself in this Time of the End with the same great and striking manifestations as He then gave, except that, the world being so much better prepared for them, His holy mediums do not suffer persecution to such violent extremes as then. But then God's Spirit could support them under every trial and give peace to every torture. The entire course of a suffering martyr could be one continued triumph over the ordinary laws of matter because the Spirit of God within him would control the sensations of his body and maintain such an equilibrium of odic force or fluid or essence as would obviate all suffering and prevent the manifestation or existence of any pain.

But how did the other Apostles preach? Much as Paul did after he had set them the example. But before that they confined themselves mostly to the Jewish nation and endeavored unavailingly to induce its authorized heads to believe in Jesus as the Messiah. Afterwards they dispersed themselves to various countries, some of which received the new doctrines with joy and gratitude while others rejected and persecuted them. The traditions respecting their deaths in the records of the church are, in general, correct. But the only fact about which you are now deeply interested is how they died. Did they seal their testimony with their blood? Did they endure to _____ter end the pangs of death? Or did they, when they reached _____alley, hesitate to encounter its shades and try by re-_____rolong a life which they had been using to promulgate _____majority of mankind pronounced delusion and a _____mous attempt to betray Jews into the power

206

of demons and lead the learned polytheist to damnable Deism? They died as they had lived declaring that of a truth they had seen the miracles they had told of; that they had experienced the Divine Spirit operating in various ways upon them precisely as they had declared in their most secure moments. They sealed their testimony by the sacrifice of the body, but they were able to do this only by the power and assistance of God given them because they had sacrificed their will to His in at least a high degree, and because, having delighted to serve God, He made even the torture a pleasure and the cross a victory over pain. May you, O reader, be qualified for the same happiness and inherit the same recompense.

Amen.

APPENDIX

CALL BY REASON: AN APPEAL

Having prayed for you and made forms of prayer for you to adopt for your own if you can, and if you desire to obtain response to such petitions, I will now return to the consideration of the imperative call now being made to men, particularly to the inhabitants of the United States and Canada, or to all in America speaking the English language.

It is to you, O Anglo-Saxons, such either by descent of by adoption of the laws, language, and country belonging to the race or ruled by Anglo-Saxon institutions and partaking of their blood in its people—it is to you, O Anglo-Saxons, that this loud call is particularly made, for you are the chosen people, called to rule in righteousness and peace the world of mankind. It is you whom I have chosen to be My subjects, because your laws, customs, and energies are all directed to the enlightenment and reformation and improvement of mankind, not merely of a particular rank or class, but of all; not that all may contribute to the happiness of a few, but that all may enjoy happiness here and glory hereafter. Be then, all ye people, ready and willing and desirous to hear the word of God addressed to you, calling you to inquiry, to investigation, to all possible efforts to learn the truth.

Awake then, ye leaders of public opinion! On you first falls the responsibility of receiving or rejecting this call. You can influence the minds of others, and you must answer in the day of judgment for the talents you possess. A strict account will be required for every opportunity of investigation you heedlessly passed by, for every manifestation wasted upon you, for every inattentive observation, for every idle word or hasty conclusion, for all you have said or done in relation to this call or these manifestations, and for all you have left undone because love of ease, love of self in some way, either of home or affection or popularity or reputation or wealth or power, prevented your doing or desiring to do the work which every man is obliged to perform; that is, his duty. You very well know what is your duty if these manifestations and this revelation are of and from God, as they claim to be.

Often have you declared or thought that if you had lived in Judea in the time of the ministry of Jesus Christ you would have left all and followed Him because He had the words of eternal life and because through Him the Father worked. But then, as now, the learned doctors, the leading politicians, the priestly order, the wealth, the power, and the wisdom of the world were opposed to the claims of the holy medium or transmission who taught that He was the Messiah promised to them from the time of Moses. Wondering crowds witnessed the works and listened to words of God's wisdom, but generally returned to his home declaring that he did not know what to think! It was wonderful, extraordinary, beyond comprehension, but the scribes did not declare His sentiments to be those

208

of Moses; the priests did not declare Him to be the Messiah; the Sanhedrin deemed Him a preacher of dangerous doctrines, and a claimer of authority and right that would bring them into collision with an irresistible power. All of the past that was in authority and respect among the people rejected the man without sin who was indeed come and present to save them from their sins, to lead them to a more perfect knowledge of God and their relations to Him and to each other, and to purer and more perfect forms of worship and greater morality, higher principles, nobler motives.

Now here is a similar manifestation of God's love for men, His willingness to teach them, and His desire to raise them from ignorance and misconception to knowledge of Him and of their duties to each other. Will you listen now, or will you now, as the leaders of public opinion did in those days, reject Christ, deny His Divine Authority, ridicule His pretensions, deride His miracles, condemn His doctrines, oppose His teachings, and imprison or injure His followers? In this day of progress when you have seen so many despised movements control society or modify all the relations of life, when you have seen the arts of yesterday forgotten and superceded by the improvements of today, if you, who have seen such spiritual manifestations, incomprehensible and unreliable as they were, usher in the dawn of greater knowledge and the revelation here made—if you also shall reject without examination or judge before trying, what punishment shall be justly declared as your desert? What depth of misfortune or condemnation ought not to fall upon you? Remember your station, your power, your responsibility! Look well to your acts; view well the foundation on which you stand. It will be no excuse for you in the day of judgment to say: Father, one condemned it and I believed him; Judge, another said it was not worth looking into and I thought he ought to know; the priest said it was dangerous to read the book or listen to the manifestations, for they might draw me away from the old landmarks, and I was paying him for good advice; and, last but surest of all, nearly every editor of a newspaper warned all his readers of its delusive character and pronounced it the contrivance of rogues or the delusions and folly of simple-minded victims, and if these editors did not know surely you will say I ought not to be blamed for ignorance! You will only condemn yourself by your excuses, for they will show that you knew how, from such small beginnings as a little leaven or a grain of mustard seed, the thing spread itself until the great mass was leavened, or fowls of the air rested in its branches. You knew what it claimed to be and what it professed to be, and yet you were willing to take the opinion of others for your guide in a matter affecting your eternal welfare when you would not have taken the opinions of those same men respecting your temporal affairs!

Go to, thou pitiful specimen of love of ease, of self-complacency; go to, thou false worshipper of mammon. California could tempt you to traverse the world to see if a fortune might be obtained, but the riches of God's kingdom and the glorious manifestations of his love could not call you away from the concerns of time, the thoughts of earth and the care of the body. How many have said: Go Thy way for this time; we will hear Thee again at a more convenient season? *I mean to read the book but now I am too busy!* Tomorrow, tomorrow, tomorrow! Thus have men in all ages and in every clime put off the work of today. Death at last comes like a thief in the night at a most unexpected time when the soul, sunk in profound repose, has forgotten its intention to see that all was made secure against the time of his coming, and for that man the earth has no more tomorrow. The grave receives the lifeless remains, buried with great honor and care, and parade perhaps, but the soul, the man himself, is where there is no more repentance.

In the spirit-world he may, however, work out his own salvation, but as the tree falls, so it lies. It falls in procrastination; it lies so. It fell saying: A little more sleep, a little more slumber; it lies so. It fell with desires to enjoy the delights of its own will till tomorrow; it lies so. Long is today to that soul, for there is no light. Long does the impulse to seek God lie dormant, overwhelmed by all the thoughts of self-gratification and all the contrivances made use of to keep the peace of the earth uppermost. God wills that even this man shall be saved, yet how hardly shall such enter the kingdom of heaven for they are rich in temporalities, but so poor in spiritualities. How soon shall the regret of remorse overtake the departed soul of this procrastinating man? All but heavenly patience shall be exhausted; all but angelic wisdom will despair of it.

But God triumphs at last. You will submit, and glorify Him who is beyond praise and love Him who is beyond being affected by your love. Be wise today, O man; be wise and understand! Today only can you work effectually; today only is it easy to sacrifice your will and begin your submission to God. He offers the riches of His grace if you will dispose of all that wealth you have already laid up, consisting of perishable earthly productions. Come, buy of Me, and you need no silver or gold; what I furnish you may obtain without money and without price!

Let us once more pray for the humble-minded seeker, for the high-minded rejector, and for the would-be servant of God. They all with one accord make excuses; they all have some very important business that may not wait for God's work to commence in them this day, and tomorrow will find them less at leisure than now.

Let Us Pray.

O Holy and Incomprehensible God, Thou art so powerful to save that Thou wilt not be disappointed. Oh, be pleased to look with favor and with blessing upon My appeals to this man who desires to know Thee but has not found Thee, who desires to seek Thee but feels unworthy to try. He is, O Lord God, humble and of such is the kingdom of Heaven. Be pleased then, O Father, to raise him to Thy feet and bring him before Thy Holy Will in subjection so that he may join Me in giving Thee the praise, honor, and glory, now and hereafter evermore. Amen.

O Lord of Heaven and Earth, O Thou who art above all, in all, and creator of all, be Thou pleased to bow down the majesty of Thy nature and let this man see how very little he is, how much smaller than the dust his eyes cannot see in his form to Thee, compared with the dust to himself. Be pleased, O God, when he is humbled by Thy power and Thy condescension, to raise him to Thy Right Hand of Power, to equality with Me, to be Thy High, Holy, and Perfect Son. O God, help him very much for he has pride and thinketh too much of himself. Be not fierce against him in Thine anger, nor severe upon him in Thy pity, but bow down his knowledge before Thine and his nature before Mine so that he may receive with humility the counsel of the spirits and the words of Thy revelation, and being so brought to a sense of his littleness and dustiness, as it were, before Thee, he may serve Thee as may be pleasing to Thee, sacrificing his own will and acting in Thine, and leading others to come down from the high places and to sacrifice in the depths of the valley of their humility.

O God, be kind and affectionate and loving to him though he despises Me, Thy Son, and refuses to listen to the words of revelation; but O God, remember it not against him for he knows not what he does. Be Thou pleased to be His God and to make him Thy servant, willing to work for Thee and to take the same equal penny which is the reward of every laborer in Thy field whether his work continueth a long or short time and whether he receives the promise of it or not. Make him, O God, humble like Me, self-sacrificing like Me, and raise him to Thy power and Sonship as Thou wast pleased to raise Me; and to Thee shall be forevermore Praise, Honor, Glory, Thanksgiving and great Renown, world without end.

Lastly, let us pray for him who would be God's servant.

O Father, I love this man. He desires not to work for himself but for Thee. Be pleased, O God, Father Almighty, to receive him into the number of those who continually wait to do Thy Will and stand before the throne of Thy Glory shouting for Joy! Amen.

Thus, O Reader, do I pray for you. Choose for yourself which prayer you are worthy of and which you desire Me to help make for you. I have already made it for you; it now wants and waits for your sanction. It can avail you nothing until you make it your own by adopting and fervently repeating it before God's Holy Throne, which has eternally existed and is now near you as it is near every man.

Heaven is My Throne, said God by His inspired holy medium. Heaven is near you, at hand as I long since told you. Turn, then, within yourself and seek God; turn within and you may find Heaven there, and the altar before the Throne all ready for the sacrifice of your free-will. Be, then, no longer fearful, but believing; be no longer confident, but believing and humble; be no longer led by men, but submit to God and believe in His Revelation.

Amen.

HISTORY OF
THE ORIGIN OF
ALL THINGS

GIVEN BY
THE LORD OUR GOD

Through Levi M. Arnold

1 8 5 2

REVISED BY HIM

Through Anna A. MacDonald

1 9 3 6

Edited by Robert T. Newcomb

VOLUME TWO

PREFACE TO THE SECOND VOLUME

In the beginning of Volume One I have shown the end and aim of spirit-life. In this volume I shall show the object of the physical relations of bodies to spirits and souls. In the last part of this series I shall declare the future course of spiritual life more fully than I ventured to declare it before.

Volume One was in three parts, originally published at different times and separately delivered and directed to the world. This Book I write and prepare for the press Myself with more particularity and more evidence of orderly arrangement because I am going to give outward proof to readers, though not to My holy medium, that I am the author. Volume One is equally Mine expect for the punctuation, but this volume is more fully Mine because the holy medium has been more attentive and more careful, more faithful and more desirous to be led and guided by Me in all parts and at all times than previously. But he will yet improve.

Still, as I have before declared, the errors in Volume One are trifling and merely verbal and will not harm you either to believe or act upon. I have left some few in this, but they are still more slight and unworthy of thought or apprehension. Let us all try to find good in everything instead of looking for weaknesses or evils. For it ever remains true, that he who seeks shall find, and he finds generally what he seeks for. If truth be the object, look for that and not for error. All who will may come and partake of the waters of life freely, without money and without price. The needy shall, upon application to My holy medium or to any other servant of My will, receive the loan of a copy of this or any former publication. Those who have superfluities shall pay only the cost of printing and circulating the books. *My holy medium is prevented by My control from realizing anything from the sale of these books that is not diverted from himself and his temporal affairs to replacing the cost and expenses incurred for printing and advertising the works given through him. He has no return as yet for his outlay and will not be permitted to retain it when he has except for the purposes named, and for his time or personal service he shall receive no outward compensation whatever. I shall take care that he obeys; if not willingly he knows and believes I can by circumstances take from him sufficient to make My declarations true. Let no one, then, accuse him of a desire to profit by his mission

*This is also true of the present edition—Ed.

in any outward way or manner. For to Me he gives the glory, honor, and praise of all that he has and he looks to God or to Me as one for all that he desires or hopes to attain.

Let each reader who is brought to know the truth of these books try by all his means and influence to induce the circulation and perusal of them by others as well as his own mind so that all mankind may be benefitted by his exertions and by the spread of Truth and Love of God manifested in Love of Man.

With the completion of the second volume, I have finished My work through this holy medium for this year, a year of importance in the history of the world of Earth because it has witnessed the reception and publication of the *History of the Origin of all Things*. My servant Emanuel Swedenborg looked forward to this time and announced this year as the very period of the production of a revelation which should establish the NEW JERUSALEM CHURCH on its proper foundation. Other foundation can no man lay than is laid, which is a foundation on the ROCK spoken of by Me on a memorable occasion when Peter, a Rock himself, confessed or acknowledged or declared in answer to My call upon him that I was known to him to be THE CHRIST, THE SON OF THE LIVING GOD. Flesh and blood did not reveal this to Peter. It was *internal revelation* which declared it to him, and on this Rock of Internal Revelation will I always lay the foundation of My CHURCH. So it was, so it is, so it will be. Amen.

August, 1852.

CONTENTS

PART I. A History of Spirit-Life and of Paradise.

History Of The Origin Of All Things

PART I

A History of

Spirit-Life and

of Paradise

INTRODUCTION

Be not afraid to read. Error hurts no man who is on a sure foundation. But if a man's belief is on a sandy foundation it can not withstand a discussion which will cause storms of winds and rivers of words. Let every man search for truth and in doing so fear not to descend for it into deeps below the surface or reach to airy heights of heaven-born aspirations. Truth must and will prevail if free discussion takes place. The Creator of All Things has

"In binding Nature fast in fate,
Left free the human will."

If man, too, leaves his fellow man free to choose good or evil he leaves him to work out his own soul's salvation, and God has before declared: No one can give to God a ransom for his brother's soul.

Let freedom, then, prevail not only in blessed America but in benighted Europe and dismally dark Asia and darker Africa. Let each one endeavor to profit by the light that now shines into the darkness and thus far is not comprehended by it. But the dawn is near. The morning of the Great Day of the Lord will soon be unmistakably evidenced and as the lightning shineth from the east even unto the west, so shall be My coming into the hearts of men. Be ye ready then to welcome the Son of God whom you have often rejected, and place yourselves under the glorious government of the King of Kings, whose kingdom and dominion is an everlasting one and shall endure when time shall be no more and when the past and the present and the future shall be blended in one great eternal infinite *now*.

Reader, dare you pray that this great and glorious day shall dawn? Do you want to see it or do you prefer the flesh-pots of Egypt, the ceremonies of the law, the dogmas of the church, to the glorious liberty of Christ and His Saints reigning on earth as they reign in heaven, which is with all power and yet in subjection to God who makes all with one mind give to Him glory, honor, thanksgiving, praise and high renown, both now and always, forever and forevermore. Amen.

Reader, if you dare to pray for this time so long gloriously expected, so long pretendedly prayed and hoped for, accompany Me in the following prayer. If not, prayer will be useless and this book will only make your sins more scarlet and your desires more unChristian. If you cannot make this prayer, do not waste your time reading this book, but proceed to your merchandise, to your cattle, or to whatever you have wedded.

Let us pray

Almighty God, who dost from Thy throne behold all the desires of all men, look down with favor upon my desire to know the truth of the things declared in this book professing to come secondarily from Thee, primarily from Thy Son the Lord Jesus Christ, to whom with Thee be ascribed all honor, praise, glory, and thanksgiving both here and hereafter, now and forever, in this world and the world which is to come to all men.

Amen.

PREFACE

Let all that is within me praise the Lord,
For He is God, and His Mercy endureth forever;
Yea, let all the earth praise God and give Him glory,
For His great deliverances, and glorious revelations!

Lest any should mistake the object of this publication I will state that all that the holy medium knew of this book before I commenced it was that it should be a Book of Hymns and that it should be commenced on the twenty-seventh of June, 1852. His surprise to find it prose was great, and his surprise to find that the hymns were one only to each chapter was still greater. But he continued passive and believing and his faith has been strengthened by what would have been to some a retarding circumstance and a confusing influence.

To those who are passive all is well; to those who are patient all will be manifest; and to those who are submissive to God will be the reward of promotion to more arduous tasks or, as it is symbolized, by having more talents or cities to control. Seek first the kingdom of God and all else shall be added to you. Seek first to be reconciled to God and you shall find His kingdom; seek reconciliation by offering the only acceptable sacrifice, that of your heart, properly called your free-will. When you do that you begin to serve God; until you do it you serve yourself, that is, your own will. When you do that you seek God's direction; when you seek that you find it; when you find it you find the pearl of great price; and you, like any other finding that pearl, will sell all that you have to buy it. You will sell your houses and lands, your wife and children, your present comfort and even your present life for that joy which only those devoted to God, those desirous of serving Him, can enjoy. The peace of God which passes all understanding is in fact a part of heaven, and the establishment of it in the heart or self-will of man gives a home to God and to His Son the Lord Jesus Christ and to the Comforter, the spirit of God which is ever at His right hand and seeking to enter every heart that opens a door for its entrance. The blessed trinity is then with him and he enjoys the Holy Communion; he becomes baptized with the fire of God's love and lies down in the promised land of contentment and peace where there is no more sorrow and the weary are at rest.

Amen.

221

CHAPTER I

THE DESCRIPTION OF PARADISE

The Time of the Establishment of Paradise

When all creation was accomplished by the production of the body of man upon the earth, I was in paradise; so were you. But the situation of paradise and the nature and extent of it have never been revealed. That subject I leave for another book. This I write now because the holy medium has time and the world has need. Yet it will be to many a pearl unappreciated. There are, however, spirits who need the book as well as men in the body. There are men who will appreciate the book and spirits shall declare its truth. But is this book, then, to be the first declaration of some things, or some part at least of its contents, to spirits? Do they derive their first knowledge through books addressed in outward form to mankind? They do, or at least some of them do. I stated in the third part of Volume One that the knowledge possessed by spirits of the fourth sphere was to be revealed to man. It is evident that if the knowledge is that which properly belonged to the fourth sphere, it could not be possessed by spirits in a lower circle or sphere.

But if it is to be revealed to spirits, is there not a simpler way than to reveal it through this outward or material holy medium? There is, but if I choose to give it in this way, shall you say that it is not the wisest way? Shall you say the simplest way is always the best to produce the greatest result? The wisdom of man does not so act. Neither does the wisdom of God, which is far simpler in its manifestations than man's wisdom, but which is so far above man's as to be incomprehensible in its simplicity. It is made complex or manifested in a round-about manner to induce and secure its reception by some. It is also evident that if spirits in the lower circles of the second sphere are so gross or sensual as I have described them, so ignorant of their situation and duties and yet possessing power, as has been described by others, to make themselves acquainted with the thoughts and knowledge and desires and hopes of men in the body—it is evident that such spirits may more easily be reached by knowledge proceeding from below upward instead of from the higher spirits downward to them only. I say only, not because the effort has not been made already to instruct them thus, but because we shall cause the procedure of this revelation to be presented to them both from above and from below.

Having thus explained briefly the reasons for this revelation and having shadowed forth what it will be, I will proceed at once to relate the history of paradise.

The Course of the Inhabitants of Paradise

Let all the people praise the Lord,
For He is great, above all the heavens, and glorious
Beyond all the creation of His hands; praise Him for all His works,
Yea, let all people praise the Lord!

Let all the people praise their God,
Who made all, and sustains all that is made, the High and Mighty
Ruler of all, illimitable in creation, and infinite in power;
Yea, let them praise the Most High God!

Let every nation, kindred, tongue, and people
Praise Him for His mighty works, His loving deeds, and His
glorious creation.
All we have proceeded from Him; yea, let every man who lives
below,
Or spirit high in heaven, praise Him!

Let every tongue, and every heart—let every soul
Praise the Most High, Eternal God, for His excellent revelation.
Let all who dwell in earth below, and they that dwell in spirit-
world—
Yea, let every creature praise the Lord!

Let all that is, and was, and will be
Praise Him, for He is good, and His mercy endureth forever.
Bless the Lord, O my soul, and all that is within me, praise Him;
Yea, let all bless and praise the Lord!

Ye spirits who dwell above the earth,
And all ye servants of His, who do His will, and every procedure,
Bless the Lord, the Most High God, Eternal Creator and Ruler;
Yea, bless Him, for His holy love!

Let us proceed to relate how the paradise, or first state of
existence of the soul of man, exists, and where it is. Let us call to
mind the revelations made previously by Me in My *History of the
Origin of All Things.* By that you were informed that it was one
of quiet happiness, that it was in such a place as pleased God to
have it, that the spirits or souls dwelt there in pairs without government other than immediate dependence upon God required by their
relation to Him as a part of Him. You were informed that spirits
leave paradise when it is no longer their choice to remain, and that
they have faint ideas of the other six circles of the sphere of experience which they are, or at least may be required, to pass through
on earth. For all do not pass through the remaining six circles; some
die too young; some never attain to the higher ones.
Where is paradise?
It is a state or condition of ethereal matter invisible to men or
to spirits who have not reached the fourth sphere. It pervades creation but not every part of it. It exists in space but is not, like space,
unlimited. It has various worlds or divisions, yet it is one in its
nature and essence and conjoined by the thoughts or desires of its
inhabitants. The more it is emptied of its inhabitants, the more is
it withdrawn from the spheres of the future life or spirit-worlds.
For these, too, have place and existence appertaining to them, though

the nature of their form and material can not be comprehended by men in the body. The nature of paradise is, however, still more refined, more ethereal than the heavens of spirit-worlds, because in paradise reside pure emanations from God.

The place of paradise is where the sun leaves its boundary of atmosphere or attracted or attractable particles of matter. But you supposed the sun's attraction extended to the most remote portion of creation by the law of gravitation and that at any conceivable or even an inconceivable distance an attractive force was manifested, though it must, of course, be almost infinitely small at an almost infinitely great distance! This however, though a proper deduction from Newton's theory of gravitation, is not the case. Each system attracts only its own matter and its boundaries are those which God, by the Word secondarily, and by His law, primarily, established. These boundaries do not far exceed the outermost planet of any system. How, then, does the spirit-world encroach upon this state or field or world of paradise? Its location is outside of it. Then the spirits of the more advanced circles pass through paradise to reach the earth or any similar globe of matter? They do, but they are not visible to the spirits in paradise; neither are the spirits in paradise or the paradise itself visible to them until they have been advanced to the fourth sphere.

How this is practicable can scarcely be explained to men even were they willing and desirous to receive the revelation, but when they are carping and desirous not to believe but to find objections or what they call oppositions to reason, it is impossible. So I shall not attempt it now, but may hereafter explain fully for the benefit or gratification and instruction of those who are true and earnest seekers after pure truth. Those who seek from vile motives such as curiosity, or from such profane motives as a desire to find herein something opposed to itself, or to what they call reason which is only their pre-established opinion, will be gratified also, but it will be by having their carnal and unworthy desires gratified to their own injury. Am I then not responsible for their being injured? Yes, if God be responsible for the evil that men in the body commit. He gave them powers to use and materials to use them upon. They abuse the powers and neglect the materials, or use them in a perverted manner, which is abuse of them. So it is with the revelations I make. Men condemn or injure themselves by reading and rejecting them because they do not properly use the powers God has given them and because they pervert the materials or subjects of revelations I furnish. Who then is responsible, they or I?

The Circumstances of Existence in Paradise

The last part of this chapter I shall devote to the consideration of the light of paradise.

Paradise is, as I have said, on the border or boundary of the solar system. Not this system only but all other solar systems. Its light can not proceed from the sun because it is beyond the boundary of the sun's influence. How, then, does the light from other suns reach the earth? Evidently it can do so only by passing beyond or through the paradise surrounding not only the sun from which it proceeds, but also that surrounding the system in which the observer is stationed. This objection is a proper and reasonable one, and as such deserves a reasonable and proper answer. I design to answer every such objection that can be made reasonably, but were I to answer others I should find no stopping place.

The light of the sun proceeds not merely as a ray falling upon or impinged against an object, but as an emanation of action, a wave or vibration of the matter upon which it acts. This matter is *caloric,* as I have previously explained. Caloric is a substance as much as air or water and is, like them, a compound. It contains the two principles or elements of light and heat and has, besides, combinations with other substances or elements which cannot be explained now. Men are, however, on the eve of their discovery, in part; they have already been ascertained to exist by the researches of Reichenbach and others who preceded him, as well as of some others whose results or theories and experiments have not yet been made public. Why do I not rob them of their reward in announcing through this holy medium the whole mystery or history of the formation of caloric? Because I am not a man in the body but a Spirit of the seventh circle of the seventh sphere.

Caloric is a compound of light, heat, od, magnetism, electricity, and other substances or elements unnamed for men in the body. The action of the rays of caloric proceeding from the central body or sun, where it exists surrounding its atmosphere as described in a former work, causes such vibration in the surrounding aura as to produce corresponding action in the caloric surrounding the planet belonging to it, or the still more distant sun. This aura pervades paradise as well as all the spirit-worlds and all illimitable space, all the infinite whole. Aura, then, is infinite? As far as creation is infinite, so far is aura; for it has no other limit than creation. Like everything else, God excepted, it was created, but, like everything else, God excepted, it is in reality circumscribed in extent. To men it is the same as infinite. To God it is different.

What, then, exists beyond aura and beyond all creation, beyond the Great Universal Whole? GOD. He is everywhere, He is infinite. No other being or substance has or can possess this quality. But man can not conceive of the infinite. Nothing finite can. The angels of the highest circle of the highest sphere approach nearer its comprehension than any others because they are more like God, being one with Him in power, will, glory, action, intention, and motive. But as I declared in the first book, already published but not yet understood by any reader, not even by My holy medium, these spirits continue to progress to all eternity and become nearer and nearer, more and more like God, without ever reaching perfection, without ever *being* God. So they will to all eternity approach nearer and nearer to a conception and knowledge of infinity without ever attaining it.

My holy medium has again and again, since I wrote the last part of Volume One, examined himself and tried the spirit by which he received those revelations. Moved by the judgments and denunciations of fallible men, and knowing his own fallibility and weakness, his entire want of power in God's hands or against His will, he has apprehended that all this proceeding might, after all, be the work of spirits acting in God's will to purify him and lead others, perhaps, into purification by misdirection, or, as it were, leading them first away from the truth in order hereafter to have them advance in the true path, even as a traveler desirous to reach the end of his journey as quickly as possible, but who has gone astray, is sometimes compelled either to retrace his steps or to take another road which also will lead him first to a greater distance from the object he desires to reach, though eventually rejoining the great highway from which he can not wander again without willfulness and which will conduct him by a direct course to the end of his journey. But My holy medium has wisely concluded that his only course is to follow his guide, to be passive to My influence, to disregard man's opinion, sectarian or interested opposition, ignorant denunciation, or even well-meaning criticism. He does what the reader should do. But the reader says: I have not the internal evidence he may have that the communication is spiritual; I know not as he may know what prayers he has made, what reliance on God he makes! No, but you may have the same internal evidence, the same outward evidence; the same prayers may be made; the same reliance on God may exist in you as in him!

I have told you before how you may have an outward evidence I do not furnish him, but you must remember the condition, that you must ask with a sincere desire to receive an answer in truth and in God's will, not from vain curiosity or desire to establish a disagreement in the spiritual manifestations. These you ought to desire to reconcile, for what greater blessing can you expect than to have revelation from God or from superior beings? This ought to be sufficient to induce you to be willing to discard pride of opinion, prejudice of education, motives of pecuniary interest, or loss of popularity. This will then produce for you a truthful answer from even an imperfect or low medium.

You will have thus a proof My holy medium has not had and will not have by inquiry. I require of him more than I do of you because he knows he has a direction from a foreign intelligence and he knows that reason and revelation assure him that God rules all, sees all, knows all, and will have all men come to Him through Me, or through Christ, which is any spirit of God sent to men. He is aware by experience that I teach only good actions, that I urge upon him good motives, and that I lead him to form good intentions. Why, then, should he believe the influence to be evil? A tree is known by its fruits. But you say the fruit of this revelation is discord? So it was when I was in the body, for I then declared that My coming and what I preached was not to bring peace but a sword; that I came to declare what would break up families and divide the nearest relatives. But was that because men received My teachings then or is this because men receive My revelations now, or is it because they reject them and war upon those who receive or received them? Answer from your own reason, O reader!

But, to return from this long digression, the aura then receives the action of the caloric of the central globe of matter, or others that may possess caloric, and transmits it by waves to every part within the limits of its own extent. Wherever these waves reach bodies capable of receiving their action, a manifestation of it takes place. If the body or substance is the caloric of a globe, it receives the impulse so as to form again a ray similar to the one which produced the wave, and that ray proceeds on its course to the earth or other globe as if it had proceeded and continued uninterrupted from the sun, or whatever source it had. But when passing through the paradisical substance it exerts no influence upon it and does not make itself in any way manifest to the senses of the beings there existing.

So I have shown you how it is that spirits in paradise and in the spirit-world do not or may not perceive the light of the sun or of other such bodies. But there is a light from God that permeates the aura which in turn pervades paradise and every other part of creation manifest to spirits before and after their connection with the body. The light thus pervading the aura manifests itself differently to different beings and maintains with them such relations as carry out the purposes of God. It is not so clearly visible to low spirits as to high ones, nor is it so visible to spirits in paradise as to those in the spirit-world. But the spirits in paradise have light from it, and are visible by its light to the angels of the fourth and higher spheres, even though they themselves can not see by its light either the grosser or finer particles or substances. They can not see God, nor men, nor the spirit-world. The adams of paradise are, as I have said, souls of men, and the souls of men are emanations from God of a part of Himself. What the nature of these adams, of these parts of God, is, I shall show in the next chapter.

CHAPTER II

THE LOCATION OF PARADISE

The Plan and Time of Establishing Paradise

God established paradise after He had formed matter. When matter had progressed to have form, which it did under the supervision or government of the Word, paradise was spoken into existence. God said: Let us make man in Our own image. Why use the plural? Because God is one but all; because He spoke to the Word which is also the image of God. So man was made by a separation from the infinite of a portion of its substance or essence which had a form and was endowed with qualities and attributes like unto God, but infinite. God and men can therefore be one. A glorious trinity can and does exist between God and the Word and man. But though they are one in their ultimate or essence, they are various in their manifestations or procedures, in their duties or attributes. This oneness can not exist with man in the body except through Christ, the sent spirit of a deceased man or a spirit which has passed through the first and second spheres. Then there must have been a time when no trinity existed! There was a time when God was all in all—one being unseparated into the Word and spirits or angels. But that time is too far distant even for angels to comprehend. Suffice it, then, for man to know God alone is eternal and from eternity, and that inasmuch as He is the Father of all it is impossible that anything else should have been eternal or co-equal with Him. Let us see, then, how man was formed.

The End for which Paradise was Established

The souls of men entered paradise at their creation. When man had been decreed to exist, the Word undertook to act in God's will upon spirit, or more properly upon God's substance, which is as much above what we have before called spirit as spirit is above matter. The innumerable creatures of God have existed as men either on this globe or in this system or universe or association of universes I have called a coelum—in this association of coelums I will call a super-coelum; in this association of super-coelums I will call a whole of coelums or universe-coelum; or in this Universal Whole, as I will call the combination of all the super-universe-coelums which combine to form it—in all these men have existed and do exist. For all these who have existed, who do exist, or who will exist or be in bodily form, spirits or adams or souls were formed at one and the same time by the Word from God's substance or essence, which was before and continued afterward to be infinite in extent and quantity as well as in every other attribute.

All these spirits, to men an infinite number because to men all inconceivable quantities are in effect infinite—all these spirits were given places of residence as I have already described. These places were in paradise. When they would, they were united; when they would, they were separated. Male and female in their existence and relationship, each pair was a community, a family, a unit. Each pair was what it pleased to be for they were possessed of God's attributes in a finite degree. But those finite powers, though very strictly confined by laws, were very extensive in reality. Quiet, though, was the great characteristic which distinguishes them from spirits who have made progress. Action or will to act was the one thing forbidden to them as inhabitants of this blissful paradise.

Action, without which men die and in which angels have all their enjoyments, was denied to the adams except at the penalty of a change of existence and position, a change and fate they looked upon as a degradation. It was in one sense a degradation; it was a fall from that peace which God had bestowed, from that happy parital existence which they possessed, to a state in which they must toil for the support of an earthly covering which was also a prison-house to them, shutting them out from that light in which they had dwelt before and that knowledge of God which they had before possessed and enjoyed.

They knew—for we may speak in the past or present tense indifferently—that they must encounter trouble, trial, difficulty, probation. But they knew also that beyond the sphere of experience was that of reconciliation; beyond that of reconciliation, memory; beyond that of memory, knowledge; and that when they should have arrived at that fourth sphere they would again be as happy as before, and that God designed them for eternal happiness. They knew not the length of time or the degree of suffering which might be their lot or that of their fellows. They knew not that the knowledge of good and evil, or the passing through the other six circles of the sphere of experience, would be attended by pain and suffering. They knew not the nature of pain and suffering because they had not experienced it, but they did know that paradise was to be theirs until they desired to leave it and that every enjoyment that they could conceive of was and should be theirs so long as they chose to remain. But of the tree of knowledge of good and evil they could not taste without dying to God's immediate presence to their senses. How they conducted themselves and how they resolved to act shall be the next subject.

The Condition of the Souls of Men in Paradise

Souls in paradise were in pairs of two emanations or adams, or, as called in Genesis, Adam and Eve. Eve was a separation from Adam after Adam had been separated from God. The two beings thus formed from one, which one was formed from God—His image as to attributes but not as to form—were united by affinities which caused them to depend and rely upon each other for society and happiness and for completion of any desire. They were inseparable without a loss of happiness; not an absolute pain, but a deprivation of that peace which in general possessed the being of each part. Each, when separate, felt the want of a part of itself; the other feeling the same want, they were no longer separate.

Still, in the long period of time which elapsed after their creation or production from God, they began to form desires for experiencing that forbidden fruit of good and evil. Long arguments would be held with the temptation; the one would call up every beauty and variety paradise could afford, the other would wish still for more. All in paradise was known, but there was something yet to know. True, it was understood that it could only be purchased by separation from the intimate half or parital part and by unknown experience of evil and by being deprived of God's presence to their consciousness. It may well be supposed that the serpent, as it is called in Genesis, the reasoning faculty as it might be called, or free-will as it should be called—it may well be supposed that this part of themselves would long and earnestly pursue the subject and examine from every possible point of view the bearing and probable result and possible experience the adam must encounter in thus expelling itself from so blissful a condition forever. For the knowledge that, once having left the state, it would return no more was fully impressed upon it.

At last a time arrives when one at least of the parital parts resolves to dare the encounter of the ills it knows not of, to escape from what has become the monotony of bliss, the ennui of pleasure. God is ready to hear its appeal or to call for its reason if in shamefacedness it makes no appeal. The change is evident to God, for His knowledge is infinite. Then the call is heard: Adam, where art thou? and the adam answers: Here am I, O Lord, desirous of knowledge of good and evil, desirous to proceed from paradise because I would rather encounter what I have no knowledge of than remain longer where I know all.

This is the general reason. There are exceptions, it is true. I have told you of one who desired to leave because, being so happy, He felt it due to God to make some return, to be of some service, to persuade some other spirits to be as grateful to God for all His blessings as He himself was. Having tried in vain to incite other adams of paradise to feel as He did, He next desired to follow those who had left paradise and try to persuade them to feel that gratitude which the goodness of God and all His benefits, conferred on the just and on the unjust, ought to inspire in every soul. This desire God granted also and, coming with such motives, He was what He was, a being so far in advance of other men as to suffer an ignominious death with prayer to God to forgive His enemies and, though left almost alone in His dying moments so far as human sympathy was concerned, despairing not of God's power and love. He ceased not to believe that He should be permitted to help men to be grateful, that He should in the spirit-world go on with that work so sacrificingly begun in the body, or rather in paradise, and that had as yet produced so little fruit. God blessed and aided Him. God raised and elevated Him. God has prospered His work since He has been in the spirit-world, and God now sends Him to reveal to mankind in the body the heretofore hidden things, in order that men may be induced to give thanks, glory, honor, and praise, thanksgiving and power, to God in the highest, in order that they may practice here the precepts He delivered to their forefathers and reiterates now through chosen holy mediums by various manifestations.

Let us pray

Look down, Most Merciful God, on this deed!
Look with mercy upon Thy servants, O God.
Look down, Eternal, Unchangeable God, on Thy Son!
Look with mercy upon all Thy adamic children.
Look down, O God, Thou King of Saints, from Thy glory!
Look with mercy upon the immortal part of mortal man.

Let all the people praise Thee, yea, let every one!
    Great and marvelous are Thy works.
Let each who dwells on earth give thanks to Thee!
    For just and true are Thy ways.
Let every creature praise and magnify Thy holy name!
    And be grateful for all Thy blessings!
O Lord God, Most Holy, High, and Loving!
    Look upon My work with favor.
O God, whose love and mercy endureth forever!
    Bless it with Thy eternal blessings.
Oh, let all who dwell beyond the grave, in earth or heaven,
    Swell the note of praise to Thee!

Thine is the glory and praise forever!
    Let all the people praise Thy name.
Thou art the Merciful, Bounteous Giver of good!
    Let those who have not yet known Thee, praise.
To Thee be honor, thanksgiving, and glory;
    Thou Ruler of All, forever and ever!

Amen.

The Manner in which Men have Passed from Paradise

When the spirits or adams left their blissful abode in paradise they usually left singly. One part would desire to remain at least a little longer, hoping that perhaps the separation would be endurable and the enjoyments heretofore experienced might yet satisfy the longing soul or being which was itself. Vain hope! If separation for a time had been unbearable without loss of happiness, or deficiency of it creating new desires, the separation declared and believed to be eternal except by the remaining part's following the first half is soon found unbearable indeed. A considerable hesitation may take place and the doubt or period of temptation may be prolonged for a considerable space of time, because it not infrequently happens that the one part was far from being discontented in its first estate. In this case the only motive for leaving is to find and love the parital part that preceded it into the unknown vale of experience. This motive makes a most affectionate disposition manifest in the body. The nature of such a one feels always the want of a being to love, and so it is one of God's wise provisions that in general the male part precedes the female, and thus it is that woman's love is distinguished from man's, though occasional exceptions mar the uniformity of this manifestation and some men love like women and some women like men, though each is perceived to be incongruous and unsexlike.

But are there, by chance, any meetings of these parital parts in the body and are they there again united? The adam or eve that leaves paradise leaves all recollection of that state behind it or, more exactly speaking, the body or earthly covering of the soul prevents it from having memory. By a law of progress or by nature of the

231

earthly body, there is also formed a spiritual body which so envelopes the soul that, even when the grosser part is left upon the earth and the spirit soars beyond paradise to that outer circle in which commences the spirit-life, the spiritual body is of such a nature that even yet it obscures or in most cases entirely hides the memory of paradise and, of course, a recognition of the spirit-body to which its parital half is assigned. As I have before stated, it is only in the highest circle of the third sphere that this knowledge or memory returns to the spirit, when it is enabled to search for and find that part of itself necessary to its perfect harmony and full enjoyment even of heaven. So that, if by a most unlikely chance two parital halves should meet on earth or in the lower circles of the spirit-world, they would not know each other, and even should they be married on earth in the body the result might not be any greater harmony than usually falls to the lot of husband and wife. For the bodies of men greatly affect character, and though this does not reduce the responsibility of man, it is true that the sins of the fathers are visited upon the children. It is a fact that the cerebral formation is largely inherited and though it may be modified by education and controlled by reason in most cases and; by God's help, in all, yet its influence is so great as to establish the character and fondness for particular enjoyments or passions, as they are usually called, so as to leave nothing more visible of the soul or adam than the great pervading desire which impelled it to come to earth, and that only faintly perceptible in most cases. The parital or adamic nature is then so obscured that the natural partner not only is unrecognizable but very often would be a most unfit husband or wife. Besides this, the interval of their respective appearance is frequently more than a life-time, so that their meeting on earth is altogether most unlikely and for no reason desirable. But in the spirit-world do they not have some idea of their relationship, or could they not give the one or the other some assistance? They do not until they have reached the circle I have named for it, unless the knowledge is communicated to them for some wise purpose, which is often done in order that the one or the other may be assisted, sometimes in quite a low circle but only when one has been advanced at least to the sixth circle of the second sphere, which is the circle of good-works.

CHAPTER III

THE LOCATION OF THE SPIRIT-WORLD

What Becomes of the Unfinished Men or Children

There is often slight experience of good and evil in the body, as is well known from its frequent early death. Sometimes the soul or adam barely enters the body, and in a few moments leaves for the spirit-world. These cases form, then, exceptions to the general rule, for they can not be said to experience good or evil on earth; where are they to pass through or acquire that which I have declared necessary to their full enjoyment of the bliss of heaven?

In entering upon this subject I feel and know that I shall give your faith a severe trial. I can scarcely expect even My holy medium to receive with perfect confidence and entire conviction what I am about to unfold. Yet this is one of the most striking objections to the plan I have previously revealed and one that is well taken by a reasonable and inquiring mind. The first case we will consider is that of a body taken from the soul at an early age, say, in adolescence. The faculties of the mind have not been fully matured. The experiences of life in the body are so far from exhausted that the soul lacks the knowledge of all the more important relations of life and of that independence which can not be thoroughly developed until separation from the parental hearth takes place. Man indeed should be a husband and a father, as I have shown in the recapitulation of his gifts which I gave in Volume One, Part II, of the *History of the Origin of All Things.* But of all the parts of life in the body, the married is the most developing to the soul or adam of man. Of all the experiences there are no others that so fully and urgently press him to ask God for help. The trials and the discords, the love and the joy, the toils and the consolations of this state are more vivifying to the soul, more elevating to the intellect, more absorbing to the passions, more reconciling to God, and more a foretaste of heaven than all the other conditions of life combined. Indeed, every character must be incomplete until it shall have passed through this experience. How, then, is this deficiency to be restored or filled up? How shall it be shown that the majority of mankind, dying ununited by the conjugal relation, are fitted by experience to stand in their places at the Last Day? You know that I Myself did not have this experience in the body and therefore, according to My own showing thus far, I must be an inexperienced and imperfectly developed spirit.

The goodness of God is such that He provides for every want; His wisdom is such that all is fully taken care of. He has provided for this deficiency. First, He leads the man in the body to have such relations to others in the body who have a different experience from his own as to make him, by the knowledge he will have by memory of the others becoming fully possessed by him in the third, or memory sphere, and in the fifth circle of it, possess within himself the experience as if it were his own. Second, He provides that after leaving the body a man may have a kind of experience by an association directly with a man still in the body having all those relations and experiences. This association is that of a guardian angel and is a part of the duty and pleasure of spirits in the sixth circle of the second sphere. Thus, by this double provision, the soul attains that knowledge and experience which is necessary for its own joy and bliss. But, then, if I entered the spirit-world under such favorable auspices as I have described, I must have progressed so rapidly as not to have remained in the sixth or other circles of the second, or even the third sphere, long enough to have been a guardian angel to a man during his bodily life This is a good and proper doubt or objection. But it must be remembered that I, like all other spirits in the higher circles and spheres, possess and include all the lower experiences and powers and can exercise the duties of any one of the lower, though it is not often done because there are enough to perform the lower duties in lower circles and the higher spirits must not and do not desire to neglect their proper and peculiar responsibilities.

The other case to be elucidated is that of an infant body which receives a soul and immediately, or very soon afterward, expires. The process here varies, inasmuch as the want of experience is greater and the knowledge or experience of good and evil is too slight to allow the soul so fresh from paradise to be placed in the usual course of progress in the circles and spheres of the spirit-world. This inexperienced spirit must receive in heaven or the spirit-world an education, and the course of instruction is so uniform that I may describe it to you and is so simple that you can easily understand it if you desire to know the truth.

The child-spirit receives its instruction from such spirits as desire to experience in themselves the relation or the semblance of the relation of parent. They give it instruction in such elementary knowledge as is usually given in the life of which it was so early deprived, for although the residence was brief it was long enough to clothe the soul or adam with a spiritual body as fully as if it had reached the age of three-score and ten. Then the child becomes initiated into such studies as the parents would have chosen for it. For though the child is thus early removed the parent is not then relieved of responsibility; his works or intentions attend the child-spirit to the spirit-world and there measurably affect its development. Still, the mercy of God tempers the injuries that the child's spirit might otherwise receive from its parents' unintended neglect or incompetence, for no serious deprivation is experienced from such a cause. The intention or desire of the parent to benefit the child is more surely carried out, so that the child-spirit shall not lose anything by its removal thus early, but may gain. The case of barbarous and savage nations is different and will be hereafter explained. The child-spirit then receives an education comparable to what it would have had on earth and, being passed through this course, enlarges in development of its spiritual body until it reaches the adolescent form and state of mental or spiritual cultivation; it is passed through the same process as if it had lived on earth until it had arrived at that age or period of bodily life and so becomes fitted for the enjoyment of the higher spheres.

The Explanation of Paradisical Discrepancies

The disparity of social condition and the difference of education of those in various conditions of bodily life might be supposed to influence the future of spirits. It does influence it but not always, nor generally, as is fondly supposed by those who consider themselves fortunate in possessing knowledge or mental cultivation, or in living in the more cultivated or polished circles of society. Before God all are equal. Yet He requires a full account from every servant of His, or from every spirit that ought to have been His servant, of the deeds done in the body, and the responsibility of that servant or spirit is precisely in proportion to the talents or opportunities given to him for proper employment. He who lays up his talent in a napkin, designing to restore it as he received it, unimproved, will find no receipt in full prepared for him. Improvement and progress are required as much as retention of good, and to whom much is given from him much is required. So, ye learned, and ye rich or polished people, think not that you will be rulers over many in the life to come unless ye are worthy to hear: Well done, good and faithful servant, thou hast been faithful so far as I have trusted thee; enter thou into the joy of thy Lord!

Every soul must give an account of the deeds done in the body and if, as I have explained, the body did not continue long enough to let the soul work out its experience, it must receive it in the world to come. Not that in the future life there is probation or acting of evil to such spirits, but that they witness and sorrow for the evil and improve well the opportunities afforded them of gathering from the experience of others, and on that depends their progress. But very often the poor man, having had few opportunities to serve God and fewer, perhaps, to serve his fellow-men, occupies the place of Lazarus in the bosom of Abraham, while he who had been regarded by men as high in sanctity may be where Dives was, a place into which Lazarus could not come. And where was that, if the higher spirits can place themselves in the powers of the lower ones? The higher spirits exercise only the good of the lower ones, and some are so low that their manifestations are mostly of bad or evil. With such the good can not associate. There is a great gulf between them, the gulf that is impassable except from the lower to the higher. That gulf is separation from God. Reconciliation with Him, submission to Him, enables a spirit or a man to cross it on the pinions of His mercy and love. This is the instruction to be drawn from that parable. Yet this is not all. There are many other things I have to tell you respecting it, but ye can not bear them now. The last shall be first and the first, last. That is, the end of all things shall be when the first state of man and the last state of man shall be one. When all shall know God from the least to the greatest, and all who know Him shall serve Him, and all who serve shall love Him and all who love Him shall be one with Him, and shout again in one united song of praise:

Great and marvelous are Thy works,
Lord God Almighty!
Just and true are all Thy ways,
Thou King of Saints!

Let us proceed to establish the knowledge of God's works and ways among men by singing to Him a new song:

Great and marvelous are Thy works, Lord God Almighty!
Just and true are all Thy ways, Thou King of Saints!
Great works are the manifestations of great love;
Just ways are the manifestations of great power;
Power and love are the highest manifestations of Deity.

Let all the people praise the Lord,
Yea, let all the people praise Him,

For all His mighty works and for all His glorious mercies.
Establish Thy power, O God Almighty!
Manifest Thy mercy, O Lord of Heaven!
Rule us with mercy and establish us in Thy power, Thou King
    of Saints.

Let all the people praise Thy holy name;
Let the earth raise its voice and the heavens rejoice,
For God reigns and of His government there shall be no end.

Let every thing that dwells on earth,
Let all who make their hope in heaven,
Let every knee bow and every tongue praise the Lord.
For unto Him be evermore love and praise;
Unto Him be evermore honor and glory,
For He loves us and will have us all to love Him forever.

Let all the earth and the people of it—
Let all the floods and the waters thereof,
Praise and honor the name of God who alone exists forever.

Let all that lives love and praise the Lord;
Let all within and all that dwells without,
Give Him glory, both now and forever and forevermore.
                   Amen.

The Head of the Corner

Let us all endeavor to find beauties in this book and we shall be
gratified. Let us all try to derive instruction from it and we will
be benefitted. Let us all seek to become its believers and we shall
be established. Let us all desire to practice its precepts and we will
be purified, and not only purified but glorified, and not only glorified but raised to God's right hand. For God's powers are no more
restricted than formerly, neither is His love nor His mercy lessened
by its long exercise, but the works I call you to do you cannot
do except as the Father is in you by His spirit the Comforter, that
will lead you and guide you into all truth. And I have heretofore
told you how you may obtain the presence of this Comforter. It
is by submission to God, by prayers to Him to enlighten and to help
you, by earnest seeking, constant desire, unceasing prayer to God

who is, though so far above you, ever present; though so infinite in all His attributes, yet ever feeling pity for your short-comings or your transgression.

God is everywhere and you need not fear that He does not know your slightest thought and heed your weakest prayer. But prayer of the lips He does not accept. If you would be benefitted by prayer it must be a prayer of the heart. If you would submit to God you must believe His actions ever-right, His dispensations ever-merciful, His afflictions the means of bringing your mind or heart into a better state; though reason may convince a man that all these statements are true, reason can not induce a man to act as if he believed them—the only power that can do that is God's, and he who sincerely asks help to believe and act thus will have it. The Comforter will come to him; the peace of God which passes all understanding will be experienced in proportion to the completeness of the submission to God; and the everlasting courts of Heaven will swell with anthems of thanksgiving that a child is born into reconciliation with God.

Would you, O reader, like to experience this joyful life and prepare yourself to enter those harmonious courts? Then pray to God in deep humiliation, in earnest desire to be His servant and to do His will, and your prayer will be heard, and not only heard but answered and granted, and not only heard and granted but you will know that it is so. You will *know* for you will find within your own internals, in the depth of the mind's consciousness, a conviction or impression of God's answer to your prayer. Do you want more than this; do you want an outward sign? Then are you in outer darkness. You are in Egyptian darkness and your wanderings may be led by a pillar of cloud by night, or what may be compared to it, but they will never lead you to the promised land of heavenly bliss. Your wanderings may serve to purify you if you profit by the dispensation you will receive, but the joy of thy Lord can only be found inwardly where no priest can enter and no preaching can avail or excite, in that sanctuary of the soul of which the holy of holies was an emblem and where only the angels of God can dwell in His presence. Then let it be your earnest endeavor, your daily effort, your nightly prayer, that you may seek God as He would be sought and serve Him as He would be served, so that you may do your part toward having His will done in earth as in heaven. Then will you find your mind stayed on God, and your reliance on His mercy and goodness and love can not be too strong or too perfect. He will lead you to that living fountain of His love at which, if you drink, you will be filled to thirst no more. He will give you that peace which the world can not give, neither can it take away, for that peace is bliss from heaven, a direct emanation of God's love, and nothing that is earthly can perceive or touch or have any knowledge of it, much less oppose its existence or destroy its completeness.

237

CHAPTER IV

THE FORMATION OF MATTER

The Sons of God Shouted for Joy

When the foundations of Earth were laid the Sons of God shouted for joy!

This has puzzled many sincere inquirers and the very brief explanation I gave in another book will not satisfy the doubts of all. I will therefore explain still further that the creation of matter was an act of the Word under the law, or will-declared, of God. But this law and action, as I have shown before, was not instantaneously executed but only commenced immediately its progressive action and execution so that the law spoken accomplished the result, but the action continued during a very extensive period over illimitable space. By this law matter was called into existence. From what did God create or cause matter to be made? Did the Word find materials already existing to make matter from, or did the Word make the materials or matter itself from something or nothing? The primitive tradition does not enlighten you on this point but I will endeavor to gratify your desire to know, your desire to see the completeness and the order and the beauty of God's creation.

The Word proceeded to form matter under God's law or decree by generation from nothing. That is, nothing as yet having existence except God and the Word, and the Word being an emanation from God, the Word had nothing to form matter from. For God's essence is above matter, being indeed above spirit which is a high manifestation or development of matter. There was then only itself and God, and both were of a nature or essence so superior to matter that matter could not be formed from either or both without so much modification as would make them different from themselves. God, then, gave the Word His power for the execution of His will, and the Word by the power of God created matter. By the Word all things were made that are made and without it was not anything made that was made.

But of what was matter made? If from any other thing, God excepted, then what was that other thing made from? For whatever exists as made was the work of the Word. The Word made matter, for matter exists, and if the Word did not make it it was not made. If it was not made it must have been God or a part of Him, which I have already said it was not. It was made, then, from nothing. It was in fact created, for creation implies production from what was not. Creation and reproduction are different; a mechanic makes by reproduction, by changing matter to another form or combination. God or the Word makes by creation as well as by re-formation. Creation of itself implies production from nothing, a calling into existence or into a state of being that which had previously no existence or state of being in any form. The Word, then, created all things from nothing because it formed matter from nothing. And from matter all things are formed or reformed or reproduced. Thus God is the great Creator because His fiat decreed and His power, delegated to the Word, produced or created all things.

The Word generated matter from nothing, as I have stated. But the Word had no other substance by which to produce the generation than the essence of itself and God. What difference was there between the Word and God except that the Word was a part and finite, though to mortal comprehension infinite? None. God made the Word in His own image from Himself, and gave it such power and attributes and extent of power and attributes as it pleased Him to bestow. The Word then, possessing certain powers, exercised them in the generation from itself of matter! The Word made matter from nothing but it was made by action, and the action was reliance on God's power and production from itself of principles or elements which were capable of infinite combination with each other. But is not the essence of God's substance *one* substance or essence or element? It is God. It is unexplainable to finite beings. It is. It has ever been. It ever will be. That is all we can know of it. The nature of it is infinite and therefore cannot be conceived of by finite beings.

So I will leave for you merely the statement that the Word generated matter from itself and produced by that generation different elements which, when infinitely combined, show themselves in all the shapes and combinations or agglomeration of matter which men or spirits can see, hear, touch, or in any way can become cognizant of. Having brought forth from itself these elements or first principles of matter, the Word breathed its prayer to God for a continuance of His aid and power to form them into such shapes as the law or declared will of God required. The processes of change, these infinite combinations, proceeded with constant action and uninterrupted force by laws of progress, carrying out the design of God, until they became gases; then from the combinations were produced fluids, and from still further combinations, solids. It is well known that of these there is a great variety, of which the highest form is metal, and the highest form of metal is gold. Thus was matter formed. Some of the processes of its development, by which the solar systems were formed, I have detailed in the second part of Volume One. But the order or law by which universes were formed is different. There is also another law for each higher association, such as coelums and universe-coelums. But these I will leave for the present and devote to the next section the consideration or explanation of the creation or formation of spirit.

The Formation of Spirit-Matter

Spirit, too, had to be formed, for there are bodies terrestrial and bodies celestial. There is a spiritual body and a natural body. Flesh and blood can not enter the kingdom of heaven yet bodies are there, and various kinds of bodies in various degrees of glory, as Paul declared. There are in the spirit-world principalities, powers, dominions, thrones, angels, and ministering spirits. By using these terms Paul meant to convey to the believers of that and succeeding ages the knowledge that there were different orders of spiritual beings and that these different orders of spirits had different degrees of glory, some of which might be compared to the refulgent sun, some to the reflective moon, and that certainly they differed as one star differs from another in glory. But it is evident that unless heaven is a state of progress these irregularities could not exist unless God failed to secure the reformation or reconciliation of every soul, or else He capriciously or with partiality made them differ in glory from no fault of theirs. But I have shown you that heaven, or the spirit-world, is a state of progress and that though all will eventually be joint heirs with Christ and the other sons of God, yet at present the inhabitants of each sphere and each circle differ from the others in glory. The bodies celestial, then, are of various kinds according to their period or state of progress.

The bodies of these spirits or adams or souls of men are spiritual. John speaks of the three kinds of substance: the spirit, the water, and the blood. The last is the body, the first the soul, the other is what we have in accordance with the custom of men called spirit; yet the idea which men receive from the term spirit is something immaterial. The only thing about man truly immaterial is the soul, the emanation from God and the part of Himself which He formed into an image of Himself.

This soul or adam, as it would be proper to call it, is after all the real man. The body of earthy matter is a covering which invests it during its sojourn in the last six circles of the first sphere; the body of spirit is the covering which invests it until it reaches the seventh circle of the seventh sphere. Here it returns to its original form and appearance which it possessed in paradise. Here it is again male and female as it was in paradise. Here it again possesses perfect happiness, but here it is conjoined with *action* which is necessary to make perfect or imperfect happiness lasting.

The spiritual body is a production from the earthly one, or rather it is a consequence of the production or investment of the earthly. The last state of the spiritual body is one of great refinement and attenuation. The first state is more gross and less different from earthly bodies. But the change is very gradual and almost imperceptible to the consciousness of the soul or adam. At last it reaches the seventh circle of the seventh sphere, as the earthly body approaches death, except that knowledge leaves no apprehension or doubt connected with the change. Few have yet reached high enough to know what this change is by experience; few spirits indeed, compared, as of course I mean, with the whole number, know that such a change must take place. They view the spiritual as immortal while it is only the adam or soul that is immortal.

What, then, is this dissolution of the spirit form and substance which thus takes place? It is merely a dispersion into space, or unorganized spirit-matter or essence or elements or principles, of this refined mass of spirit. It is not corruptible for it is the incorruptible body which takes the place of the corruptible; it is the last remains of that fall from paradise which the adam or soul was taught to consider as a degradation but to which it submitted because action was so necessary to its happiness.

Does the soul or adam fear this change? Oh, no, perfect love casteth out fear and the love of God is perfect according to the nature and capacity of the adam for perfection. The change, then, is looked forward to as a release from an incumbrance, for by this time the soul is educated to act without the spiritual body even as experience in the earthly body enables the spiritual body to act without it. And I have shown you that when this experience in the body is wanting, it has to be supplied in the next state by education. But in the sixth circle of the seventh sphere all are prepared to dispense at last and forever with that which for such a great length of time, in some cases it might almost be said a great length of eternity, has served them and been looked upon as an inherent part of itself or of the being called Adam.

Let all that is within Me praise the Lord,
For He has fearfully and wonderfully made Me.
Let all the souls of men glorify God
For He divests them of all imperfections.

Let every Adam praise the Lord
For His most wonderful providences,
Yea, let each soul magnify His power
Who is Eternal and Everlasting and Ever-Acting.

Behold the workmanship of Thy Word
And the results of Thy holy law,
Established from the beginning and continuing to the end
O God, behold them with Thy undying, undiminishing love.

Be merciful, O Lord, to them who are ignorant
And receive not Thy glorious revelations.
O God, let not Thy remembrance afflict them very sore
But be pleased to let them know Thy everlasting mercy.

Let every tongue of Earth and every voice of Heaven
Let every adam shout for joy;
For unto Earth and Heaven has God revealed His glory,
And made known His inexhaustible love.

Let no man take Thine honor for his;
Let no man presume to speak Thy will;
Let no man kneel to aught but Thee
Or bow down his soul before any idol.

Let all who will come to God be watchful;
Let all who desire to serve Him be attentive;
Let all who will be instructed teach others;
And let no man turn back who desires the Lord for his portion.

Let all that is within Me praise the Lord,
For He is merciful and loving and kind.
Let all that seek Him follow Me
And praise Him whose mercy endureth forever.
                              Amen.

The Formation of Earthy Matter

The formation of spirit, as we have called, in compliance with usage, the refined material or substance that exists in the lower heavens. was like that of all other things formed by the Word. But in all cases the Word acts in God's will, and acting in His will consequently has His power, for God's power always executes His will in such manner as His will chooses to be executed.

Spirit is the more refined elements of matter which the Word produced at the time of the formation or generation of matter. Spirit is matter, but still it is such matter as earthly matter cannot take any note of—cannot see nor feel. Spirit-matter pervades the earthy even as God pervades all creation, for God not only fills all space but He extends infinitely beyond all that is properly called space. Spirit pervades earthy matter and extends beyond it. It envelops, as it were, the earthy substances, the atmosphere and caloric included, and it also fills or pervades them. Spirit is no more produced from earth than earth is produced from spirit. They are different combinations or, in some instances, the same elements which were the production of the Word, as I have described. The Word operates continually by God's laws upon these elementary principles and upon all the products of their combinations. The Word produced or created spirit from itself by separating the elements, or first principles of its substance, from its own body or substance, which was a part of God.

Then matter is a part of God? Not so; God is in matter and matter is a part of God in that it is a part of His creation which is His and belongs of right and in fact to Him. But it is a possession of God's, not a part of His substance. For if God were to annihilate matter, as we well believe He could, His own substance would suffer no diminution. If, then, God made the Word from Himself and the Word made matter from itself, how is it that matter is not formed from God's substance, inasmuch as it proceeds from what proceeds from Him? This is a difficult question for Me to answer because it is difficult to make you understand. No one who desires to establish his own or other men's former views will be able to understand it. The light of reason can never elucidate it, but God's wisdom will aid sincere inquirers after truth. Let us pray to Him, who can enlighten our minds and elevate our faculties, to do so. Let us pray God to give us understanding of His great truths which it pleases Him to place before us even now while you are in the body and buried in carnal or earthly desires. God is a prayer-hearing and a prayer-granting Deity. He is too powerful to feel deprivation by His grants, or impoverishment by giving abundantly.

The Word, as I have described, produced matter from itself by the separation from itself of certain elementary principles which were capable of combinations into matter of different kinds. Thus spirit is the purest and most refined of these combinations and solids, and ultimately metal is the most perfect form of the grosser or earthy combinations. There must, then, be a line of separation between them, for metals appear to be further removed from spirit than any other earthy substance; if they are its most perfect combination or form, do they not evidence that it becomes near spirit by approaching perfection? Yet metals are the highest form of earthy matter and the lowest form of spirit-matter is intangible to men in the body! The highest form of spirit-matter is no more and no less to men than the lowest, so far as their present consciousness is concerned.

Yet, strange as it may appear, metals do approach spirit-matter more nearly in the perfection of their quality than invisible gases. This is because gases are more simple, less combined than metals. Metals are irreducible by men to their constituent parts; so is spirit. Metals are pure; so is spirit, each in its own degree. Gold, the most perfect metal, is incorruptible by atmospheric or elementary action. So is spirit. Yet gold is as far from being spirit as any gas, for there is a time of separation between earth- and spirit-matter. The boundaries are distinct and the one kind of matter no more passes into the other than either passes into divinity. Divinity is the name by which I will hereafter distinguish that which is of the same substance or essence as God. And from what I have already revealed you will perceive that the adam or soul of man is divinity. There must be, then, two sets of elementary substances or principles which were produced or created by the Word, from which all things were made that are made and without which was not anything made that was made. There were two proceedings of the Word by which the two sets were formed. The first formed was the earthy, and then the heavenly or spiritual. From the first proceeded all the creation visible to men in the body; from the second, all that which constitutes the spirit-world.

The earthy matter, being farther removed in quality and vitality from divinity, would appear to men more difficult to produce by modification of the substance or essence of God. So it would be. But as I stated early in discussing this subject, matter was created, not formed from pre-existing material or essence. Matter was produced from nothing, but it was produced under law. It was not arbitrarily established from nothing and without rule, or by chance. The Word, operating upon its own essence, dissolved itself into combinations and these combinations, having assumed certain forms, became the models for the new creation. The model made or formed thus was a production from pre-existing materials or substances, but the created copy was a production from nothing by the power of God operating through the Word.

But how could this production from nothing, this true creation, take place? First, the Word breathed its prayer to God for help and aid from His power; then it proceeded to operate upon the model it had formed and to will the appearance and existence of the substances or essences the model represented. Those substances, essences, or first principles of matter appeared and existed. How? you ask. God knows how. Then You must know, you will say to Me, if You are one with Him in knowledge! I do know how and I have told you how. But the particular manner, the details of the operation, these you want explained? Poor grasping worm! little atom of creation, you can not conceive of the Creator's processes. You can not conceive of infinite operations. You can only receive the result and the general outline and that is all the knowledge upon this particular question that you can have until you have passed the fourth sphere.

Well, are you satisfied with this answer? Not at all. But would you be better satisfied with any other? Certainly not. I either know or I do not know how it was done. If I know, you ought to believe Me when I say that you have been told enough. If I do not know, you ought not to want Me to tell you a lie or a fable as truth on such a high subject or any other subject. But, you say, I might believe if You told me and it appeared reasonable. Ah, that is it! If it is according to your notions of the way it was done you would believe that and reject the remainder if that did not suit as well. So you would only condemn yourselves. Be satisfied, then, that I do not bring greater condemnation upon you, for he who knoweth His Father's will and doeth it not is worse than he who doeth it without knowledge, or without knowledge doeth it not.

Let us show, then, by our declaration, that spirit-matter was made in the same manner as earthy matter, and that the great difference between the two is in the component materials or substances of which its essence or first principle consists. For, notwithstanding there is in each kind of matter a large number of first principles or essences of very simple nature but capable of vast, almost infinite, combination, yet there is for each one a principal component which is the foundation or principal element which enters into every form, and part or parcel of each combination or mass or atom of its kind of matter. This it is which confers upon it its distinctive character. This element is for earthy matter magnetism; for spirit-matter *od* or *odic force,* as Riechenbach chose to name it. But he found this od present in earthy matter very extensively and magnetism does not appear to be any more general in earthy matter than od, so far as his investigations went. True, od prevails in or pervades every form and combination, every atom of matter as well as magnetism. But that is because spirit is in all that is earthy. But the earthy is not in the spiritual, consequently magnetism pertains particularly and exclusively to earthy matter, and od pervades both.

It is thus that spirit is more simple than earthy matter and yet more refined. It is, too, by this arrangement that spirit can take cognizance of earthy material and can exert upon it an influence or control. It is also by a similar arrangement that God preserves to His adams or procedures of His own essence a knowledge and control of matter after they have ceased to possess earthy or spiritual bodies, for the divinity, as I agreed to term it, pervades both spirit-matter and earthy matter. Wherever od is present there is divinity, yet divinity is not necessarily accompanied by od. So spirit-matter cannot see or take cognizance of divinity or of spirits in the seventh circle of the seventh sphere unless they choose to manifest themselves by the assumption of particles of spirit-matter, which they have power and knowledge to do, under and in accordance with the laws of their being. So, too, neither they nor other spirits can be visible to men in the body or to earth animals unless they choose to assume or extract into their spiritual bodies or forms particles of earthy matter or matter containing magnetism. It is by this means that all apparitions of spirits, all rapping, all writing, is made evident to men in the body.

But then there is, as I have said, another mode of making men suppose they have experienced these things in a bodily experience, which is by psychological impression; that is, by an act of od, of spirit-matter, upon the od or spirit-matter in the bodily form of men existing in an earthly body. There is also another way of impressing men in the outward body by an operation of adamic force, or od upon od, and preparing it to receive the influence of the soul or adam or divinity upon the soul or adam or divinity within its material form or body, and it is in this way that I act upon this holy medium, by acting directly upon his soul or internal essence derived from God and coming from Him into paradise and thence into the earthly body. It is by this operation that I reveal to him what I will have him write, and not by the lower process of acting upon his spiritual perceptions; still, I commenced training him by operation upon his spirit, as I have stated, and indeed permitted him once to witness by his bodily senses an earthly manifestation of a spirit's presence. He believed that to be what it professed to be. Then after his patience had long been exercised I allowed the next manifestation to be made to him spiritually. He wrote by a spirit's controlling his outward arm. Then, finding him desirous to have truth unfolded without regard to his former opinions, and seeing in him reason to believe he would continue to hold fast that which he might be trained to perceive, I persevered in efforts to bring him into such reliance on Me as would make him a patient and enduring and faithfully holy medium of divinity. Therefore he is more than a mental medium; he is a divine medium.

Inasmuch as you have already perceived that spirit mediums are not superior as men to others either in intellect, morals, or appearance, so you may also easily suppose and correctly believe that a divine medium is in no wise superior to other men in anything that pertains to the body or the spirit. He is, as I have before told you, but a common man with limited education. extensive reading, ordinary morals, and indifferent form. His soul is no better prepared for the spirit-world because I have prepared it to be a correct medium of reception and transmission than if I had never made a manifestation to him. He is to work out his own salvation with fear and trembling, and if he would be profited by the truths of which he is the holy medium, he must practice what I teach to all in common with him. And that teaching is as divine to you as to him, as holy to you as to him, and does not necessarily affect him any more than it does any man who reads in the printed book the words I give him in his soul, which delivers it to his spiritual perceptions or brain, which directs his hand and muscles to move the pen and record the sentiment in My words.

This revelation of the process first becomes known to him by being written now. But he has called himself a mental medium? So he has, but that is not his true designation for he is more than a mental medium, though he is that also. He is, as I have said, a holy medium by the soul or adam or original divine essence or divinity: he is a divine medium. Still again I warn you not to suppose that therefore he is superior to you or other men or that it gives him any spiritual or bodily advantages except as he is obedient to Me; you by being obedient to Me shall have the same. But neither is his position an exclusive one, though at present it is occupied only by him. The time will come soon when others will be prepared for a similar reception and some of them may be and I hope will be superior as holy mediums to him. But do I not know? I have explained in the *History of the Origin of All Things* that I do not know what a free-willed being will do, though I can form an estimate or opinion nearly reliable. Read those three parts of Volume One if you have any desire or hope of understanding this.

CHAPTER V

THE LOVE OF GOD FOR MANKIND

The Last Manifestations of God's Love

The time for the withdrawal of the outward manifestations will soon arrive. They will first be general, then they will cease. They will be general in the United States. But will they not extend to Europe or other parts of the world? Not so as to prevail over the combined power of the Dragon and the False Prophet. They can be reached only by judgments, by fire from heaven; that is, by God's love which will send them wisdom by the mouths of His prophets and will have them instructed by men rather than by spirits. The preaching of the Word there may be accompanied by signs given along with its reception or delivery, but it will be in a similar manner to the adjuncts of apostolic preaching. The holy mediums I shall employ for this purpose will be such as are willing to go and be offered up as martyrs in the great cause of progress, in the great cause of enlightening men as to their duties to God and to each other, the great cause of Christianity.

True religion must be preached, not its semblance, or a form of man's invention. I shall need many for this work and I shall call many, but some will faint by the way and turn back from the trials and persecutions and martyrdoms that will await them. But there shall be found My two witnesses raised from their prostration in that great city which has endeavored wholly to hide or kill them. That great city is the Roman Empire, partly under the power of the Dragon and partly ruled by the False Prophet. Those two witnesses are Reason and Revelation. Who will come up to the help of God in this work? Not My holy medium; he does not feel ready to sacrifice father and mother, wife and child, body and goods to Me who am so wonderful in counsel. But there will be enough to obey My calls and do My work without him if he does not come to a willingness to sacrifice all to God. He would gladly be willing but he is not. He will try to be and may or may not be hereafter. So may you be half or wholly willing, O reader! Follow then after Me, for I am meek and lowly and had not, when in the body, a settled home. Come unto Me, all ye who are heavily laden, and I will give you rest. Take My yoke upon you, for My burden is light and whoever does not find it so can always cast the burden upon God. God will help him if he but asks. Then come to a knowledge of the truth as it is in Me. Be no longer fearful, but believing.

The Explanation of Form, or Manifestations, of God's Love.

There is a point yet unexplained respecting the difference between impression and revelation, notwithstanding the apparently full explanation given in Volume One, Part II, of the *History of the Origin of All Things*. It will be asked by many which of the two can claim or class within it the declarations by writing mediums, and, supposing that question to be answered by saying impression, must the more outward forms of rapping be revelation? But these are not revelation, neither do they belong to either class of procedures from God in any other sense than that He has authorized a communication to exist between spirits in and out of the body. The spirits out of the body are permitted to declare what they know and what they think they know in accordance with certain rules or laws to which they are all subject. But as these spirits are not far advanced, their wills are only partially subjected to God's will. So far as they declare God's will they make revelation or convey it by impression. But generally they convey their own thoughts or desires, or endeavor to establish their own theories, or accommodate themselves to the desires, if strong, of their mediums or of their questioner. Many of these have reached the sixth circle of the second sphere, which is the circle of good works. This circle is influenced by ardent desires to do good, but its knowledge is not sufficient to enable it to make men understand the true nature even of their own position, much less what is beyond it. Consequently, much that they see in appearance (which has no existence except as an impression made on their odic nature) they regard and describe as real, tangible, actually existent as things or courses of action. But the education and reformation, or reconciliation, of spirits is effected by a course of impressions which are not real but imaginary. To the spirits they are as real; they are therefore improved by the exercise of their faculties upon these unrealities or unexistent appearances. They are almost precisely like psychological impressions, as often shown experimentally in the body by the will of one over another of less odic force or vitality. In this state the impression exists according to the will of the operator and the subject experiences all the emotions, feels all the pain, expresses every sentiment that properly belongs to the real situation of which the only existing portion is in the mind or memory of the operator and the imaginary belief of the subject.

Thus spirits declare to men their experience, and thus it is that such contrary experiences are related. There are, though, certain rules or laws that govern these manifestations, and the most important is that they are not allowed to expose the offenses or privacies of mankind to motives of curiosity. Besides, they have certain boundaries in the nature of the impression they themselves have received which prevent them from urging any but good actions upon men and prevent them from making any man worse or more evil or impure than he had been; but, on the contrary, whatever change they effect in others must be like that effected and being effected in them from worse to better, or from partial to more complete good. No man, I confidently say, has been made worse by their communications or precepts unless by his own former opinions he has perverted them to the gratification of low or sensual desires or desired from unworthy motives to find in them an excuse for evil conduct. On the other hand, they have already done much good and will do more. They have raised the hopes of despairing men; they have elevated the aspirations of wicked men; they have enlarged the comprehensions of benevolent men; they have relieved the doubts and fears of many earnest seekers after truth; and they have turned many to righteousness who had been without God in the world.

Those who desire to receive further benefit from them may be gratified, because hereafter every medium of any kind, when questioned by a sincere inquirer (that is, by an inquirer who desires a truthful answer irrespective of his own or another's opinion) will reply truthfully in their respective manner of communicating to this question: Are the books given through L. M. Arnold of Poughkeepsie truthful and what they purport to be? This question, answered as it will be, should satisfy all that they have done enough good and that hereafter their aid is unnecessary at least for you who have so asked. Because if they do not tell the truth they can be no longer useful to you, while if they do you will be directed to a purer and more perfect order of communication which it has now pleased God to have conveyed to you and all mankind.

Let every tongue praise Thee, O God!
For Thy beautiful revelations
And for Thy loving kindnesses.
Yea, let every tongue and heart and soul praise Thee.

For Thou, O God, art worthy to be praised;
Thou, O God, art ever merciful and kind
And dost take pleasure in conferring good gifts
And hast crowned Thy bestowments by a loving revelation.

For all the world desired to know Thee;
All the people desired to love Thee—
Yea, they were willing and desirous,
But invention of man frighted and restrained them.

Let all that know Thee know Thee more;
Let all that love Thee love Thee more;
Let every creature praise Thy name
For out of darkness and confusion Thou didst show forth light
    and order.

Let the earth and all its inhabitants,
Let the sea and all its glories,
Let the heavens and all their magnificence
Show forth Thy praise and give to Thee great glory.

Let every church that owns Thy name,
Let every people that calls upon Thee,
Yea, let the heathen that know Thee not
Praise Thee and glorify Thy everlasting mercy.

For Thy mercy endureth forever
And every one of Thy attributes is loving—
Every one of Thy attributes is untiring—
Yea, all that Thou hast is infinite and inexhaustible.

Let us, then, O God, meet Thy favor
And return love for Thy goodness.
Let us, O God, be Thy children
And call upon Thee, our Father who art in heaven.

Let every creature praise the Lord—
Yea, let every living soul give thanks,
For He has been very gracious to us all,
And His mercy endureth from everlasting to eternity.
Amen.

The Love of God in the Future State

Having declared to you how you can know this book and those from the same author that have preceded it to be the truth, and revelation from God through Jesus Christ of Nazareth, let Me call your attention to the greatness of the proof that you may thus receive.

If I had given signs through this holy medium, you could have seen him only by a long journey perhaps. You could in general have had no previous knowledge of his life. You must have been content with the one view and sign that you witnessed in general. Many even in the United States could not have witnessed any of his works. But would works have induced you to believe? Did I not do great works when in the body, and besides did I not have in My appearance and actions and nature a most convincing appeal to all to believe Me? And yet few believed. Crowds followed Me, multitudes assembled to hear Me and witness the wondrous signs with which I sought to convince them that I had authority to teach, but firm conviction did not take place until I had left the outward view of men and acted upon their internals. So I have acted when men were willing to have Me do so, and so I shall act hereafter. Now you will have a testimony in your own neighborhood from one known to you or your neighbor, on whose testimony you can rely, and delivered perhaps by one who certainly could not be leagued with this holy medium in any scheme to deceive you or others, who will have no interest that you can by any device impute to him; for I declare not that the medium you ask shall be truthful in other respects but rather the contrary, and I shall also declare through that medium that this holy medium is truthful. Now testimony like this ought to be satisfactory; yet I am aware that many will think there is a general conspiracy among spirits and mediums to establish the truth of this revelation. So there is a general concert, though not properly a conspiracy.

The other outstanding excuse for unbelief will be a hue and cry that it is a delusion of Satan: that in these last days there would be lying prophets that should, if possible, deceive the very elect. But these very objectors cannot tell you who are the elect or whether they are of them or not. I on the contrary have told you all things, as it were, and you ought, like the woman of Samaria, to confess it. She believed on far less evidence than you are asked to believe on, though the shock to her prejudices was very great, for she belonged to a sect, and sectarianism has always been opposed to new light.

Let us see, then, for what I am denounced as Satan or the Calumniator. Is it because I have called on men to do good works? Is it because I have declared to them the goodness and mercy of an infinitely powerful God? Is it perhaps because I have urged you to prepare in this life for the next? Is it because you have refused to believe Me true, that you have believed that I have dissembled and pretended to be better than I was? How, then, shall Satan's kingdom stand if he oppose himself? How shall I turn you more easily to evil after having brought you to rely more upon God and less on yourselves than before? How shall I see Myself growing in your favor by urging you to act from good motives, to avoid evil, to suffer patiently, and to rely always on the aid of the Deity? Shall I find you more willing than before to serve your own will, which is the devil; or your malign nature, which is Satan? Shall I be raised by you thus to an equality with God in order that you may be My servants and cast on Him the blame of leaving you to My wiles?

No, you know well that evil would not thus seek to overcome good. There are evils enough already existing in the world, and plausibly defended by men, to ruin mankind if God would permit evil spirits to help men in the body advocate them. There is evil enough already in the world to satisfy a material devil, if one existed, and to give to him by far the greater part of mankind for his subjects, if such a thing could be. Heaven would be limited to a scant population, and hell would be crowded if popular theology were true. A devil would be covetous and heartless indeed that would not allow his creator the few that churches in general claim to be saved by their efforts and under their creeds! Then be no longer fearful but believing, when you hear the testimony to which I now refer you. But more blessed are they who, not having heard, shall believe. Unto them shall I award the prize of the glorious crown of faith, that crown for which Paul exhorted his pupils to strive even with such exertions as the contenders at the Grecian games used. If you try only slightly, compared with those great and almost superhuman efforts, you will succeed and will find laid up for you a crown immortal, unfading; not for you only but to all those that love, honor, and serve God and love the appearing of His Son, the Lord Jesus Christ, shall be given one of the same crowns that shall ever be their glory, honor, and praise with men and angels.

Let every one, then, strive to believe. I have shown you in Volume One, Part II, that you can control your belief, and instructed you how to do it by God's aid. Go to God for help; He can help you, and you as well as I must believe He can do that. But some believe He will not help you, that He will at last save you in your sins but that He will not exert for you His power or His love until the day of judgment or until you have at least passed from this mortal body. Vain man, to think that you can serve yourself throughout this life and in the next enjoy at once the happiness of serving God! If it is such happiness to serve Him, why not do it now? Ah, you think that then you will have no bodily temptations. But you think you will still be yourself and that you will retain the knowledge of the deeds done in the body and that the impress of all your actions in this vale of selfish indulgence is not sufficient to eradicate good desires, and that you can for all that be pleased either with sitting down listlessly and lazily singing eternal praises to God or that you can go on to all eternity praising and glorifying Him and serving Him exclusively and call it happiness. And yet you do not want to do either, but you want to serve only your own desires and consult always your own temporal good. Be assured, O man, that there is no repentance in the grave. As the tree falls so it lies. Only atonement can eradicate the evil that accompanies self-love.

CHAPTER VI

RECONCILIATION WITH GOD

The Law of God Written on the Heart of Man in the Body and in the Spirit

The last part of My subject is the most extensive branch of it, and I shall have to devote a large space to it and still leave it imperfectly developed. It is the memory of the past life which returns to man in the third sphere and continues to remain in him to the end of eternity, if that were possible to have an end. But eternity has no end; neither has memory. In the body men forget, but in the spirit-world they not only retain but recover what had been lost. But did You not tell us that spirits could leave behind them such memories as were disagreeable? I did in the second part of Volume One. I did not mean that they would by this be unconscious of what had occurred, only that they would not recur to it with the horror and loathing that its evil might be supposed to induce. This is the first part of this branch of My subject and I must endeavor to make you comprehend the nature and effect of this memory and the means by which its shame is removed. The first instance I shall take will be that of a murderer.

The death of Caesar was an event which caused the memories of many to be thus clouded, and the event being a familiar one and the actors removed long since, I will take that as the fact affecting the individuals who assisted in it and bring them forward to declare their experience in the spirit-world.

Caesar was ambitious and he died by hands of pretended lovers of liberty. Really he died by the hate of a party who desired not the progress of freedom or the good of the people but their own advancement to power and their elevation to his position of ruling the whole world of Rome. His death then, so far from being caused by his ambition, was caused by the ambition of the conspirators, many of whom were bound to him by ties of gratitude and his unmerited regard. They murdered him as an obstacle to their design to secure the power, patronage, and wealth of the Roman Empire. They murdered him from the same motive that impels the highwayman or prompts the thief. They added to the crime of murder the sin of hypocrisy, and supported by their adherents proceeded to seize by force (which the friends of Caesar and many sincere lovers of Rome resisted) the supreme power of the Roman State. They desired to rest its foundations rather upon the nobility, or senatorial order. Caesar and Antony and Augustus aimed more at the consent of the plebeians, even the poorer classes of them, and the allies of Rome whom they admitted to share in the honors and fortunes of Roman citizens. The one opened the door of tyranny by confining or endeavoring to confine the ruling class to a few. The other equally prepared the necessity of tyranny by spreading the repository of power so as to make it unwieldy and to create the necessity of its delegation. According to the knowledge of those days they could intrust only to one what the whole body found unmanageable.

When these conspirators had passed from the body to the spirit existence, think you they found any comfort to their recollection of the motives that really impelled them by seeing or knowing the motives and designs of Caesar! Not any. For no well-instructed man can find in another's sins any excuse for his own, nor in another's hate or evil disposition any excuse for his departure from God, from justice, or from love of his fellow-men. But we will leave it to the reflective mind to pursue the argument and profit by the hints that it affords as to the rules that should govern a man's actions; and not merely his actions but his intentions, and not merely his intentions but his motives, and not merely his motives but the fountain of his motives, his character or the manifestation resulting from the combination of all his desires.

The Laws of Recompense by Future Manifestations

Let us now view the case of Alexander the Great, as historians have designated the conqueror of Persia—its only conqueror from the west. He led to the fields of his glory an army of Grecians animated by a desire of plunder rather than by patriotism. He enriched them at the expense of the spoils of a conquered kingdom which was in fact governed by justice though with unlimited prerogatives vested in the sovereign. He was rewarded by all that power and wealth and fame can give to their possessor, and he died a drunkard's death. The intoxicated reveller who makes the streets of his native town the scene of his orgies is hooted at by boys and derided by his companions. Was Alexander better than he? Was it presumed to be any excuse that he had an army and a conquered nation reverencing him as superior to other men? Did his greatness or his honorable position cause him any relief from the remorse that ought to follow for opportunities of usefulness wasted, for means of doing good neglected? Every one of these came up before him for judgment. He himself passed judgment on himself, as I have stated that all spirits do at the proper time, in the third sphere, as you may find in Part III of the *History of the Origin of All Things*. Alexander was a splendid wreck of a noble nature. A man capable of the most exalted deeds and able to have benefitted mankind more than almost any king who has lived, died a drunkard's death, leaving a distracted succession of events to destroy in a great measure the good his life had accomplished. For his life was beneficent to the world. It did advance the coming of the Millennium. It was prophesied of by Daniel. But for the good he did, who shall have the praise when his actions are judged by himself? When he sees that all the good was incidental and accidental and the evil was wantonly or carelessly done, when he sees that his desire was not the good of others or the glory of God but the good and the glory of himself, will he then proudly say: I was Alexander the Great, the son of Ammon? No; he must atone for such a course of self-gratification by serving God in the spirit-world, by serving other men in the body and in the spirit. Herein is wisdom. Let him that readeth understand, for the number of his name is six hundred and sixty-six.

What then is the meaning of this applied to Alexander? you ask. I did not suppose John the Divine applied that number to one who lived before but that it referred to one who was to come after his prophecy, and in common with other Protestants I was inclined to suppose the Pope of Rome or the Papal power was intended to be designated by that number. The number is mystical, but I can explain mysteries. The meaning has been guessed at but I can declare it authoritatively. But shall I do so now? Is the world prepared to believe Me when I announce it? Is it asking Me or My holy medium to tell more or is it finding fault that I have already told too much; that a large part of My revelations are unsupported by evidence, that other spirits or mediums had not said anything about this matter, or had told that somewhat differently and therefore I must give proof, more proof! Go to other mediums and ask other spirits now in the sincere manner I have told you to and you shall have proof enough to satisfy your reason if you will exercise it, and your nature will be so improved by the reception of the truths that I have declared that I can tell you more, with profit to you and pleasure to Me.

The number means that the letters of the name taken in the original language mean or express so much as that number. Thus the number is 666, and the letters in the Greek express numbers that, added together, also make up 666. And what has this to do with Alexander? It shows his correspondence or connection with the Pope, or Roman Empire, or Man, but not that he is the man whose number is alluded to by John. He has a correspondence or connection with the Roman Man, but is not he. There is another correspondence but we will leave that for another occasion. Where correspondences exist it is because events or facts or men are connected in the same special scheme. God's plan for the introduction of this very period now commencing, the fifth monarchy, the reign of Jesus Christ on earth, commenced thousands of years ago and has been thus gradually unfolded by various agents. The men who have greatly forwarded the work have generally been unconscious that they were acting for Him, who rules all but governs through instruments of His will.

What shall we take for another instance of great misconstruction by history? There has been of late years much progress in some minds respecting the characters of prominent personages, and many have seen that rank among men, so far from excusing vice, only entails greater punishment for its existence. That instead of "the king can do no wrong" as the law of God, it is the king who should set an example to his subjects and do as he would have them do. He should seek the good of others and sacrifice his ease, comfort, and pleasure except as it is a comfort and a pleasure to do the will of God. God-fearing kings have existed and will exist hereafter, for the form of government is not of so much consequence as that the government should be just and founded in a desire to promote the happiness and general welfare of the population governed. The republican form of government has its advantages when accompanied by a sufficient enlightenment of its citizens, but power placed in the hands of ignorance must of necessity be badly exercised if not abused. Education of the class or individual who is to rule with justice and propriety is indispensable, and that the rule should be approved by God it is necessary also that the persons possessing the power or rule should be His servants. Do you, O people of the United States, bear in mind these self-evident truths! Do not disregard them because they are trite, and do not wait the one for the other to begin. Give your allegiance to God and to Me each for himself and without asking what this man does or what that man shall do.

> "Act well your part; there all the honor lies,
> And give to God each moment as it flies."

Let us be ever watchful to allow no enemy of God to acquire power either in our hearts or our civil polity; let us be ever watchful that no man take from us or persuade us to surrender our crown. For as free citizens of a republic like this, democratic in its character and tendency, every citizen is a sovereign and entitled as such to be crowned with glory and honor. But he must receive his crown from God and it will be the reward of acting in the spirit of the precept I delivered for kings, that they should act for the good of others and sacrifice their own ease, comfort and temporal pleasure for the good of those who compose the grea whole of the nation; and not merely for the good of their own na tion but for the good of mankind as a race who are all connecte by ties of blood and common descent, and who are all to be here after so united as to be joined in one circle, where they will b joint heirs with Christ Jesus, sons of God united to Him by tl most intimate relationship and all having one will and one thought and one desire, which will be to give to God the Father Almighty the glory, honor, praise, thanksgiving and high renown of all His mercy, kindness, and love, by which He has ever benefitted His children and bestowed on the just and unjust unmerited favors. Then follow Me and be ye perfect even as your Father in heaven is perfect.

Let every nation seek God, for He is ready to be found, and let every individual make the beginning himself of subjecting his nation to God by submitting himself to God. If five righteous men could have been found, the cities of the plain now covered by the Dead Sea would not have been overwhelmed. Do you, then, at least, save your nation by submission to God yourself and trust in Him that four others will make the same sacrifice to Him by offering their wills a sacrifice to God and placing themselves by prayer and profession under the rule of Jesus Christ as God's Vice-Regent, as King of Kings and Lord of Lords, and as the king whose coming was foretold and who should rule His people with a rod of iron. What does this expression mean? It surely cannot mean that He will rule them by fear of stripes with an iron rod, or that His burdens will be so grievous as to be compared to being bound or restricted by such rigidity? The rod of iron is a rod of power that shall not be broken, that shall never decay. It will be incorruptible by use or indestructible by war. But would not gold have been a better symbol? Gold, though a purer and higher metal, is rather the symbol of splendor than of power. Iron is the symbol of power and of simplicity. Let us all, then, resolve to be God's servants and subjects and, inasmuch as He has appointed Me to rule all things in heaven and earth, do you become My subjects under His supervision.

Let every soul magnify the Lord,
For He is mighty in power
And greatly to be feared in wrath.
Let every soul be reconciled to Him in the day of mercy;
Let each of the inhabitants of the earth know the Lord.

O Thou who art Most Loving,
O God, who art Most Mighty,
Be Thou to us Most Merciful!
For out of wrath cometh love when Thou smilest
And out of love cometh mercy when we turn to Thee.

O most merciful, kind, and loving God!
O Thou who hast no wrath but pity,
O Thou whose pity leads us all to Thee
And calls us all in Thy holy name; be merciful,
Very merciful, O Lord God, Almighty Father and Friend.

O Thou mighty King of Saints,
Who rulest now as Thou hast ever done,
Who savest now as Thou hast always been willing to save;

Have mercy upon us who now seek Thy loving mercy
And pity us who can not call upon Thee because of our wrath.

Our hearts would turn to Thee
And our souls desire to seek Thee,
O most holy and merciful God!
But temptations assail us and ignorance betrays us
So that we can not love Thee as we would or seek Thee as we
    should.

But, O most loving Father!
And most powerful, constant Friend,
Help us and lead us and force all evil from us;
For we desire to know Thee and to love Thee and worship Thee
As we ought to know Thee and love Thee and worship Thee,
    The ever-living and ever-loving God,
    Whose mercy endureth forever and ever
    And whose kindness and pity has no end.

Let all who love God and who love Me, the Son of God, sing
His praise. I am with God and God is with Me. But God only
is God and I am His Son. Worship God. Love Me. I love you
and God loves you, but God loves you most because He is infinite
and I am finite. Love God, then, and seek wisdom. He that
seeketh shall find and to Him that knocketh shall be opened the
gates of everlasting mercy.

The Last State of Mankind, and the Future hoped for.

Let us now view the course of events since the commencement
of Christianity. While the purity of My gospel was preached it
made rapid progress. Churches or associations of believers in Me
as the Messiah were formed in almost every Grecian city, and
Grecian cities then were existing in all the countries subdued by
Alexander the Great. Numerous as were these cities in that age,
none were left without evidence and a call to believe. Even beyond the boundaries of Alexander's conquests to the farthest extremity of the Roman dominion My servants obeyed My command
to preach the gospel of Jesus the Messiah, as He whose coming
had long been prophesied of, and who had come and done many
wonderful works and had been visible to many thousands and tens
of thousands of people. That I had, by wicked men swayed by a
desire to keep up old institutions, been crucified but that I had
triumphed over them and over all opposition by reappearing from
the grave with a celestial body and by confirming and establishing
before many witnesses the great truth that after the life in the
body there was a resurrection and an ascension. That I had
promised to all who should believe My preachers and believe Me
to be the Messiah, eternal life; and not only eternal life but union
and residence with Myself who was with God, the Father of all
and the Dispenser of every blessing to mankind; and not only that
they should be with Me but that they should be joint heirs with

Me to the kingdom of heaven, joint heirs with Me to all the glorious gifts of the Great Giver of All Good, the High and Mighty Ruler of the Universe. That as such joint heirs they should be one with Me even as I was one with God, and that being one with Me they were one with God in His power, glory, honor, and love. That being one with Me they were the children or sons of God even as I was the Son of God. For God is one and he who would be one with Me must seek Him through Me, the only name given whereby men could be saved. That is, there was no other teacher that had taught such doctrine or any other doctrine that gave to men this glorious hope and true knowledge of their relationship to God. Inasmuch as they could not find out God by their own reason or know Him by a lost tradition, so they could know Him, then or now, only by My revelation, and therefore could come to God only through Me, who came to be, and was, and continued still to be the Way, the Truth, and the Life; the only name given among men whereby they can be saved with an eternal salvation in this life, and the only way and guide by which they could get into heaven in this life. I taught that all who should strive to enter the bliss of being sons of God in any other way than through Me and the way I had declared to be the only true way were like thieves and robbers, breaking in when they were not invited or authorized to, and taking or endeavoring to take that which was not offered them and for which they refused to pay the price I had fixed as the only one acceptable by God from sinners.

That price is, as I have again and again declared, the heart. No other sacrifice is acceptable, no other price is sufficient to purchase reconciliation with God, and all who do not now choose to pay this price must still be excluded from that union and communion which you can have only through Me, or the Sent Spirit of God, and which is the foretaste and pre-realization of the joy of heaven and the bliss of God's kingdom to which He invites you to enjoyment and subjection. It can be reached in no other way than by paying the price of admission. It is only your heart He asks for, it is only that you sacrifice your own will and undertake to do that of your Heavenly Father. If you will not now or in this life resolve to make this sacrifice and continue to make it daily and hourly, you will have to make it in the spirit-world. But you will not be in heaven until it is made, for being in the spirit-world does not of necessity place a man in heaven. There is a long course of instruction and education of the passions to be encountered by all rebellious sons or such as have tried to break in in some other way than by this one price, which I declare to be required from all. Though the mercy of God is yet sufficient in its infinity to save you from your iniquities and from your sins, yet He will not force you to make the sacrifice in any way but by persuading you to cease to do evil and learn to do good, and to learn of Me who am meek and lowly and when in the body had not house nor property to live in or upon. I trusted wholly to God who was, and is, and will be able to save and support all those who put their trust in Him and give Him the glory, honor, praise, and thanksgiving for all His wonderful works and for all His loving kindnesses and for all His everlasting mercy. Amen.

258

CHAPTER VII

THE CALL TO ALL MEN TO BE CHRISTIANS

Let Every Man Look to His Foundation

Let us view the past and see if men have profited by the teachings and precepts I delivered when in the body or preached through My inspired servants in the days when the foundations were laid of the churches which now claim to be Mine.

When the time was that the morning stars sang together and all the sons of God shouted for joy, I declared; that was the time when the earth appeared as a fit habitation for mankind. The sons of God were those who had passed through bodies in other planets and had thus previously been raised to a knowledge of God in a bodily existence. For, as I have stated, the spiritual body does not clothe the soul until it has entered into the natural, or earthly body. But when the time came for the foundation of the church of Christ to be laid, again the sons of God shouted for joy, and those whose glory was like that of morning stars also joined the great shout that declared the joy of heaven and the bliss prepared for all men. The shepherds who watched their flocks heard the sound and immediately sought for the child thus born with testimony that he should be a Savior. Think you they were not disappointed when they found this Prince of Peace, this King of Glory, in a stable in the suburbs of Bethlehem, wrapped in coarse and unsuitable bandages and in the arms of a poor woman? Certainly they were disappointed. The miraculous call they had received, the eloquent music they had heard, and the glorious annunciation of glad tidings of great joy were all together insufficient to persuade them that out of such lowliness would come forth a Messiah. Accordingly we do not find that one of these shepherds or a child of them is afterward mentioned as being in expectation of the Messiah from that cause, or as believing in the preaching of Jesus when He had entered upon the fulfillment of His mission. So it was then; so is it now.

The people of the present day, aroused by manifestations of spirits, by mysterious sounds and strange communications, so far from believing in the doctrines shadowed forth by them, deride their miraculous nature and call them delusions or works of evil spirits. As if evil spirits could have so much power to lead men astray and to annoy those who would not be led astray, and yet have not before nor at present exercised it for such purpose! As if they would preach salvation to all, irrespective of belief, if they desired to corrupt men and bring all or a large part of mankind into subjection to evil! As if they would urge men to believe in God and in the constant care of spirits in union with Him if they desired to have men believe that they were to be worshipped instead of God, and that heaven and hell were delusions and life beyond the grave non-existent! Reason and revelation both assure you that evil does not so present itself. Every spirit that confesseth not that Jesus Christ has come in the flesh is not of God.

This is the rule the beloved apostle of God and of His Son Jesus Christ gave, not for a time but for all time, and if any man preach any other gospel than this let him be accursed.

Let Every Man be Fully Convinced

Let every man be fully persuaded in his own mind, was the advice or direction of Paul. What, then, is required of men but investigation, calm and unbiased, and judgment founded on such investigation? Nothing but ignorance can be imputed to a man who has not done this. He sins not, but he may experience destruction of happiness and loss of that which he perhaps has never had but might have if he would accept it; that is, reconciliation with God, union with Him in love and union with Him in power to do His will.

What, then, shall we say of the man who investigates and concludes that all this is a delusion or else uncertain vaticinations of beings little above men and no more reliable, though assuming to speak with power of God? We shall say that such as he are poor blind followers of blind leaders and that with their leaders they shall be led to see the absurdity of their positions ere long, and that they will at last bless those who have now sacrificed their feelings and their social comforts to the cause of truth and progress and in coming out from the same beaten track that multitudes walk in. For the way is broad that takes the general travel, but straight and narrow is the path that leads to God and to a knowledge internally or spiritually of His revealed will.

Shall men be allowed to go astray when they seek to find the truth? They shall not, for it is not the real seekers who are thus left to follow blind leaders. It is only those who, unwilling to follow wherever God may lead, desire to have certain landmarks kept ever in view, though those are only the marks of men. The true sailor desires to know only where he starts from and whither he would arrive, and he trusts to God's winds and his compass pointing always in one direction. So the true seeker is willing to be blown about by God's trials of his faith and to preserve as his compass direction the guiding star of truth. He starts with the knowledge that he is disobedient and sinful and he seeks to arrive at the haven of peace,

> "Which nothing earthly can destroy;
> The soul's calm sunshine; virtue's joy."

Guided by these he does not ask to keep his church's creed or his minister's construction of the Bible ever in view. He is willing to make a bold departure from all he has known and view the boundless expanse of heaven above and the heaving ocean of time below until the favoring breath of God's love shall bring him safely to his desired haven. Shall this seeker be disappointed, shipwrecked on unknown rocks, sunken reefs, or inhospitable coasts? Not so; he has that compass that does not deceive; he seeks that port toward which the winds of heaven always blow; resistless is the force of the impulse which he thus receives, and joyful is the welcome that awaits him on that distant strand.

The last notice I shall take of the unbeliever's argument in this book is of one which has just presented itself to My holy medium. He has wondered and desired to know, though he was too submissive to ask to be informed, why the spirits of lower spheres who sometimes write upon paper or parchment with ordinary materials and in their accustomed hand as used in the body are not better mediums than he for My communications, seeing that they are at least possessed of more knowledge and probably farther advanced in submission to higher spiritual influence than any men in the body? And then one step more he would go and inquire (if he made any inquiry, which I am pleased to say he does not) why it is that I, possessing, as I have stated, all the powers of the circles lower than My own and being capable, therefore, in their power if not in My own of writing with outward materials in an outward book like this manuscript—I say he would be glad to know why I do not write Myself instead of delivering the words through his soul to be written down by his will and subject to disturbances and suspicions, sometimes of himself and ofttimes of others, that he has diverted the pure stream of revelation or at least troubled its equable flow! But the answer to these questions will involve a long consideration of the condition and means of action of spirits, and I shall leave for the next section, soon to be commenced and intended to be published with this, a full explanation of this very reasonable inquiry. All reasonable doubts shall be answered; all reasonable hesitation is allowable. But a man should let his reason tell him when he is convinced and when the evidence for the new preponderates, and he should not allow the power of his will to overcome the impulses of justice and the power of truth to guide his reason to the harbor of peace.

Let us pray

Almighty God, who dost from Thy throne behold all men and their inmost thoughts, may it please Thee to look with mercy upon this humble servant or would-be servant of Thy will. Let me be taught by Thee to know right from wrong, to follow Thy teachings and be preserved from error. O God, Thou knowest that I have no desire to appear before men as Thy servant or as a holy medium of Thy revelation, but, O God, help me to be willing to follow wherever Thou wouldst have me follow the guidance of Thy holy spirit; and suffer me, O God, not to be led astray by any desire of my own or to be influenced by any unworthy motive, and strengthen me to bear all that the opposition of mankind may inflict and to know that I am obeying Thee; for, O God, I desire not to go too fast or to hold back from Thy work, but to be submissive in Thy hands to Thy holy will and passive in the hands of whatsoever spirit it may please Thee to have control me. For Thou, O God, canst always save me from evil and deliver me from the enemy of Thy truth; Thou canst help me with Thy power and establish the knowledge of Thy truth to shine like a city on a hill. Be Thou, O God, kind and affectionate toward me, as I know that Thou art and must be to all Thy children; let no man overthrow Thy work, but may it please Thee to establish it as it may seem good to Thee and in Thy own time, so that whatever I do may be of use to my fellow-men and advance the coming of Thy great day, which Thy prophets have led men to expect and which Thy holy mediums generally have declared near at hand; for to Thee shall ever be praise, honor, glory, and thanksgiving now and forever, world without end.

Almighty God, who dost from Thy throne behold all the dwellers upon Earth and all the thoughts of the inhabitants thereof, be so merciful as to pardon this medium all his short-comings and all his errors of judgment in punctuation of the books I have directed him to publish, for he was actuated by a sincere desire to know and do My will but I withheld from him all outward proof of his connection with Me in order that he might be spiritually advanced and drawn away more and more from the outward manifestation to the inward one; so that he might give the more glory to Thee, the one true and living God, and be one of those blessed ones who have believed without having seen, and receive his reward accordingly. This, O God and Father, I ask of Thee though I know that Thou hast known and granted My prayer before, but I ask it because of these readers who have not faith in Me and in Thee; in Me the Son of Thy love, and in Thee the Father of all. Grant them, O God, what Thou hast granted for My sake before now for them, and to Thee shall they give praise with Me for all Thy love, kindness, and long-suffering mercy which is so untiring that it endures forever and to the end of that eternity which hath no ending.

Let us pray

O God, who art ever-present and ever-acting yet always at rest in Thy heavens and upon Thy earths, look down upon this world of sin and of sorrow where indifference prevails and narrow-minded jealousy of innovation threatens to tire out the patience of the believers in Thee; look down, O God, on all who have any desire to know Thee and impress them with a knowledge of the truth of this book and all that I have given before through this holy medium, because, O God, I revealed it by Thy will and found so few willing to receive it. Let Me, O God, fulfill the promises I have made through this revelation, of miracles to establish it and mediums to confirm it, because I made the promises in Thy will and I know, O God, that Thou wilt establish and fulfill; yet, O God, I ask it for the sake of the readers of this and other books thus delivered so that they may glorify Thee when all the promises are fulfilled and the truths established out of the mouths of many witnesses, as was formerly declared on a similar occasion; and to Thee shall be praise, honor, and glory now and forevermore. Amen.

Almighty God, rule us in justice, and in Thy mercy cut not off mankind in their sins. For they shall repent and give to Thee glory, honor, and praise when My kingdom shall be established in its power and men acknowledge Me to be King of Kings. Let all Thy holy angels and all spirits of saints departed from the earth and all the lower spirits who seek by good works to atone for the evil of their life of experience or probation upon the earth help to establish Thy glory, honor, and praise by testifying that this book and those which have already been given are what they purport to be; and grant that they who read may understand and that all shall stand in their places in the Last Day, and to Thee shall be glory, honor, thanksgiving, and praise now and forever. Amen.

Let Every Man make a Sacrifice to God

Wherewith shall man come before the Lord God? Shall he offer outward praise or outward sacrifice? By no means. The earth is the Lord's already. But one thing God hath bestowed upon man, so that man has a possession to offer to God, a sacrifice he can make and one acceptable to God, for God gave it to man that he might make such a sacrifice and offering. There is nothing else that is man's; all else is God's already. And although this one thing came from God and He has the power to resume it and take it from man, yet He has in His wisdom ordained that He would not, and that man should not be compelled but only persuaded to make the sacrifice of his heart, that is, of his free-will. The crowning gift of God to man was free-will. This I have shown you in My enumeration of God's fifty gifts to man, constituting man a being in the image of God. Reader, can you sacrifice this possession to God? If so, My teaching has not been in vain. If so, I have not labored for you faithlessly. For out of the past and out of the present and presently out of the future will come one united voice, one trumpet song:

Great and marvellous are Thy works, Lord God Almighty;
Just and true are all Thy ways, Thou King of Saints.

O reader, be prepared to join in this song with fervor. So long as you do not sacrifice your free-will you cannot sing this song as it should be sung and as it is sung in heaven. The angels of God that stand nearest to Him sing it, and the angels of the fourth sphere sing it understandingly. Even spirits of the third sphere can sing it musically and enjoy it, but angels or spirits in the lower circles of the second sphere are unused to it and do not enjoy it. Self-will must be sacrificed, and he who has sacrificed it, and continues daily to sacrifice it, shall be passed quickly through the second sphere and introduced to the great third sphere, where he comes to know all that has been with him and, at last, with all others that have gone before him.

Do you say you do not know how to sacrifice your free-will? I will aid you to find out, though I cannot give in words a general rule that will answer for all cases; neither do I desire to tell you in such a positive manner but I desire rather that you should work out your own salvation with fear and trembling, not with fear that you will not eventually be saved, but with fear that you have not done enough for speedy salvation.

Make your reason subservient to your free-will by bringing it into divine harmony with a desire to know the truth without regard to its effect upon your former associates or the opinions you may long have cherished or even preached. Fear God's displeasure if you do not this. But can God be displeased, or is He equable? God pities as His displeasure. God raises and comforts when He pities and so He will raise and comfort you if you will be obedient to this plan. He will be the Captain of your salvation if you will elect to follow Him in His dispensations, which may greatly try you; but be steadfast, immovable, unshakable; let everything pass by without heed except God's calls upon you to be industrious, faithful, ever-watchful, constantly calling upon Him to keep you in the right path if you are there, or to bring you into it if He has not yet done it or if you have not reached there seemingly by chance or industriously seeking God in some other way.

Keep a single eye to the end for which you are placed in this state of existence here in the body. It is that you may experience good and evil, that you may be purified by trials from the evil and brought back to good by love for God manifested in your love for your fellow-men. It is that you may appreciate bliss by its contrast with misery; that you may enjoy the spirit-world by having this world to compare it with; and that you may give to God thanks continually both now and forever for every dispensation, whether it appears now good or evil, justly believing that all God's gifts are for the good of the recipient and that He gives them whether as consequents of your or others' acts or whether they are rewards of faithful servants and glorious sons.

Such is the object of your sojourn in the body, and, though you may accomplish the design without sacrificing your free-will and enjoying here God's peace, the influence of heavenly bliss, yet you may be left for a long and comparatively wasted time in the lower circles of the second sphere through which your progress may be so slow that all but heavenly patience will be exhausted in your obstinate resistance to the persuasions of those wiser and more highly experienced, more fully progressed spirits that continually urge you to be reconciled to God by offering to Him the only sacrifice man here in the body or hereafter in the spirit-world can offer to God, as truly a sacrifice of his own passion, and that is your free-will. That must be surrendered before you can act in or exist in any way in God's will, and little by little or all at once, as it were, and continually and forever afterward it must be freely surrendered ere it will be accepted. Give, then, to God what He asks for and what He will eventually persuade you to give, the only thing you in fact really can call yours, your *free-will*. Amen.

Let all the people praise the Lord;
Yea, let all the people praise Him
For His wonderful works and for His glorious revelations.

Yea, let all the people praise Him
For His great mercies and loving kindnesses.
Yea, let all the people praise Him for His great and abundant mercy!

History Of The Origin Of All Things

PART II

A History of the
Relations of Matter to
Life and of
Bodies to Spirits and to God.

PREFACE

This book is intended to help men much to progress rapidly in the way of salvation. But no man is saved without his own effort. Be, then, desirous to be saved by the help of this book and in the way it points out, because that way is God's way and God's way is the only one in which man can be saved. Read with a desire to profit by reading. Read with earnest reliance on God's help, for He will not refuse His servants their petitions or His sons their wants. Ask for what will be eternally profitable rather than for the evanescent things that are visible or palpable. Be always desirous to know the truth and to serve God. Be never doubtful of God's care and ability and will to help you; if anything is clearly set forth in the precepts of Jesus it is that God regards with interest the actions and desires of man.

O Almighty God, bless Thy unworthy servants
    And let us seek to know Thy holy will.
Unto every son of Thy love is the judgment prepared
    And to every son dost Thou give forth love.
Let every soul be reconciled to Thy holy dispensations!

Lo, every one that thirsteth shall be refreshed;
    To every one that hungers shall be meat.
For God's riches are not limited, but infinite,
    And all that love Him shall have full return,
For the mercy of God endureth forever, yea, forevermore.

Lo, here we are, O God, offering to praise Thee with our lips;
    Let us also join in with our hearts!
And let us never cease to praise Thee here or hereafter;
    For out of Thee proceeded always good and loving mercy
And unto Thee, O God, shall be praise and honor evermore.

Lo, let all that is within man testify of Thee
    And let all know Thee, the one True Guide.
And let none go astray who desire to reach unto Thee
    But save us all with everlasting salvation,
For Thy mercy, O God, is unbounded and endureth forever.

Then let every creature here and in the holy heavens know Thee;
    Let every life and soul magnify Thy holy name!
Yea, let all that is within and all the outward praise Thee with glory,
    For Thy mercy and love and bountiful favor
Are ever present and will bless all men forevermore.

Amen.

CHAPTER VIII

OF LIFE IN GENERAL

The Creation of Bodies for the Reception of Life

When the world of matter was made, the Word placed it where it was directed to be placed by the Will of God. The place was not one place, but all was as one because it was in unison or harmony wherever it was. But wherever it was it was without form and void; that is, void of form and of life or action. It was passive, and gaseous in a most attenuated degree. By the Word was imparted to it another quality, that of action by condensation. The cause or process was the addition of a law respecting it by which it was impelled to act upon itself so as to combine in various ways its various elements, or original or ultimate particles of different essences. There are not many of these ultimate essences, but their combinations with others, being first united in pairs of various proportions, are numerous. These pairs of first union are held in connection by so strong an affinity that science will never be able to separate them. Nothing less than the wisdom of God can do it. High spirits have this power but do not exercise it except in His will. The first principles or ultimate essences of matter are four, of which three belong to earthy matter, while spirit-matter has, as I have already stated, another called od.

Od is the distinguishing feature of spirit-matter. It is intangible to men except by its effects. Sensitive persons can feel its effects and witness some of them also, but in general mankind is unable to perceive its primary effect and can ascertain or know its existence only from the evidence of others, or from the secondary effects which are manifested from its primary actions upon earthy matter. *Magnetism* is a quality or essence of earthy matter which does not exist in pure spirit-matter. Spirits are not influenced at all by magnetism, though they know what it is because they can perceive it by the powers they possess and can witness all the phenomena which it produces by its various combinations. *Caloric* is another ultimate essence or base. It belongs exclusively to earthy matter and does not affect spirits. If it affected spirits they would feel its loss as men do by sensations of cold or the excess or increase above its average quantity by the feeling of warmth or heat. Heat is not caloric any more than cold is; it is only an effect of caloric; so is cold an effect of caloric because it is such a condition of surrounding matter as abstracts caloric from the human system. Caloric pervades all bodies as fully as magnetism or od, but it also exists uncombined, as do magnetism and od. Od, indeed, does not combine with magnetism or caloric or with the third unknown and unnamed substance of earthy matter. Od pervades all earthy matter but does not combine with any of it. The third substance I will name *body of man* or, as sufficiently explicit, body. By this I do not mean flesh and blood or the form of earthy matter that conceals or confines the spirit or the soul of man. I mean that fourth ultimate essence which is the foundation, as it were, the great base of all the solids or visibilities of earthy matter and which, though never seen uncombined as magnetism or caloric are, is yet so combined with these other two substances that it may be ultimately appreciated by men as a substance. It is not capable of separation even into its original first or ultimate combinations, which are, as far as the powers of men are yet awakened, unappreciable. These first or ultimate combinations are many, and all were originally gaseous. The whole of this substance or essence was at once combined thus inseparably with magnetism and caloric in various proportions, yet definite and by order, not by chance. The affinity for these other essenses is so great that the combinations are irresolvable by any power less than that which caused the affinity and combination.

Having now given you a brief outline of the manner in which earthy matter is formed, I will state in the same brief manner that spirit-matter contains three substances or essences, of which od is the substance that pervades all earthy matter and is also in every combination of spirit-matter as one of its constituents. Od is indeed the grand base of spirit-matter, as body is of earthy. The other two substances or essences of spirit-matter are in the same relation to od that magnetism and caloric are to body. But the different nature of these substances from anything of earthy nature, which alone is cognizable by men in the body, will prevent a further disclosure at present respecting them and their combinations with od, except that the past furnishes a proof that man may advance in capacity to receive by receiving. The ignorant man must receive little by little the most familiar kinds of knowledge. Those who have much also receive in small quantities compared to the whole they have received, but inasmuch as they have already received a hundred or a thousand-fold more than the ignorant man, they can add to what they have a hundred or a thousand times as much at once. To whom much is given much is required, and to him that hath shall be given, is true in both physical and spiritual things as well as in mental.

What shall we do to make you acquainted with the spirit substance or matter at this time? We will leave you to digest what you have been told. If you cannot receive this you cannot have more because more could not add to your knowledge unless you have faith in what is already given. Such is the law of progress here and hereafter. So is one law sufficient for all mental or intellectual progress and for all divine progress. This one law is divine in its origin, though promulgated through the Word. It is the law which gives progress to soul and spiritual intellect (which is a different thing from soul) and to mental or earthly intellect (which is a different kind from the spiritual). But in the earthly intellect magnetism is the active agent, the essence which is its base and which never leaves it inactive. In the spiritual intellect it is a higher or more refined substance or essence which has the same office to fill. So there is a correspondence between the earthy and the spiritual.

In the soul, or attached to the divine essence or substance, is also an intellect or perception which is indeed a kind of mind, but it is the divine mind which pervades all divinity. It is wisdom of God which, by virtue of the unity of the sons of God with His divine nature, becomes a part of the son, or a possession which is shared by him. So the divine mind becomes the fountain of wisdom for all His procedures or adams. Again, adam or adamic force is the substance or essence which, as I have stated, pervades all earthy and all spiritual matter. It is a substance possessed by God even as He possesses all things created and uncreated, but the adam is a procedure from Him of a part of His divinity combined with a created substance which is properly the distinguishing feature of the soul of man, and which in paradise and in the seventh circle of the seventh sphere forms the body of the soul or procedure.

Though this base, as it might be termed, of the soul is created, it is not perishable and needs no renewal. It is that part of the soul, or of man, which distinguishes it from Deity, and it is this essence or created substance that God can annihilate. If He did annihilate it, it would leave unexistent memory of the past and, what is inseparable from that memory, consciousness of individual existence. *Individuality* is its feature and quality, which I will use hereafter as its name. Individuality, then, as well as *divinity,* pervades all matter and it is by this pervasion of individuality existing in all creation or all combination of created things that sons of God in the highest circle act upon and control all created matter. But divinity also pervades them and all created matter, and thus divinity is that quality or essence of the Deity which is one and unchangeable, which is the means of His control of all His creation. God is All in all. He is Creator of all and the Pervader of all. It is by pervading all things that He controls all, and it is by controlling all that He sustains all and maintains all in harmony and order. Such is Creation.

But where is the Word in this scheme! It is not assigned a place and yet by it all things are made that have been made. It is the Word which pervades all things created or proceeding from God, for the Word is God. Yet though the Word is God, the Word is not the whole of God, as I have explained in Volume One, Part I. God is not the Word in any other sense than that He is everything. The Word is God though, because it is uncombined with anything else and is a procedure from God without combination, as the soul of man is a procedure combined with individuality. Then does not the Word possess individuality? Not as a part of itself. The Word has memory of the past by pervading all it has created, and when any part of its creation is annihilated, the Word no longer possesses a memory of its having been.

The Word is that part of God, then, which pervades every part of creation, every created thing. The Word of God is God, but not the whole of God. The Word is divine; it is divinity. The soul of man is divine in part; all that is immortal is divine. The soul is divinity and individuality combined in that intimate, irreducible combination which I have stated to belong to all primary combinations. Nothing but the will of God can separate the two essences which form the soul. But besides the combination of the two essences forming the soul, there is a pervasion of the soul by divinity or the Word of God in an uncombined state. You are requested to bear in mind continually the exact meaning of words now, for otherwise you will not obtain an understanding of the high and holy revelations I am making. Combination and pervasion are different, as you know or may ascertain. Divinity is combined with individuality, forming the soul of man or the real man. Divinity also pervades the soul in an uncombined state and it is by this uncombined divinity that the Word, which is divinity, acts upon man's soul. And although God in His fullness also pervades all things, it is chiefly through the Word that He acts upon men. He may act directly because His power is unlimited. Should He act directly it would be by means of His pervasion of the soul— and of all other things—by His fullness.

God, then, is All in all. God is all, but all is not God. So God is the Word but the Word is not God. This now explains the darkness that you thought was in My first description of the Word in the first part of Volume One. Then the light shone into the darkness but the darkness comprehended it not. Now the light is comprehended by those who have desired to receive it, and read attentively with that desire to understand, or receive.

There is then, first: *God eternal, infinite, unchangeable, and incomprehensible,* except to Himself. He is One. He is All. There is nothing but God, except as He wills it to be. There is nothing that He has not and does not continue to will to be. If He willed it not to be it would cease to exist, and if He willed it to cease utterly, even all memory of it would go out of existence. He is the One True God whom the antediluvians worshipped. He is the God of Adam, of Noah, of Abraham, and of Christ. He is the God in whom you and I live, move, and have our being. By Him we exist and without Him we become more than dead; we become as if we had not been. For though the divinity in us is immortal and therefore imperishable, yet when its *individuality* is separated and non-existent, the divinity is purely God and is *as united to Him as if it had never been separated.* It is thus that man is immortal. But he has also a kind of immortality in the unchangeableness of God, in consequence of which he will never be annihilated but will ever be maintained as a glorious manifestation of the power, wisdom, and love of Deity.

In the beginning was the Word, and the Word was with God, and the Word was God. There was a time when the Word, now separated from God by being a procedure from Him, was not only in God but with God and was God. But now the Word is with God only as co-existent and as being pervaded by Him. The Word is quick and powerful, sharper than a two-edged sword to the dividing asunder the joints and marrow, the body and spirit and soul of man. But the Word is acting ever in the will of God and

270

never had free-will or a will of its own. It has therefore always done the will of God and will always continue to do so.

The Word, then, is the highest procedure from God and the most like God, being indeed a part of God without having any combination with any created substance. The Word is of one substance or essence uncombined, which substance is divinity, which in turn is a part of God. Is God, then, composed of combined essences or is He all divinity? He is *One* and He is *incomprehensible* to man and to spirits. God is One, yet besides being divinity He is love, power, wisdom. These are His attributes besides others. *The Word is divinity,* without these attributes. What then is divinity? Divinity is deity, but deity is not merely divinity. Divinity is power, but power is not merely divinity. Divinity is love, but love is not merely divinity. Divinity is wisdom, but yet wisdom is not merely divinity.

Divinity is that part of God which by His will became the Word. It is the Word, and is that quality or essence by which the world and all else was made, though not unassisted by God, the One Universal One—*the one whole, uncombined, unseparate, inseparable Being* who is and was and ever will be *infinite in everything* and that is thousands of qualities or faculties which, in the first sphere, never have been and never can be imagined or conceived of. God is *One Infinite.* The Word is *one finite.* The Word is God, but *God is more than the Word.*

From this brief recapitulation of what these beings are and are not, you may get in your mind some idea of them, but, though the Word is finite, your nature is now so obscured by the fetters of earthy matter and also the prison garb of spirit-matter, that you cannot conceive of the high nature and attributes of the finite Word. Let us therefore leave that, and briefly recapitulate the relations of the different parts of creation to each other, to the Word, and to God.

The Deity caused the procedure of the Word from Himself in that very remote and incomprehensible time called the beginning. The Word existed in the will of God a long time inactive. But when God resolved to have a creation of matter He willed to have it, and the Word proceeded to execute God's will. It created matter, as I have described in the first book of this series, creating by God's power from nothing certain substances or elements or essences with which it should act ever after as long as God is pleased to have it exist. These substances or primary essences I have described to you in this chapter.

The essence that, combined with the Word or divinity, forms the soul is individuality. It is a combination of the two essences having by the law of the creation of individuality an extraordinary affinity to combine in certain relative proportions, definite and when combined irresolvable except by the action of the Word of God. When combined, there can be no change of the combination by an addition of the one or the other part without a resolution of the combination. But the combination may be pervaded by either

271

substance in a greater or less degree, except that individuality has no uncombined existence because of its affinity for divinity, and therefore soul or adam is pervaded only by God and the Word, or divinity, except that other separate or determinate combinations or souls or adams may also pervade the soul of man. Now it is by this pervasion that I act upon the soul of this holy medium. It is also by pervasion that the Word of God acts upon the soul of any man. It is by pervasion that God acts upon the soul of any man, or spirit, or upon the Word. The Word pervades the soul always. I pervade it only under special conditions. The action of the Word is above My action and the action of God is above the action of the Word.

Why, then, should you listen to Me when you have the Word ever present in you? will be one objector's cavil; and why should you listen to Me or the Word when you have God ever within you? will be another's. You should listen to Me because I would not ask you to unless it were God's will that I should speak to you thus, and because He chooses to have His revelation made in this way. He never has revealed Himself directly to His creatures and is not resolved to do so now. He has never acted unnecessarily, and He ever acts by the best means He possesses, and He always has such means as He chooses to have. Let us all endeavor to be willing to hear Him through His chosen servant, whoever or whatever it may be. If we cannot succeed in this endeavor, it is because we do not try, and we shall in consequence be left without the blessed knowledge of His revealed will. Be, then, desirous to know God and to know His will and to do it. So shall His will be done on earth as in heaven. So shall you be reconciled to God, and until you are reconciled you can not be at peace with Him. When you are at peace with Him you will enjoy that peace which the world can not give nor take away. My prayer is that you may strive earnestly and seek perseveringly to obtain this heavenly blessing. Amen.

Next lower in the order of creation, as to equality or dignity, is spirit-matter. Individuality might be called matter, as it is created and destructible, but inasmuch as it is unchangeable in its relation or combination with divinity, it may well be called by another name and regarded as immortal or eternal. If called matter, then it should be called immortal or eternal matter, while spirit-matter should be called incorruptible though changeable matter; and earthy matter corruptible, or perishable matter. Not that earthy matter is truly perishable by decay, but that it changes its form by decay and by change of combination effected by action of itself upon itself. Spirit-matter changes only by action upon it of divinity or adam or soul.

Spirit-matter is a combination of od and two other essences, or primary substances, which exist in various relations and combinations and is pervaded by *adam*, by *divinity*, and by *God*. It is pervaded by adam specially, and by divinity and deity generally or always. Spirit-bodies are formed by action of the Word upon od, causing od to combine with the other substances forming spirit-matter in definite but variable proportions; as od pervades all earthy matter, whenever a spirit is formed in a body, it is by the Word's having established the law under which the od in the earthy body combined with the two undescribed substances and formed a spiritual body in the form of the outward or earthy type. Not precisely in the form of the outward body, but in such form as that body should have to be perfect.* Thus the spiritual body has none of the imperfections or deformities of the earthy body, though it sometimes chooses in its separate existence to manifest. its identity by assuming or manifesting an assumption of them. So are fears of men acted upon by apparitions, and their apprehensions excited that they may retain in the world or state to come the defects of bodily organization.

Let us return to the organization of spirit-matter. Od is the base, but though it is always this which at least is present, it has not that affinity for other spirit-substances which causes it to seek them and combine with them unless impelled by adam or soul or some kind of life. Spirit is no more life or alive than earthy matter, and it requires the impulse of life to combine and sustain it. Without life's being in a sustaining relation to it, it dissolves into its constituent materials which, however, do not thereby perish though they cease to act upon each other. The two unnamed substances are extremely subtile in their nature and, having no relations, connections, or combinations with earthy matter, they cannot be discovered by earthy perceptions. They are to od what electricity or magnetism is to body, or the gross or visible combinations and forms of it. When set free, they mingle with the spiritual atmosphere, which has existence in the spirit-world, in a manner analogous to the atmosphere of the earth; and as oxygen and nitrogen form the atmosphere by their combination, so these two substances form the spirit-atmosphere by their combination in a certain definite proportion.

One of these substances combines also with aura, and it is by that affinity and power or susceptibility of being combined that spiritual bodies can pass from planet to planet, from system to system, and from coelum to coelum, and so on through all the illimitable creation. Aura is only another name for Word, which, as I have said, pervades all things and exists throughout the whole infinite creation and even beyond the creation as far as God extends, which is infinite and beyond anything that finite beings can understand or imagine or suppose, in the most remote approach toward conception of it. Aura, being the Word, when combined with the spiritual substance, can act with and upon the spiritual substance so as to be its controlling medium in a double manner. First, by pervading od and all substances and forms; second, by combining with what we will call magnetic-od, that is, one of the substances that combines with od and the one that is capable of certain kinds or forms of combination with Word, or aura.

*In modern terminology this is often referred to as the "etheric". Ed.

273

Word or aura thus possesses a double relation to spirit-matter: the one general, the other particular; the one derived from pervasion, the other from combination. The one it has of itself; the other it has by the will of spirit-intelligence. For the combination with aura is always made, like all other spirit combinations, under the influence of life and this life can be derived only from God and it exists only in His procedures or adams or souls. The other substance might be called magnetic-od too, but that it is grosser, wanting in the power of combining with aura, and therefore we will name it, for men, electric-od. For electricity is to magnetism as magnetism it to od. The electric-od is to magnetic-od as magnetic-od is to aura. Electricity may combine with magnetism but not with od, but magnetism may combine with od under the influence of spiritual-life or mind or intelligence, which in fact is sustained and established in being by adam, or primarily by divinity, which is Word or aura. For these three are one, being different names for one substance manifested in different ways, yet having always one nature.

Spirit-matter, then, exists generally uncombined with od or itself, except that magnetic-od and electric-od are usually combined into a spiritual atmosphere when not in combination with od. Od is the base of all spiritual bodies, or what may be compared to earthy fluids or solids or visibilities, while the other two are like earthy gases or atmosphere compared to solids or visibilities. Od, pervading all earthy matter, is ever ready to be acted upon by life, and when the soul or adam enters the newly-born child, the action of its entrance causes the immediate affinity of od for magnetic and electric-od to take place, and the moment this affinity exists the aura, also pervading all substances and all space, transmits to the affinity or desire of od the desired or required quantity of the other two substances, which are as readily transmitted as electricity by an insulated wire and far more rapidly. The body is then formed immediately in the perfection to which it is entitled and is retained so long as the laws of its sustaining life require. There is no possibility of its disruption or dissolution because the law of its existence depends upon the affinity caused by the life, which, in the case of a spiritual body, is an involuntary action of the life and beyond the power of the life to pretermit or interrupt. The *spiritual body,* then, is formed of a combination of *od, magnetic-od,* and *electric-od, united* by an affinity produced by the contact of individuality, or its combination with divinity called adam, and pervaded by *adam* or *soul,* by *aura* or *Word,* and by *Deity* or *God.*

So, too, *earthy matter* is composed of combined *body, caloric,* and *magnetism,* pervaded by *od,* by *aura* or *Word,* and by *Deity,* and in special cases like human bodies by *life* (or the combination of aura, or divinity, and individuality). But how is the life of animals distinguished from that of the human form? This question I will answer in My next division.

The Qualities of Matter, in Relation to Life

Animals are well known to be of various degrees of development. The chain of existence has been perceived by man's reason to extend from him to the lowest discoverable attribute of life in regular gradation. From this many have supposed that life was spontaneously produced from inanimate matter or that all matter was endued with life and that consequently life was only a manifestation of a property common to all matter and always existing, never ceasing, only changing from one object to another. The belief in metempsychosis of the soul is only a consequent or analogy drawn from the assumption that life belongs to matter and therefore does not perish as long as matter remains for it to combine with.

Animals differ in their life from each other, and they all differ from vegetables, while vegetable life can scarcely be said to include all that animal does not, for there is a kind of mineral life which really exists, difficult to discover and prove by reason, yet suspected and believed in by some. It is mineral life. The popular, or once popular, opinion that rocks grow is not altogether without foundation, though it is very different in detail from what was supposed.

The difference between animal and human life is that the latter has added to it a soul or adam, and I have explained that the soul causes the existence of the spiritual body. Animals have only earthy bodies, yet animals have life and the life of the animal body is the same as that of the human body. It is only a limited intelligence— unreasoning it can not be called, yet reasoning loosely and incorrectly on most subjects and guided far more by instinct than by reason. Instinct, too, is a species of intelligence which animals and men both possess. The instinct of some animals is superior to that of man, yet man has a large development of instinct. It is instinct that prompts many actions and opinions under which men reason and act, besides those of infancy with which all are familiar. But instinct is a peculiarity of animal life. There is an intelligence or definite course of action in vegetables which resembles instinct as much as instinct resembles reason, but it is no more instinct than instinct is reason, or reason the true intelligence or mind of man.

Life in its lowest form is a quality of matter dependent upon electric action. It is inherent to a certain arrangement of particles, or ultimate atoms and combinations, of the magnetic and electric elements with their base or body. It is produced, like all other things, by the Word. The Word produces it by certain laws which ever exist and continue to act to produce it or develop the circumstances by which or in which it is manifested. The form of life which men have most investigated is the vegetable, by which matter assumes before their observation such extraordinary changes as extension, erection, and variation in function. First the seed is formed; the tree appears to result from it; the flower and the fruit follow in due time. The seed is cast to earth, the tree itself falls and decays—all the investigations of science but declare these well-known facts, though expressed sometimes with so many details of the process that the ignorant are induced to believe them more knowing than they really are. Whether life is a principle or an effect of a principle is yet unknown. Whether life is an end or a means is also unknown. But I will reveal to My readers what life is.

Life, as I said before, is a quality of matter dependent upon electric action in its lowest form; in its higher form, dependent upon magnetic action; in a still higher form, dependent upon odic force; and in its highest form, a procedure from God Himself through the Word. The first manifestation of life in earthy matter is that of compression, or gravitation, or condensation, for the inherent laws or properties of earthy matter cause it to make continual progress toward solidity, and when solid to become more dense. Matter first exists as gas of a very attenuated nature. Its condensation goes on by changes and various combinations caused by its laws primarily, by its action on itself immediately, until it reaches the state in which we see it or know it, of solid, fluid, and gas. It is also evident to men that the addition of caloric changes many solids to liquids and a further addition makes the fluid gaseous. So, too, it has been ascertained that gases can be reduced to solids and that when this result is accomplished caloric is disengaged or set free, as it is usually termed. Caloric is indeed the chief agent in producing these changes at the will of intelligence, but laws of matter impel it to act continually without the interference of any intelligence and by those laws continual condensation goes on. By these laws matter will at last be brought into a solid and very dense mass.* But the Word can set other laws in operation that will cause it to expand and continually recede from solidity. Matter will never need to be re-created, because the Word can cause it to assume any shape it may desire and it can as well employ this present matter as any other which it might produce. For matter does not lessen with all its changes; neither does it receive injury by any combination or dissolution.

How it is that life is the result of electric action I can hardly make known to you for you are unwilling to believe, and what we do unwillingly we generally do with difficulty. A law of the Word relative to matter is that its action must be constant, unintermittent. By that law, motion is produced and motion is regarded—and justly too—as a sign of life. Electricity is the agent of motion and of change. As it continually passes from one body or combination of earthy matter to another it acts on the substance it passes through or upon. It loses some of itself in one place, it gathers in another. But in such change of combination an average of condensation is increased. Thus motion is the element of progress and the immediate cause of life.

This seed of a vegetable we have instanced, being placed in the earthy matter where earthy combinations of solids, fluids, and gases have free access to it, receives by electric action an impulse

*Cf. "dark stars", discovered after this book was written.—Ed.

to combine with the earthy substances which surround it; and, having had laws of the Word impressed upon it governing the course of its agglomeration of matter, it proceeds or grows in a certain determinate form manifesting life indeed, but life under great restraint or confined to one position and a similarity of form to its parent stock. The germ of the vegetable or seed contains only a certain combination of earthy particles or arrangement of ultimate atoms of body with certain combinations of caloric and magnetism which are peculiar to a particular vegetable form or manifestation, and that combination exposed in this way to electric action collects and agglomerates in a certain definite manner other combinations of earthy matter by which the plant is produced, by which each seed produces or expands by agglomeration into a vegetable whole of its own kind. Having passed all the stages of its being, if not interrupted by some deficiency of electric action or foreign influence, it dies or decays and returns to a more solid form than it before possessed, leaving behind it in general parts of itself as seeds which contain again that certain form or manner of combination peculiar to its species.

The principal difference between vegetable and animal life is the power of locomotion. The same action of electricity upon the same seed or small part of itself, containing, however, its peculiar combination of matter, exists in animals. This seed must in a similar manner be placed where the action of various earthy substances in all their different solid, fluid, and gaseous forms can have access to it, and by the incessant electric action of the germ or seed they are joined to or combined with it, causing growth and development until at last the locomotive animal appears from the germ or seed, or ova or egg, which contained its constituent, determinate, and peculiar combination. This combination evinces instinct if it belongs to it, for instinct is unreasoning and is only a law of affinity expanded to suit the circumstances of its action or relation to the mass upon which it is required to operate.

We have, however, higher animals which appear to possess powers of memory and reason, or actions founded on previous actions without being consequent upon them. These reasoning powers are the result of magnetic action combined with electric, and acting under the laws impressed upon the combination peculiar to the species or individual. The law comes from the Word, and the Word, pervading all things, is ever present to sustain the action of its law if such necessity should exist. But the laws of the Word in relation to matter are self-acting and self-sustaining, as they were ordained to be.

Reason then, as we call the actions of animals founded on past experience of themselves or others, is a quality of matter, of earthy matter! Yes, but that is a low animal form of it. The reasoning powers of man are dependent on further combination with spirit-matter or od. It is this combination which enables him to carry his mind, with all its attendant attributes, to the spirit-world where of all that aided his animal reason only od can enter. The soul or adam has intelligence before it is united with the body, but intelligence is thwarted in its effort to act through the bodily senses, or perceptive faculties, and exists in a latent or slumbering state while within the prison-house of the body. Its first action on entering the body was to cause the production of the spiritual body, but that was accomplished by affinity, under a law requiring or admitting of no exercise of reason.

The next perception we can have of its action or existence is the aspiration for something better than the bodily or earthly life affords, a desire for a higher state of existence and an instinctive—or rather intuitive—assurance of immortality. This is all the soul generally gives to the intellect in the body but sometimes it is made a channel of communication with God, or the Word, or another soul, or a high spirit of God by which truth or revelation is conveyed to the mind of the earthy body conjoined to the spiritual intelligence. The soul receives and transmits to the odic part of the mind by the magnetic-od of the spiritual body. The odic force, or od, then transmits the communication to the magnetic substance of the earthy body, or mind, and it directs the action of the obedient muscles to execute its will. This is the process of writing this book, as I before explained in the first part of the series.

The spiritual intelligence, though, acts with the earthy intelligence and, though located in different portions of the brain in greater or less predominance, the two intermingle with each other and act in concert, by which the earthy and the spiritual bodies are connected and made one man or being instead of two separate beings, as they are separate existences. The nature of man is dual: first, by being parital in paradise and in the seventh circle of the seventh sphere; second, by being double-bodied on the earth; and third, by having a spiritual body and mind acting in harmony with the adam or soul in the spirit-world until it arrives at the last circle of it. This is an explanation of that doubleness which so influences men in this life and in the future. A striving to do two incompatible things—to follow two contradictory objects or courses of action, and this manifestation of man's duality is the remote cause of the doctrine of a principle of evil wholly or partially independent of God, of the doctrine of human depravity and the fallen nature of the soul of man.

This it is which has been so much in the way of man's progress that it has now pleased God to make new manifestation of His power and love in this period of time. It is this doctrine I am most anxious to eradicate, and to elevate your ideas of God and of your own immortal nature so that you shall be brought to a willingness to submit to God, the Giver of all Good. Then let me **entreat every man to investigate calmly, fully, impartially, remembering on the one hand that he must discard every desire but that of arriving at truth, and on the other that God will require a strict account of the manner in which he reached his conclusion and that his reward will be found according to his works. Let no man say: Thou art an austere master and I fear to investigate,** for if I believe I shall have more to do, more responsibility. Remember that you already possess the responsibility, and that to wrap your talent in a napkin is by no means satisfactory to Him who gave you the talent. He asks of you not that which He gave, but its proper results, the improvement of every opportunity to enlarge the sphere of your usefulness and responsibility. Remember, then, that you are called and invited to the marriage feast and that the master of the feast will have His tables filled with those who have on wedding garments; that you must, if you do not procure the garment, be cast into darkness where there is weeping and wailing and gnashing of teeth, or in other words, deep and humiliating disappointment.

How, then, shall you obtain satisfaction? How shall you be held excused if you do not believe after you investigate? By your fervent prayer to God to help you to know the truth, by your sincere desire to receive the truth, by your ardent pursuit of the inquiry after truth, by reading faithfully, carefully and repeatedly the works I have delivered through this holy medium. If you have followed My directions you will be brought to a knowledge of that peace which cometh down from heaven. More blessed are they who give than they who receive, and most blessed are they who believe without having seen. Give to others all the light you can, and lead them to all the truth you can give them. Fear not to sacrifice your friends, your opinions, your political prospects, your business qualifications, even your family ties, if such to human reason appears to be the inevitable consequence of your avowal of and public support of the truth. Be assured that not a hair of your head shall be touched without God's knowledge and that none of these consequences shall fall upon you, nor the most trifling approach to them, without His notice, and be assured that He is an abundant rewarder and a sure paymaster. He will abundantly compensate to you every loss, though they should equal Job's losses, and the first payment you will receive will be that peace which none but God can give and of which no one but yourself can deprive you. Of the nature and extent of this peace I have before written, but I have not, and cannot describe its fullness. Man in the body seldom enjoys much of it because he is too outward in his views. He seeks and finds other enjoyments, while if he were wise he would seek only this and so seeking would surely find it.

CHAPTER IX

THE RELATIONS OF MAN TO GOD

God's Requirements of Man

There is a sure and perfect salvation provided for all mankind. It is not by works but by mercy. It is not of themselves attainable but by the assistance of the Deity who has prepared it for them. But the Deity does not act Himself when He has agents or sons, high spirits desiring to act for Him, benefitted by acting for Him, and under them an innumerable company of aspiring spirits all desirous of doing the will of God and of serving God in any way most pleasing to Him. These lower but aspiring spirits, having various degrees of advancement and station, all advance themselves by attending to duties suitable to their capacity. Having, then, this inducement to serve God, which is to serve their fellow-beings, it is reasonable to suppose that they will be desirous of doing so, and that they will enjoy not only the action or work which is pleasurable but the reward which is so glorious and so sure.

You no doubt think that when you are one of this joyful company you will be one of the willing and obedient ones and that nothing will divert your attention from your duties or cause you to suspend for a moment of time those pursuits which are thus the cause and the means of your progress and the reward of their own performance. But, O reader, could you know as I do that even in the spirit-world the sacrifice of your free-will is required and that it is made generally with reluctance and always by degrees; could you know how powerful is the influence of bodily experience or habit upon the spiritual intelligence which in that world to come enchains the soul and prevents it from soaring at once to the feet of God; could you know that you will find so many clogs, so many difficulties thwarting your desires of progress—you would, after all, doubt your future happiness.

Yet you will be happy at last, and comparatively happy at first, in the spirit-world. For compared with the spirit-world the earth is dark and miserable. Yet here on this earth, here in this state of existence, dark and miserable as it is, you might enjoy the same progress toward heavenly bliss, you might practice the same means which in like manner would cause you the same progress in loving God and approaching His throne and His unity. You might begin now to do those works of good to your fellow-creatures, to sacrifice that free-will and to soar upward toward the feet of God in obedience to the aspirations of your soul if you only would resolve to do so. Not only might you do it here, but you could accomplish more here in a year than there in a hundred, perhaps; or there might be even a far greater, almost infinitely greater, difference. What you think is so easy there is far easier here, and all that man needs in order to be placed in advancement on the road to happiness here or hereafter is a firm resolution to do the will of God. He can do God's will only by sacrificing his own, and when he sacrifices his free-will the victory is won. He has then fought the good fight and henceforth he can say that there is laid up for him a crown of eternal glory and that crown must ultimately be his no matter how he may behave here.

Here, though, is the place to obtain the victory easily. Life in the body is short, but exceedingly great is the reward of faithfulness in it, and could you appreciate the joy of heaven or the inextinguishable rapture of the redeemed son of man who sacrifices on earth his free-will, you would not hesitate though that sacrifice should entail destruction of every earthly hope and prospect, destruction to every earthly possession of wealth power, fame, or even family and friends or social position. Great is the reward and trifling is the requirement, in general. There are times when the sacrifice seems too great to make, but then the faithful servant of God can cast his burden upon his Master, for God is ever ready to hear the cry of His servants and minister to their every want. Far be it from Me to urge you to seek God for the sake of receiving a reward in this present state, yet I do say that no man sacrifices houses or lands, wife or children, that doth not receive a hundred-fold in this present world, and in the world to come, life everlasting. Far be it from Me to ask you to serve God for the applauses of men, but yet who seeth the righteous man trampled under the feet of this world, or his seed degraded in the land?

Oh, Americans! you know not the blessings you enjoy, the ease in which you live, the advantages you possess. Long ages did My servants bleed and die after suffering prolonged afflictions for religious liberty. Long did the cause appear to them hopeless and the heavens as brass. But now no power threatens you, no persecution endeavors to restrain you from following the dictates of the spirit of God. Instead of persecution unto death you should rather fear homage for virtue and inflation of your pride by the well-meaning admiration of your acquaintances. It is excess of consideration that endangers your path now, rather than the opposing influences of a hostile community. I do not say that the avowal of belief in the present manifestation of God's love produces these temporal or earthly rewards, but that good works in general and the sacrifice of your own will to God has done and will do it still in this favored land. Still, the opposition to spirit believers, though somewhat acrimonious, is in general tempered by the reflection that effort to restrain or punish their credulity, as it is generally termed, is unavailing and opposed to the spirit and letter of the supreme law of the land.

In fact, then, you have great advantages and God has chosen the time well that He should now manifest His love and call you to return to Him that free-will which He conferred upon you as His highest, His crowning gift. Then, O Americans, walk humbly before God. Resolve not to oppose anything that you are unacquainted with because you think you know something else. Be careful that you be not found resisting God's work, for be assured that this work will not stand unless it cometh from God, and you all know that you can not successfully fight against God. But, you say, Mormonism is a delusion and it has made great progress. You will say Mohammed was an impostor and yet his religion stands to this day! But remember, O man, that you have opposed both these unavailingly while God shall say: Thus far shalt thou come and no farther, but there shalt thy proud progress be forever stayed.

What is Mormonism? Is it, after all, when preached in its purity, a pure religion and a calling of men to the kingdom of Jesus Christ that shall be established as the fifth monarchy? True,

Mormonism has been perverted by those who should have been its protectors, as they were its leaders, but so was Christianity, as all admit though some call one thing perversion and some another. But God will bring good out of evil, and though Mormonism has borne bad fruit and shall be cut down, it has nevertheless prepared the minds of many, not only of its members but of others, to receive the pure spiritual teachings. The evil is transient, the good is permanent. The mercy of God will pardon the ignorant and deluded, while His justice will punish every man according to his works, his intentions, and his desires. For all these, if evil, man must make atonement; for all these, if good, he shall be rewarded by advancement in that spiritual progress we all have to undertake and accomplish.

So, too, must we view the followers of Mohammed, a rank impostor but nevertheless an instrument of good. He receives his reward by having to make atonement for his sins, by having to suffer remorse for the success of his teachings and the obstinacy with which his followers resist the truth. Read history; learn the state of the church that called itself Christian in those days. See how the world and its delights prevailed over the love of God and the peace from heaven. See how its bishops contended and even fought for power and rank. See how the people became idolators and worshipped images. See how they divided the unity of God and dared to make Me His equal. Was this a religion to be defended by miracles and retained in power by My strength, or was it not well that the effeminate Greeks, slaves of lust and cringing to despots, should lose—what? Not their liberty, they had none; not their religion, for no conqueror can take that from any man, and the followers of Mohammed were uncommonly lenient to an opposing faith for that age of the world; not their wealth, for they were left no more burdened with taxes than before—it was their pride that was abased. They were robbed of their overweening pride that made them esteem themselves superior to all other nations and as the particular favorites of heaven. Who shall say, after making himself acquainted with what history has recorded of them and their conquerors, that they were better in any respect than Arabians? Who shall say even that the departure from truth was greater in Mohammedanism than in Greek religious worship? Even the statues of the profligate despots, in compliance with the old pagan custom, were almost declared sacred by the high dignitaries or authorities of the church, while the common people actually did consider them as worthy of worship. And truly they were as worthy as the images of bishops or other distinguished men called saints which were publicly established and treated with the same respect that the pagan Greeks long before bestowed upon their deified heroes. Who shall say the religion of the Greeks is now less pure than it was, or still less may it be said that it is less pure than it would have been had its corruption by unprincipled aspirants for power continued?

Was not the Roman Church equally corrupt? No, out of that church came a cloud of witnesses to purity and truth. To be sure, they met with persecution, but the blood of the martyrs was not lost; it was good seed and produced other crops of martyrs. And if they did suffer in this life, who shall say that, laying down their lives willingly and joyfully as many of them did, they were not happier than if they had exchanged places with the robber noble or the pilfering priest? Who shall say the slayer was happier here or hereafter than the slain? The tears of humanity and the power of God are both testimonies that they did not die in vain. No man sees to the end. God does. Out of evil He brings forth good, out of discord harmony, out of wickedness and disorder righteousness and peace. To Him be evermore praise, honor, and glory, world without end.

Advice and Warning to Mankind

Let us once more return to the case of Mohammed the impostor. He preached one God. He led many from the worship of numerous idols to a belief in the one true and living God. Though he joined with that profession of faith a declaration that he was the prophet of God, by which he made them believe and assert a lie, yet so long as they were deluded and sincere in endeavoring to do good they were pardoned the false belief. Though he was not excused, they were. His followers were enthusiastic—a proof of their sincerity. They were frugal and abstemious — a proof of their virtue. They were above the fear of death—a proof that they were willing to sacrifice the present to the future. All this is very well, you say, yet still they slaughtered, burned and destroyed! In all these they but did as surrounding nations did also, and were neither better nor worse for their belief in Mohammed as a prophet. They were better for believing in one true and living God and so were their enemies, for the belief in His judgment often restrained them from worse acts than they did commit and, barbarous as were the Saracens, the Turks, and the Persians after their conversion to Mohammedanism, they were no worse but rather better than the Persians when Magians, or the unconverted Tartar that soon after overwhelmed Saracens, Turks, and Greeks in one common calamity.

Look now at Mohammedanism and see where is its strength; look at its rulers or governments and see their feebleness; look at the people and see their poverty; and look at their teachers and see their ignorance. Do they now offer any obstacle to the spread of truth? Do they appear resolved to demolish all new revelations, or are they not looking for something to occur that shall change them and their faith? Not that they suppose the change will be for the better, but that it will be inevitable. It was to be or it would not have been, they will say, and they will receive My prophets hereafter with faith, and having once submitted will be strong in faith.

Great and marvelous are Thy works,
Lord God Almighty!

Just and true are all Thy ways,
Thou King of Saints!

Did not My servant John prophesy that the delusion should come and that the deluded armies of scorpions should overthrow and scatter and destroy, and that the third part of the earth should be destroyed by them! And if God foresaw it, could He not have guarded against it if He had desired to, and was He not bound to do so by His own nature if their coming and conquests were not for a good purpose? My friend, the question is not what you would have done, but what did God do? He may have arranged matters very differently from what you would have considered wise, but I doubt whether you are prepared to say He made a mistake.

Where is the day of joy and where is the time of trouble such as was never seen before? The time when men should call on the mountains to cover them and the hills to hide them from the face of the most high God? The mountains and hills are the sects in existence. In these men will try to be hidden from God's revelation of His will, which discloses His attributes under a new form or aspect. This is the time of trouble such as the world never saw, when many shall run to and fro and knowledge shall be increased. But the signs are not so evident as you supposed they would be; you expect there would be signs in heaven and mighty deliverances on earth. So there are now. Heaven sends forth its armies and the powers of the world are mustering theirs. Romanism is making a desperate effort; it has emptied Ireland of so many of its adherents that it totters there to its fall. They now begin to fear that in grasping for America they have lost Ireland. They now desire to stay the flood, the mighty river of population which so unprecedentedly rolls from that fertile island, that might be the garden of the world, to the comparative wilderness country of America. But the flood will not be stayed; the people will escape from the hands of the priests, for here they can not prevail.

Another flood now begins to rush upon this favored land, bearing with it wealth and strength; rivalry of the two floods shall be the safety of My kingdom. The fusion of the two with the Anglo-Saxon shall again restore the purity of the type or people. The combination of Irish and German will make Americans most inevitably. Power will be proportional to population where freedom prevails, and so long as freedom of discussion prevails, truth fears no overthrow. True it is that Orthodoxy, as its professors delight to call it, fears Catholicism, as the Roman believers like to call their profession. But though Orthodoxy is weak, I am strong because I work in God's will; they work in their own. My revelations shall be established and men shall know Me to be the King of Kings, the long-expected, long hoped-for, long prayed-for Prince of Peace. My kingdom shall be established and all shall join in giving to God the glory and honor of the victory He will have obtained over the powers of evil. Let no man take thy crown, reader. Thy crown is thy free-will, the crowning gift of all thou hast received from God. Let no man take it—let no church or association of men take it, for I claim it. But I claim it only as King of Kings and Lord of Lords. I ask you to be My subject and I promise you, in return for your obedience, advancement in that glorious path of progress toward the mansion of everlasting bliss. I promise you eternal life. And all these promises I am authorized to make by God Himself and to Him will I lead you if you submit to Me. But, will you perhaps say, you would rather submit to God and not submit to Me? I do not object, but no man knoweth the Father but by the Son, and he that hath seen Me hath seen the Father. I am the way, the truth, and the life, and if any man come unto Me I will be his guide, his helper, and I will lead him to fountains of bliss and mercy, to fountains of the living water of God's love which shall be manifested as His peace that shall be with the man evermore so that he shall thirst no more.

Come unto Me, all ye heavy laden, and I will give you rest. Yes, rest from your labors. The grievous burdens of the popular theology are too heavy to bear. The priests themselves dare not preach the creeds of their own sects. They are much farther from believing them than from preaching them. From preaching them they are restrained by the fear of men's opposition; from believing them by fear of God's displeasure. How can this be, you ask? Is not the last a sufficient fear for both? And if God's displeasure is feared, so as to prevent their belief, would it not also prevent their preaching what they do not believe?

No, reader. The heart of man is corrupt, and scarcely knows its own motives when it seeks to know them. But when it seeks to conceal them from others it generally begins by concealing them from itself. So the priest begins by refraining from preaching doctrines too dark and cruel for an enlightened and benevolent audience who have free inquiry in political and civil affairs and have learned to dare a little free inquiry into religious teaching. The priest fears to shock his people's benevolent ideas of God by too much dwelling upon his dark points of doctrine, such as the vast proportion who must be doomed to hell by the sect's creed; the enormity of the offense of those who do not believe what the founder or founders of his sect believed and, as a consequence, the well-merited punishment which they should and must receive, and the exclusiveness of their position by which they have secured the favor of God and can pass from this to the next state unfearing and with confidence. But alas, the priest's own fears betray themselves, for he is too apt to portray the horrors of that passage from time to eternity. He loves rather to avoid exciting the apprehensions of his hearers, by dwelling upon the goodness of God and the gratitude we owe to Him and the pleasures that God has laid up for His children. Having thus denied in effect his sect's creed for fear of offending men, he soon begins to see that such a dark doctrine must rob God of His beauty and holiness, and he finds that he cannot believe even so much as he must continue to preach, because he gets beyond his congregation in enlightenment. Yet his convictions are not strong enough to prevent him from trying to please God and his congregation: to please God by acting as if all men were to be his companions in a future state, and to please the congregation by complacently assuring them that they are the especially favored people of God. Having thus reconciled to his own satisfaction God and man, he acts in the fear of one and cultivates the good opinion of the other.

But, you may say, it is necessary that he should dissemble somewhat to his congregation, that he should imitate Paul in being all things to all men in order that he may save at least some. But I would have him also imitate Paul in that he would not overlook the short-comings of a single hearer of preaching, but would urge all to serve God by doing the works of repentance, and the works of repentance are restitution to God of the heart undefiled by any desire of doing aught but God's will. I would have him, like Paul, seek no counsel but of God. I would have him ask not man or men what he shall declare to the congregation, but seek God's guidance and follow it wherever it may lead. It may be that some will refuse to hear unwelcome truth, but the consequences must be left to God. Do His work, leave to Him the future, and if each day has its duty, each will have also its reward. Imitate the independence of Paul as well as his willingness to assume the character most agreeable to his hearers. Depend not on men for support, for bread, but rely on God and in trusting to Him feel as much confidence that all will be well here if you do your duty and obey His will as you feel or profess to feel that He will reward those who serve Him in another place or state of existence. God is not a hard master and I have told you before that the sufferings of the martyrs were apparent rather than real, inasmuch as the body can be sustained in endurance by the mind, and though health is certainly a great blessing, peace of mind is a blessing greater beyond comparison. Reader, do not apply all that you find in this book to others when I warn or threaten or expose. But take that part also to yourself and try yourself by an impartial judgment, remembering that with the judgment or measure that you establish or declare for others you yourself shall be judged. The promises are for you, too, but remember the conditions of the promise and believe that you need reformation as well as your neighbor, and that you are as sinful as many a one you regard as being as far below you in spiritual advancement as was the publican in the estimation of the Pharisee.

CHAPTER X

REVELATION TO MANKIND

There is a great work to be accomplished in this present time. All men who have desires to love and serve God are called on to help. Not because God will be benefitted, but because He will be pleased with the evidence thereby given that the man desires to serve God, and because the work is serving his fellow-men. It is nothing less than subjecting all to God. It is nothing less than bringing men to regard and obey Christ as their great Head, spiritually, temporally, politically, and socially. Christ is King of Kings and He is now ready to take the government upon His shoulders. He already has invited you to submit to His rule. He continues to invite you. I appeal to you to be My servants, for this is the day of My power. Let each man subject himself first and then enlist in the army of heaven in the legion I will lead against the powers of darkness and ignorance. On My sword shall be written the Word of God, and legions of spirits or saints from heaven shall follow Me. Walk humbly; be willing to serve as a private. No man shall be promoted to command My followers because of his desire, or his social rank, or his knowledge after the manner or education of man. I will raise up such as I will to have raised, and they shall be such as are faithful witnesses to My coming into their hearts and who give Me the glory of causing their success over the corruption they had rioted in. I will bring forth in them the works of repentance and faith and they shall be willing to lay down all earthly possessions and employments to follow Me. Yet it will not follow because they are willing to do it that I shall call on them to do so, for though I love a cheerful giver, I return many-fold all I accept. I want the will sacrificed, the willingness once accorded to Me and continued to be accorded to Me, and I accept the will for the act. I come not to destroy but to fulfill all righteousness and peace, and duties performed by My servants will be the plain requirements of common sense and such as reason, if unbiased, would direct a man to perform for his temporal good.

I desire to have all men seek to know God and Me, His Son the Lord Jesus Christ, by prayer and by receiving in answer a revelation of truth. I desire to have you seek for this revelation each for himself and each in his own heart. For if God has power to speak to man, He will exercise it under such circumstances. Many of you believe or profess to believe that revelation ceased long since, but you cannot find any authority for such belief in the Bible. You judged rather from the fact that it did not come with outward signs as formerly. But the power of God is no less, and however long the rain may be withheld it comes at last and is received by the parched ground joyfully, and a return at once is made by the springing up of plants in places barren before. Thus it should be with men spiritually. There has been a drought of revelation. For a long time it has been restricted, not by the will of God but by the perverseness of men. What men did not desire was not given.

Too many yet say, on one side, the Bible is sufficient for us; and, on the other, amplify it by a commentary. On the one hand they say inspiration has ceased; on the other, they pray to have it for their assistance in writing or speaking their sermon or conducting their affairs. They ask God's help, well knowing that He has not outwardly or visibly laid hands upon matter or men and helped in that way, well knowing that if God ever helps it must be by working upon man's internals or by influencing their own or other men's opinions or thoughts. How is it, then, that inspiration has ceased and yet you pray for it? How is it that the Holy Ghost is regarded by you as a part of God and as God, and yet that you would deprive the Spirit of God of all the power or use of the power to address you in the heart, in the mind? How is it that you would say that what is declared by the Holy Spirit is unworthy of belief, and that you must not alter any opinion derived from tradition or experience that the Spirit of God within you shall declare to be wrong, or that it would lead you gently away from? Are you not unwilling to be guided and governed by the inward light that God causes to be manifested in your heart even though it should come in answer to your own prayer for enlightenment?

O man, how hard it is to surrender your will! How easy it is for you to resist the holy influence that God in His benevolence and mercy casts about you! He asks you to surrender your free-will as an acceptable sacrifice to Him, and you reply that it is not reasonable that you should do so! That if He would have you obey Him, He must ask of you only what your reason approves! Then He can only ask you to follow your own inclinations until you find that reason cannot save you, and that it is not a sure guide, and that you get no better in your heart because you submit to reason and follow your own will. But, you say, we have the Bible for our guide and if what we receive in our hearts is anything different from what we find there it must be error, for the Bible is truth; then, as the Bible is truth and all else than it must be error, we may well dispense with looking within ourselves for a guide inasmuch as we ought not to follow it if it does not lead us as the Bible would, and we need it not if it only undertakes to lead as the Bible does!

Having thus stated your best argument in its strongest presentation, let us look at your inconsistency. You ask God to help you understand that book whose teachings are your infallible guide. You ask Him to help you preach correctly in elucidation or enforcement of its doctrines or declarations. If, now, the Bible is your sure guide, would it need reason or revelation to insure its reception by all, or its being understood by those who read it? Does not its incompleteness become apparent when you find it necessary to add so greatly to it in order that it may be understood by the laity? If incomplete, would not inspiration or revelation complete or perfect it better than reason? And if incomplete, would not prayer to God to help you understand and to enlighten your dark minds be heard? And how can He answer you but by inspiration or revelation? If He answers by the latter would it be too much for Him to have some willing man record His revelation and publish it to the world? By no means.

Will you not, then, admit the possibility of revelation at the present time? But, you say, there can be no necessity for it or God would have given it before? God has given it before and the knowledge was lost, as I have told you. God caused Moses to publish all he could collect. But much of that was lost in the captivities and idolatrous departures of the Israelites. Again by prophets God made revelations so much as or more than the people could bear; that is proved by the fact that the prophets were generally slain instead of being honored by the ruling powers. Then came the revelations I gave when in the body, only a part of which were even recorded in memories of men, and only a part of what was remembered was written down in after years, and of what was written down but a small portion has reached the present time;—yet you would persuade yourselves and your brethren that that little is enough for you. In one sense it is enough, for you do not practice the precepts recorded as Mine, and being unwilling to receive as your guide what you have, you do not deserve more. But God gives to the undeserving. If He did not, none would receive mercy.

So at this time in mercy and love God reveals to you more through this outward book, and He is also ready to reveal to each of you inwardly and specially to suit your own case and circumstances. I ask you to listen to this. It was this I asked the people to listen to when I was in the body, and it was because they were so outward as to be unable and unwilling to turn to their own internals that I declared it was expedient that I should leave My followers. After My departure they sought and found inspiration. But soon it was lost by slavish dependence on other men. Since then, the world has not been in a state to receive another written revelation with benefit. I have told you that disobedience to God's known will is sin, an unpardonable sin; this being so, God has not added to the condemnation of mankind by revealing more of His will. Whenever men did turn within with sincere desires for light they received it, sometimes knowingly and sometimes unconsciously. But because many received inspiration without appreciating it or even being aware of it, though perhaps they acted on it, you must not suppose God could not cause them to distinguish inspiration from reason or imagination. And revelation is a more perfect and distinct declaration than inspiration or impression.

Therefore I assure you that there is no difficulty in My holy medium's distinguishing what I give him from the results of his memory or reason or imagination—not only from the results of his will upon these faculties, but also from what may be called the involuntary action of his mind. How he can do it is not easily explained to those who have not experienced something analogous to it; those who have done so ought to believe that others can distinguish as easily as they what comes from a foreign source and what originates in themselves; and these ought further to believe that God has power to make some, or all, even more sure than they have ever been, by more unmistakable differences or distinguishing features, that they receive from a divine source.

I have already given you in previous books much light upon the manner in which I deliver this revelation to this holy medium and how he can and does distinguish it from his own mental action. But I propose to declare now that he is not liable to make any important error because I can call him back to it should he make one. It is a part of his education and training by Me for this purpose that he has made immaterial discrepancies and corrected some of them in the manuscript copies. But these manuscript copies are the best evidence of how few they are and of the cause being a desire on his part, perhaps unconsciously or thoughtlessly, for communications or revelations on some particular point. These copies in manuscript will be preserved as evidence, to those that seek such, of the real character of the delivery of these books.

But I will also explain why I did not in some other way deliver this revelation, as I promised in the first part of this volume. I might have it written by a medium whose hand was controlled like Hammond's when he wrote *Light from the Spirit-World* and *The Pilgrimage of Paine*. But all such mediums are physical, and have been and may be used by lower spirits. Not that I could not separate such from the influence of lower spirits, but that they in general are unwilling to be separated from lower spirits, as they are desirous of yielding control to those they have known personally or by reputation and without regard to their fitness for revealing entire or perfect truth. I could perhaps have trained them to a different course. But as I have told you, no man is forced to be a medium contrary to his will though many are selected who have not formed a desire for it. Many mediums have declared they did not want to be such! These had times when they desired manifestations and when they were willing to serve God and yield their own inclination or will to His. At such times the spirits of lower spheres operated upon them and advanced them in the work appointed for them. These mediums, though, did not possess all those qualities of mind and disposition that this one did and I deemed him the best subject that offered a willing mind for the reception of truth; and though, as I have said, a better is wanted and may arise, I shall use him as the best at My command at present. This is not because he is the best or most perfect physically, mentally, or morally but chiefly because he is the most passive and submissive to Me while he is to men the most impassive and uncontrollable. He is thus impassive to conviction by argument or persuasion of men. Some men knowing a thing can be reasoned out of it; others can be persuaded out of it; some yield to one influence, some to another. There are men who are psychologically influenced by other men. This holy medium has never been and cannot be magnetized even slightly by other men. He is therefore impassive and unsubmissive to them. These qualities do not make him a better man morally or save him from errors of judgment, mistakes in argument, or deficiencies of conduct. With these qualities he may be unpleasing in deportment and unskilled in art. But they are valuable to him for this work and, having caused him to be in a social position neither too high for usefulness as an humble instrument nor too low for respect or credit among men, neither so rich as to be independent of the world's frowns nor so poor as to be unable to command all the means of serving Me in the publication of what I deliver to him for that purpose, I chose and called him, and though a considerable period of preparation passed away before he was ready for the work, the same qualities that were most resistant to My influence and will in the beginning are now the safeguards of his convictions and the maintainers of his position as My holy medium. What next I shall have for him to do I know not but he will undoubtedly do it.

Do you see that last sentence? He wrote it to the end though he perceived it was contradictory to what he believes to be the knowledge possessed by Me. This shows his passiveness and submission. But have I told a lie to make an example of that passiveness? By no means, for I do not know what I shall have him do next. He is free-willed though voluntarily subjecting himself to God. So are other men. What other men may need and what he may resolve to do I cannot tell. I can, as I have stated, form a judgment founded on My knowledge of all the physical and social influences that have affected and that may affect them; and, knowing the future perfectly and the present in every ramification, I can tell with small probability of error or mistake. But man is a free-agent and though free-agents are accountable their acts can not be foreseen. They can be controlled by divine power; but when controlled by that power or any other they cease to be free-agents and cease to be responsible for the acts they then perform. I can control circumstances so as to cause effects and influence actions of men, as I have said, and these auxiliaries are not only at My service but they are used, and I am therefore most unlikely to err or mistake in My conclusion respecting even the course of a free-willed being like man. For all that, though, I do not know what I will next call on My holy medium to do and I am not obliged to form any plan because whatever I do will be in God's will, and His will is always the result of, or in accordance with, His perfect wisdom.

There are yet to be answered the two questions: why I do not deliver this book by spirits of lower circles writing with outward materials, as they have written, or by writing with them Myself and delivering the manuscript complete and perfect into the hands of some man or body of men.

The spirits in lower circles whose province it has been to use outward materials do not wish to act so and are not so submissive as to desire to serve God entirely. They are less controlled by circumstances than men, and are freer to act in their own wills. The spirits of the second sphere act in their own will almost entirely, though they are restrained by laws unperceived by them from doing what will harm others or make them more evil. God leaves them to have every wish gratified until they learn that their own desires when fully gratified are enough only to satiate and tire them without giving them happiness. This knowledge leads them to submit more and more to God as I have described and to receive more and more from His bounty and His mercy until they progress to the highest circle of the seventh sphere. Then they act

291

entirely and perfectly in God's will and then they may write such a book as this in such a way and have it perfect. Until then it would be filled with errors arising from the operation of their own will and imperfect opinions.

But why do I not write Myself in the usual manner and leave the finished manuscript with this or some other man or body of men? is the final and strongest question you can put as an objection to this mode of communication. I will answer you conclusively, I think, to your reason as well as to your faith. The answer to faith is simply that it is better thus.

To reason a longer course of argument is necessary. First, then, should I do so there would be the same objections to belief in it that now exist, for you would of course see no miraculous appearances in what was prepared after the manner of men. Second, you would find what you would think were errors in it then and search as successfully as now for discrepancies or contradictions. Lastly even My holy medium would be suspicious of imposition. For he could not have the same positive knowledge he now has that it is not the work of any visibility. His belief in its authenticity would then, if it existed, be founded in faith produced by some other manifestation than the mere finding of the prepared manuscript. At best it would pass with many only as an imitation of the finding of the Mormon Bible or Golden Legend.

But could I not assume a bodily form and deliver it to some body of men; for instance, to some church assembly or congress of church or sectarian delegates? Would they receive it, think you, or would they not, at least a majority of them, treat it as a work of magic or as a delusion of the arch-enemy they outwardly believe to exist—and thank God that their faith was strong enough to withstand even so imposing an effort to draw them from the faith of their fathers? They would most assuredly deliver My work to the flames rather than to the printer, and I should be left undeclared to the world as a savior. For it is in My character of Savior of Men that I desire to be viewed. I come to call men to repentance, to reconciliation with God; I preach now what I preached eighteen hundred years ago, that I and my Father are One and that I do His will and have His power, and that by His power all things are possible but in His wisdom all things are not expedient. Now as then I declare to you that the outward must cease or disappear so that you may more certainly be led to the inward, and though then I suffered crucifixion when it was perfectly in the power of God to have saved Me and permitted Me to step down from the torturing cross I was doomed to die upon; (I say, though, if it had been expedient it would have been His will that I should have done this, and I should have acted in His will) yet I did not do it because it was not expedient.

I might have pursued other methods to convert men to a belief in My mission and divine authority then, even as I might take other methods now, if it were God's will; but, O man! if you have any faith in God and in the inspiration or unity of His Son, believe that He and I are having everything provided for you in the truly best way—in the way that will soonest bring you all to reconciliation with Him and to a proper preparation for that eternal, unending, supernatural bliss which He has prepared for you and for all men who were ever or ever will be born either on this or any other of the innumerable myriads of systems of planets existing now or heretofore or hereafter in any part of the illimitable creation which is spread over infinite space. God is good and His mercy endureth forever!

Blessed be the name of the Lord God of Hosts! All His sons do bless Him and glorify His name as it is, and was, and ever shall be, world without end.

The Soul as Separate from Inclosures

Let us return to the consideration of the past, present, and future of man. In paradise I have told you he has a body formed by the Word, composed of a combination of individuality and aura, or Word, or divinity. This combination I have also informed you is pervaded by Word or divinity, which is uncombined with it and which is, as it were, the body or substance of the Word. It is also pervaded by the Deity, by God Himself, whose body, so to speak, pervades not only man and spirits or souls but also the Word and every part, parcel, and atom of all procedures and created matter or combinations of matter. God thus pervades all without being combined with any, for His procedures are uncombined with Him though they are parts of Him combined (except the Word) with created substance or essence. The Word, as I have said, is only an emanation or procedure from God of a part of Himself endowed with a separate existence and finite in its powers and perceptions. Finite in reality and as compared with real infinity, but infinite as compared with other procedures or any conception that we can form of it. Thus in one sense, and to us, the Word is infinite, though in ultimate fact there is but one Infinite, who is God.

We have traced the formation of man so far as his constituent parts are concerned. We have yet to see how much we can declare of his formation into body, or shape, in paradise. The soul or adam being composed of divinity and individuality, and the quantity of the combination existing in an individual being determinate and unchangeable, it is reasonable to suppose it has an unchangeable form. But this is not a consequence of such formation and condition, for the soul exists in an indeterminate form in paradise. It is capable of expansion and contraction. It can make itself more or less dense. It can even stretch itself by attenuation to the whole boundary of paradise, or it may be all contained in the newborn infant of the smallest size that ever lived. It is by this expansion that it traverses paradise and it is by this contraction that it enters the earth-body prepared for it. The expansion and contraction are both under the control of its will and are performed with such rapidity that it may be called instantaneous, though it is not instantaneous.

Let us endeavor to understand the precise limits of paradise as they existed at its establishment, and the same as they exist at the present time in this system. The boundaries of paradise were, in the beginning of its existence, about the surface of the only body existent in what is now a solar system. A solar system was then a vast globe of attenuated gaseous matter (which has since solidified, as I described in Volume One, and by the process of solidification become changed into primary and secondary satellites revolving about the central body, called the sun or star). Having this globe of earth-matter on one side, the thickness or outer boundary of paradise was fixed by the wants of the inhabitants of it, allowing them ample range and scope for every imaginable enjoyment. This was perhaps double the size of the central earth, or sun globe. But, no, you can not receive an idea of its size correctly; suffice it that the sun-globe extended as far as its farthest influence now extends, and that paradise extended over a space as much greater as to be illimitable by any measurement of man though not to reach or join the paradise that surrounded an independent sun, or earth-globe. Where double suns or stars existed their paradise was one; their boundaries of extent conjoined and do now conjoin, or have contact, by which they roll, as it were, upon the surface of their true extent around each other. The inhabitants of the bodies enjoy, or disregard as the case may be, an apparently confused order of movement of the stars and planets visible to them, which, nevertheless, when understood is found to reveal the true relation the suns and planets bear to each other and the real harmony and beauty of their location and arrangement. Paradise, then, exists surrounding each sun-system or double or treble or quadruple sun-system and extending therefrom a vast distance into space. Paradise though, having location and extent and limits, must have form and substance. The form I have given as a sphere inclosing a sphere; the substance I will also describe.

The substance is the surplus spirit-matter uncombined with itself or with aura. It is mixed only with itself. Soul or adam has the power of combining it in a certain degree and manner, as I have stated that od (under the influence of life which is possessed by aura and individuality combined and called soul or adam) is capable of, or rather possesses an affinity with, or for, the other combinations or substances of spirit-matter which I have called magnetic-od and electric-od. These two higher or more refined elements of spirit-matter combine in various proportions without od but not permanently. They too are combined by the influence of aura or of soul but they do not retain the organization or combination except while the will of the life maintains it. The moment the will ceases to act the combination dissolves.

In paradise, then, the soul or adam has power by its will to form any combination into any shape or semblance it may choose, and by this power it possesses whatever it wills to have. For this od-matter is as real in its existence, though so attenuated in its substance, as earth or body-matter. Having a form of its own such as it chooses to have, large or small, round or square, tall or short—in fact, just such a shape as it for the time pleases—the soul has by its command over od-matter all its desires gratified. It has only the one restriction I have before stated upon its power of will, and that is that it must not leave its residence. It must not expand itself beyond nor must it in a contracted shape leave paradise. The moment it resolves to do this it begins to partake of the forbidden tree, or fruit, called knowledge of good and evil. But if the soul does so resolve to enjoy that which is forbidden it does enjoy it and as the first step leaves paradise and enters a body prepared by the Word for it, formed by the Word from earth-matter and pervaded still by the Word so that the Word can communicate with the soul and reveal to it such intelligence as is called instinctive or intuitive, as is its true term or significance. This intuition is derived from the Word who wills thus to add to the knowledge the soul possessed in paradise.

But what did the soul know in paradise; what kind of intelligence had it? It had a mind consequent upon its possession of a portion of aura combined with its individuality. The combination possessed memory and the combination possessed what its aura had before its combination. This part of the soul had a knowledge of creation, of the acts of the Word previous to its separation from the Word, and of the thoughts of God previous to the Word's separation from Him. It was thus endowed with high intelligence, and was allowed to use its high powers in adding to its store of materials of mental action and reflection by its almost unlimited power of creation in paradise. This memory or knowledge was all lost by its entrance into body of earth. But the soul did not know that it would lose this by its departure from paradise. This loss of its memory and consciousness of previous existence was the death which was the declared penalty of transgression of the bounds of paradise.

Unconsciousness of the past is death. In this sense the soul never dies after leaving the body. The body dies because it, like animals, loses all consciousness of existence. It is thus that all flesh is grass. It is thus that there is one flesh of men and animals and that they all die alike, as is declared in the saying of Solomon the son of David—a text that has puzzled commentators and many sincere inquirers but, viewed thus, is simple and plain. You will find, O unbelievers! that I am not come to destroy but to fulfill; not to pull down but to build up; and to declare to you the Bible that you ignorantly worship.

There is in man, then, a hidden intelligence or mind, dead for the present state of existence but receiving from the past a life which is again visible in the future. This is the true man, the real soul, the mind or intelligence with which the soul was endowed at its creation, at the time the aura or Word-procedure combined with the created individuality. This intelligence or mind is hidden more or less from its entrance into body to its arrival in the seventh sphere. There its full consciousness and memory again exist, and it has also added to these a knowledge or memory of all that second intelligence, that of the spirit-body, possessed. All that the earthy body possessed, acquired, or experienced had in like manner accompanied the spirit-intelligence when the earth-body and its animal mind, or intelligence, perished, or separated from the spirit by what is usually called death. Death changes the man from earthy to spiritual, while birth changes his form from divine to earthy. The dissolution of the spiritual body changes man again to the divine nature or perfect soul. He is then united to God, and one with Him, though not one with God in all His attributes until the soul reaches the seventh circle of the seventh sphere. I have called these existences spirits heretofore, but they are not properly spirits but Sons. The soul, then, becomes a Son and all the spirits in the seventh sphere are Sons while those in the seventh circle of that sphere may be called full Sons, or Perfect Sons.

CHAPTER XI

PAST AND PRESENT MANIFESTATIONS OF GOD'S ACTION BY HIS SPIRITS

Present Manifestations of God's Action upon Men in the Body

Shall a man live then after death? was asked of old. Now the faith that he does is established. Not that the teachings of eighteen hundred years ago sufficed to establish it, but that the continued effect of what was then preached has overcome skepticism. There has been, however, a renewal of skepticism within the last hundred years, caused by the decline of religious faith in God's mercy. The church taught no doctrines satisfactory to the soul; the soul, therefore, longed for something better than the church could offer. Leaving the church, it sought to arrive at truth not by revelation but by reason. Reason is a blind guide and they who followed it fell into the ditch of atheism, if they continued under its guidance, or at least were led to skepticism, a miry slough in which they lost faith in immortality and looked for God's revealments of power rather than of mercy and love; for His indifference rather than His care; and having thus made themselves, as nearly as they could, independent lords or gods, they were easily induced to trample upon the rights of humanity, the love of fellow-men, and the happiness of love of God. Thus was the evil spread; thus did it destroy life, for without love of men and of God the soul of man is dead; that is, separated from God. For true death is not dissolution of the body, but the separation of the soul from its fountain which was immediately the Word and ultimately God.

This skepticism extended itself insensibly into the minds of many who were unconscious of having invited it, but in whom misgivings and doubts arose in their deep reflections and wasted their faith. Belief in apparitions fell before this skepticism; belief in God's love and care for men was shaken till it tottered to its fall. But signs given through holy mediums in various ways have checked the progress of skepticism and will forever overcome the doubts and disbelief of sincere inquirers. Much has been accomplished by these trifling manifestations, dubbed by the leaders of public opinion in and out of the sacerdotal order as contradictory and unworthy of attention. Though these manifestations, being the efforts of spirits not above the sixth circle of the second sphere, have been contradictory and the descriptions they gave of the future state unreal (because descriptions of unreal things must be unreal) yet they present to mankind the experience of individual spirits in the next state of existence.

The next phase of manifestation is that of inspiration of the spirit of man—his mental intelligence as distinguished from his bodily or animal instinct or intelligence. This inspiration extends to other than religious subjects when the individual has proved· by submission his willingness to be governed by it. It is thus that many preachers of the Word of God are qualified to preach, and whether orally or with notes is no matter, provided they have inspiration. Such inspiration is not easily distinguished from reason, and many have received much of it without being aware of it even when they deny the possibility of its occurrence, for they have earnest desires to be governed by God and yet they are unwilling to admit that He can govern them by inspiration. God accepts their willingness and pardons the sin, or omission to look for Him in the way He does manifest Himself, as originating in ignorance and as being a pardonable one. Still, it is evident that the man thus placing his principal dependence on reason will often be led astray by it and thus be the means of leading others away from the true path. God sometimes overrules even these errors so that they are productive of good to others. God is able to overrule even man's free-will, as I have explained, but I have also declared that He will not do it.

So the man is left in his twilight state, sometimes brightening, sometimes darkening, until presently he passes into the spirit-state, there to continue to waver, as it were, upon the verge of light without perceiving his own blindness to it. At last, finding himself left behind by those who have listened to the inspiration as inspiration, he begins to look upon it as possible that he, too, has been favored by holy communion with the Son of God. For the inspiration thus manifested to the spirit, or mind, of the spirit-body of man does not proceed directly from God but is delivered to man by the Son of God who, being sent to him, becomes for him a Christ, since, as I have explained, Christ signifies Sent of God. But if all inspiration be from such a high source, why does not the man become more truthful and reliable; why do not the different receivers of this inspiration agree in their declarations upon spiritual matters and upon our duties to each other and to our God? Because there is in all a want of dependence upon inspiration, and a leaning upon their own wisdom, learning, or experience.

Do not some who believe that inspiration may take place, and who profess to receive it and to endeavor to act in accordance with it, become truthful, and capable and worthy of being servants of God, reliable as exponents of our duties to Him and to men? They do; by thus submitting to God and having no other desire than to do His will, they become His faithful servants and, being His faithful servants, they do declare His will. But not having faith enough to depart from the landmarks or bounds which others have instilled into them or placed around them, they are not qualified to declare the will of God except in parts and for particular occasions. The followers of Fox even bind their preachers with earthly rules, though they profess to have derived those rules from the united voices of inspired men. But this is declaring that the fathers are wiser than the children and that God does not manifest Himself as in days of old. The preacher is required to submit to the counsel of men appointed by the church who are not preachers. What is this but saying that they who preach do not speak by inspiration unless we who do not speak are also inspired as they are. And yet every preacher differs somewhat from every other so far as his reason or earthly mind overrules the spiritual reception of the Son of God.

How then, if all preachers by inspiration preach more or less error, shall we distinguish the truth from it; how shall we or even they themselves know what part of their recommendations is worthy to be followed or what we should refrain from when they tell us to taste not, handle not? Go yourself to God and ask Him to direct you what to believe, what to do, what to leave undone. For He will as readily speak to you by His Christ as to any other man. But even I may just as easily be mistaken as the preacher and I may thus reject as much or more truth which he had really received or delivered as he, and be equally liable to take the errors as truth! This is so, and yet you will be held accountable only for the talents you really possessed. That you did not know the law is an excuse for you; that you desired to do right is received as if you had done right, as I have explained in the first volume. But if I am to be blessed by having my motives taken for acts, why, then, need I act; why should I not confine myself to good resolutions? Because that is impossible. The existence of this motive for resolving and refraining from action is an unpardonable sin because it is a sin against knowledge. As I have before shown, you will have to make attonement for in before you can be reconciled with God.

There is, then, much inspiration existing already in the religious community, and yet they reject not only the works of lower spirits but denounce and oppose this very work that I am delivering through this holy medium, while he, perhaps, says they who denounce are in the outer darkness which prevents them from seeing the spiritual light shining through his publications! This shall be only until the medium of inspiration meets with this book. Either he shall cease to be inspired or he shall confess that this is inspiration, and not only inspiration but revelation, and not only revelation but the Word of God, given to man as a manifestation of the mercy and love of a gracious God who now, as ever, delights to confer good gifts upon His children and to establish His truth in the minds of all earnest seekers for it.

Even yet, you will say, there seems an inconsistency in putting off the day when the inspired medium will declare the truth of the present revelation until he meets with it. Why should he not, rather, declare it by inspiration to be truth when he first hears of it? Some will go further and ask why not cause him to declare that such a book exists and should be sought for? Because this would require an interference with man's free-will which God does not choose to make. He therefore is patient, and waits until the inspired man meets with the book, and then He impresses the holy medium to say it is true and that it ought to be so regarded by all. But then even if the man is unwilling to sacrifice his reason, his prejudice, his tradition, his sect's landmarks, or the control of the sect's rulers over him—if the man is not willing to take counsel of Me only and let the dead bury their dead and let the earthly possess the earth while he, taking no counsel of man, resolves to obey God rather than man—he cannot declare it truth, and by thus putting to shame the revelation he has had and the witness given to him of its truth in his own heart, he sins against God and crucifies His Son afresh. Then shall he be cut off from His communion with God; then shall he cease to have the help of any Son of God or of any spirit of God to preach or pray or do any act or deed either spiritual or outward. Thus shall My work be established and until it be so established the dew of heaven shall cease to fall; the inspiration of mankind, or their revered guides, shall fail to appear; and the drought shall be very sore to many. But at last all shall know the Father, for he who hath seen Me spiritually hath seen the Father spiritually.

So, too, the various kinds of mediums shall no longer receive communications from the spirit-world until they declare the truth respecting this book. They shall cease to be holy mediums until they repent and do the first works, which will be to declare the gospel of glad tidings in their communications. Now many will be prepared to receive this as truth, but, as all mediums will at last unite in declaring it to be so, eventually the uninformed, the unwilling servants, the followers of past traditions and present enthrallments, will be induced to prepare themselves for the reception of knowledge of God. The demonstrations that God favors this revelation will continue for a time with those who proclaim its truth, and then will cease as the miracles which accompanied the proclamation of glad tidings by My servants eighteen hundred years ago ceased when the gospel had been preached to every nation, tongue, and people. Many have supposed this command not yet fulfilled, but they err in supposing that it was to be literally understood to refer to every nation which existed on the earth at that or some subsequent time. It was intended only to direct a general preaching of the gospel so that no nation might be supposed to be excluded from the rights and privileges of equal enjoyment of it, of equal station in the church of God, and of equal claim to the power of prophecy and communion with God.

Having thus stated the causes which will impel men to believe these books to be truth, I will also state what I shall expect of those who believe. First, they must by every possible exertion undertake to become familiar with this book's contents and with those of its predecessor, for none of the later are intended to supercede the earlier. The first part is, as it were, the foundation; the second, the basement; the third, the upper stories; and this volume is, so to speak, the covering and inclosure of the whole. Later I shall furnish it in detail but you need not know how nor when. It will be in My time, and My time will be, as always, God's time. Having made yourself completely familiar with these books, I shall then expect you to be led by their precepts, seeing that their precepts are precisely those that I taught eighteen hundred years ago. Being thus in accordance with previous revelation, you are not required to discard your former guide (if that guide was the Bible) but to interpret it correctly and be governed by what it teaches when truly understood. This I ask of your own will, but then I also ask it as being My will and the will of My Father who is in heaven, for He gives not what is useless or what shall be superseded, but all He gives is fitted for the instruction not only of the generation to whom it is first addressed but for all succeeding times. Yet the first generation will understand less of it than those who succeed; the first will act upon what they receive very differently from following generations, whose acts will also be based on the same foundation; but the revelations afterward made, being in their nature or effect explanatory of the first, lead the possessors of the whole to view the first very differently, to see new light and meaning in it. This meaning, though less outward, is no less evident, and though this inward or spiritual sense was not perceived at first, it always existed and the instruction might, through searching and seeking God's help, have been available to them. But unless progress is made no better understanding can be received, and no progress can be made except by submission to God and to His teachings through the outward declarations of His inspired servants, prophets, or holy mediums as well as attention to the voice, the still small voice, within each man.

It is worthy of remark that God said to Moses: I will write My law on their hearts and put it in their inward parts—not that He has done it or had already done it. It is too often assumed now that this is universally done, whereas it is only a promise that it will be done in a contingency to occur. This contingency depends on each man's acts, and the man who desires to have God's law within him must seek to induce God to put it there. This is easily done if the man wills to have it so, if he desires to receive the law that he may obey it; but if he desires it merely from curiosity, to know what God's law is and with a view to test its justice, judge of its perfection by his reason, or of its truth and power by his experience, or his understanding of the Bible, or any other test, he does not obtain his wish. God does not give him the condemnation he would thereby experience. He leaves him to his own desires but He does not grant them when to do so would leave him in continual sin and rebellion against God's law which would be consequent upon possession of knowledge of it without the desire to obey it. Still every man does have implanted in his spiritual intelligence a sense of right and wrong, called conscience, which is not God's law written on his heart but a power or standard of testing the accordance of his actions or intentions or the supposed actions or intentions of others by the rule or law of eternal justice. Now although education modifies this conscience, or tradition transforms it to a blind guide, after all its heavenly birth is often manifested and it shines all the brighter for the darkness that has just before enveloped it. But there is no repentance which it enforces, there is no deed that it prevents unless there is also a violation of the law of education or tradition; for this conscience is so blinded by these circumstances that it views all things through the medium of their colored glasses. Thus, though judgment is given based on eternal principles of justice and good faith, there is a leaning to support former decisions of the same tribunal, or those which we have been educated to take for our guide.

Having allowed you to see that conscience cannot lead a man to purity of faith and action, let me tell you what it can do. It can lead a man to look within for a decision upon his own conduct and to endeavor to learn how he ought to behave under various or all conceivable circumstances. By this endeavor he secures God's help, and when it is evident that the man desires to be governed by the principles of eternal justice and right, he receives God's laws as far as he is able to receive them, which is just so far as he is willing to obey them. He then has God's law written on his heart and placed in his inward parts. But it is easy to obliterate the record. A patient endurance of evil, a long-suffering of deliberate or unpremeditated transgression by others of these laws, should be willingly found in us. Otherwise we shall, like Moses, dash down to the earth and shiver into thousands of fragments the divine law which God's finger had traced on that tablet which he had prepared and placed in our inward heart, as He is represented to have traced and placed with Moses the tablets of stone upon Mount Sinai. If Moses set such an example of impatience, can we do better than follow it? You can do better than Moses did in all things if you fully obey God and act always in entire submission to Him. For Moses sometimes erred, and you would not be asked to be perfect even as I am perfect if it were not possible for you to be so. But there is other instruction to be derived from this relation of the breaking of the outward tables of God's law (which Moses did actually receive and break in the sight of His followers who had witnessed the miracles which had released them from Pharaoh's power) and that is that the tables will be re-written whenever due submission and atonement is made for the impatience or rebellion. Not only will the lost be reproduced, but the broken will be restored to its original beauty and the mercy of God shall be evident to all.

The last subject I shall touch upon in this chapter is the great and final question: What shall a man do to be saved? I replied to this in Part II of the first volume and I must once more recall to your memory the sermon that Paul preached to his jailer, when he had converted him by showing that he was acting against God in manifesting hostility to His Sent Spirit, the Christ of Jesus of Nazareth. If you will reread what is said there you will better understand this.

What shall I do to be saved? ought to be the constant inquiry of every man who has not received an answer that his heart, his reason, his conscience, his whole being, approves. So long as a misgiving exists he should ask it; so long as any doubt disturbs or any faith is wanting, let the man ask the question. Day and night, morning and evening, it should be dwelt upon and every answer tried by the proper test to know if it be the true one. But from whom shall the answer be expected? Paul is not and Jesus is not; at least, they are not visible or in any way tangible to the bodily perceptions!

The answer must come from God if it is to be conclusive, because no one but God, the Supreme Judge and the Everlasting Father, can tell on what terms you shall be saved. But God does not speak to me either, you say, and how shall my bodily faculties take note of His answer if He should; for if He speaks in some manner new to my experience how shall I know from what source it comes? These questions or objections are all proper. I do not want you to be led astray or to fancy that every loud voice is God's or every pretender is Christ. There is in every man a witness and that witness is at first his conscience, which, though it may often be blind, is yet capable of receiving the light of God's regard whenever the man is willing. The free-will of man must be sacrificed; then the conscience acts unbiased by the man's desire. When the conscience so acts it conveys its assurance to the man with unwonted strength. He appears to recognize in it a voice of God. It does speak with God's voice, for a Christ enters the man who has sacrificed his free-will, and where Christ is there is God in effect as well as in reality, for God is everywhere.

The next manifestation will be the inspeaking Word of God acting upon their hearts, placing within them the law of God. This strengthens their conviction that they hear God's voice or that of His authorized speaker, or agent, or Christ. Christ himself will speak to them when they are ready to hear Him; when they seek for His advice or counsel with a view to obey it—not to judge of it, but unhesitatingly obey it. They will feel that He is indeed a Counselor, a Guide, a Prince of Peace, because His words will be to them mighty and their effect happiness, perfect joy.

302

This should assure them that they are right thus far, and, being now so far reconciled to God and so far advanced toward the sonship as to have His Son in them, they can enjoy the Holy Communion with Christ daily. They can continually call upon Him not only when they break the outward bread by which they live but when they desire to receive that bread which comes down from heaven and sustains in spiritual health the soul of man. It is this spiritual bread that Paul speaks of when he says that the Rock which furnished the water to quench the thirst of the wanderers in the deserts about Sinai was Christ; when he says to the people of his day that their fathers ate of that manna which was Christ. For though these miracles of producing the outward fountain of water and living upon the manna or bread which fell from the atmosphere to the earth's surface really did occur, even as they are truly related to have occurred, yet there was a spiritual significance which was unperceived at the time but which is to be revealed to you for your instruction now. Miracles are intended not only to benefit those who witness their outward performance but to be instructive to following generations who receive the accounts given of them with faith in God and belief in the truth of His servants who recorded and transmitted to them the wonderful account of the extraordinary manifestation of His love and power. Miracles are not only manifestations of love but are also the demonstrations of power, and none are given as manifestations or demonstrations only but as both; as the benevolent acts of a Heavenly Father who loves His children and gives them from His abundance far more than they can reasonably ask, and far more than they can at all appreciate while in this outward habitation.

What, then, must a man do to be saved? will still be asked by some who have not read attentively and thoughtfully and prayerfully what I have written. Again I will answer you, more briefly, and then by a re-perusal perhaps you will begin to comprehend the deep things of God. First, surrender to God your free-will; that is, subject your every desire to a desire to serve Him. Let your mind appeal to your conscience when you feel a desire to do any act, to know if it be one you can ask God to help you do. You can get a correct reply to this from conscience. Not that conscience will answer every man alike to the same question, but for your state of progress it will be a proper answer, and a continuance of such questioning will rapidly advance you in the knowledge you most need—of the fact that there is in man an inspeaking voice which, if listened to, will lead him with more and more certainty to the paths of peace and communion with God's Spirit. Then, second, you will begin to hold communion with God through your internal nature and His spirit dwelling within you; and that communion, if you continue passive, perfectly submissive to it—addressing it, not familiarly as if you were its equal, much less haughtily as if you were its superior, much less dictatorially as if it were your servant, but respectfully and prayerfully as if it were your god,—will be truthful. That communion will be not only truthful but it will be distinguished as the Comforter that will lead you into all peace. May you seek for it; if you seek you shall find as surely as if you knock it shall be opened for you.

God's Action upon Spirits and their Transmission of It to Men

There is in every man a soul, an emanation from God, and a spirit, a consequent of the unity of aura and od, and a mind or intelligence consequent upon the unity or combination of aura and magnetism—for magnetism is the predominant quality or substance in earth-matter, as od is in spirit-matter, though od is a base for all spirit-combinations while magnetism does not exist in all earth-combinations except as it pervades them all. But as od is the controlling element in spirit-matter, so magnetism is of earth-matter; the former is the base of spirit-intelligence while the latter is the base of earth-intelligence. It is earth-intelligence that is possessed by all animals in a greater or less degree, and this earth-mind is no more immortal or continuous in existence than earth-body. When the body dies it also dissolves, ceases forever to act. The spirit-mind then acts upon the spirit-body precisely as the earth-mind had acted on the earth-body. But before the body dies the earth-mind conveys to the spirit-mind all its knowledge, which indeed is simultaneously acquired by both minds from the experience of the perceptions which are under either's control. Thus whatever the outward perceptions, the nerves of sensation, take notice of, is conveyed first to the earth-mind and from there is imparted to the spirit-mind. Then also whatever the spirit-mind receives from its superior perceptive qualities, which are dependent on its connection with aura, is in like manner conveyed to the earth-mind. Yet each mind can distinguish the source of its knowledge and we can, in the body, distinguish only the receptions or workings of the spirit-mind by what is thus imparted to the body or earth-mind.

It is only when the spirit is disenthralled that it perceives itself to be a complete man also, possessed of body and intelligence, of powers of reception and expression. It then perceives that it has been released from a bondage, and it soars upward toward that congenial atmosphere which I have informed you is the conglomeration of all the out-of-use particles or combinations re-separated of spirit-matter which surrounds the outside of the sphere of paradise, forming thus the spirit-sphere, a hollow globe like that of paradise. It is here that spirits dwell and have their mansions of rest. They do not, however, remain inactive because they are at rest; they rest from labor but not from action. All action, however, is enjoyment to them and they do no more than they are gratified by doing. Having reached this sphere they are at liberty to return to earth as often as they choose to, but they are not always allowed to interfere so far with earth-matter as to make men in the body cognizant of their presence of acts. For when they desire to do this it often happens that it would be inexpedient or hurtful. But when the spirit desires anything it is gratified by realizing to itself the consciousness of its possession. It does not by any means follow that the spirit has it because it has desired it and appeared to itself to have received or performed it. It is to the spirit as real as if it were actual. Thus a spirit desiring to appear to a relative and carry on a conversation with that relative is permitted to have its wish realized by supposing itself in the presence of the relative and passing through the same scenes and conversation that would ensue if the relative saw the spirit as the spirit desired to be seen. Yet all the time the relative knows not that the spirit is disposed to manifest and is no more conscious that the spirit is thus realizing its wish than if such wish had never been experienced. So, too, a spirit which had possessed on earth a favorite horse may desire in the continuation of its earthly habits and likings to enjoy a ride upon that horse. Immediately the horse is ready in the spirit-sphere and the spirit, having to its own sensation mounted the horse, they proceed to gallop untiringly over just such roads and through just such scenery as the spirit desires. But the horse, the roads, the scenery even, the whole has no existence outside of the spirit-body. It is a law of the spirit-sphere that everything the spirit desires is possessed by it, but it is only in that way. To the spirit it is as real as is any bodily experience or perception, but the horse, the road, the scene, are only impressed upon the mind, existing as the spirit desired them to exist, while in reality and to the perceptions of other and higher beings they have no substance whatever and no existence except to the perception of the spirit desiring them. But they are visible to these higher beings when looking at the spirit desiring them and appearing to itself to have them. The high beings see them as a man may see in a mirror what is not the reality but the reflection of a reality. Spirits thus enjoy a varied and gratifying experience without in any way disturbing others or the relations of others to matter. When the spirit desires to have an assemblage of other spirits about it for any purpose, such as social enjoyment or leading an army, debating in a parliament or ruling a church or other body of men, it to every perception of itself has them, and for as long a time as it pleases the scene continues and the illusion is perfect. But whenever the spirit is ready to dismiss them and desires to sleep, to view another scene or perform some other act, the former one, in a natural manner so that the illusion is not dispelled, passes away and what is desired by the spirit again appears to occur. Spirits, then, have not real experience, all is unreal. They act but produce no effects upon matter or upon other spirits. They enjoy what they prefer or wish for but they do not disturb others.

How then do spirits manifest themselves by rapping or other demonstrations that we in the body observe? By having progressed from their state of acting merely in their own wills to having a desire to do good. Having formed this desire, they progress and arrive at such a circle as qualifies them for action. Then, assisted by higher spirits, they produce real effects. But is it not declared in the second and third parts of Volume One that low spirits, having most of the earth-connections upon them, more often appear to men than higher ones and that even the first and second circle spirits have thus appeared? This is true, for that depends on another rule or law, which is that spirits of low circles having desires to benefit mankind are permitted sometimes to manifest themselves, and having so performed acts worthy of acceptance they progress to a higher circle by deserving the reward: Well done, good and faithful servant! thou hast done well in this very little matter and thou shalt now have one greater to attend to. But do spirits in these lowest circles then do good works, or are good works commenced in the sixth circle? The sixth circle is the circle of good works and there are spirits generally and actively employed in endeavoring to benefit others. But the lower ones sometimes have unselfish desires as the result of good motives, and good motives are the means of progress both in the body and in the spirit. Those in the lowest circle can become reconciled to God only in a sufficient degree to advance to the next by having an unselfish desire, by having a desire to serve God. It is this desire to be obedient and reconciled to Him which leads to advancement, and it is this desire which enables the earth-matter of the low spirit's desires to be visible or apparently visible to the man in the body.

Thus we will suppose a man has been murdered; he desires to reveal his murderer's name to some friend. He appears, as he supposes, to that friend and the friend, instead of listening patiently and quietly to whatever the spirit has to communicate, begins either to repel the spirit by his fears or to drive away his apparition by questions. The spirit, unable to act in its own will, becomes a medium of another's will and the man, desiring perhaps to take advantage of the opportunity for gaining hidden knowledge of unlawful things, receives answers according to the folly of his questions. Again the man remains quiet, composed, passive, and the spirit imparts its desire or will which is perhaps the declaration of the name of the murderer and the circumstances attending the commission of the crime. But is this a good work and a work impelled by a good motive? It may be and is sometimes. But generally this kind of desire is not allowed to manifest itself, because it is not from a good source but is much more often the result of revengeful feelings, and revenge belongs to the first circle of the second sphere. Still, when the spirit desires to appear to the man because the man desires to have an explanation of this occurrence or because the spirit desires to relieve the mind of the survivor, merely, without desiring to be revenged upon the murderer, then the two desires united have the power of manifestation outwardly or earthily. The man beholds because the spirit wills to appear to him, and the spirit appears to the man because the man desires to see it or hear it, as the case may be. The double or united will of the two induces the Word or aura to confirm their desire where no evil consequence is apprehensible. This is not yet clear to you. I will take a simpler instance and explain by it the origin of the rappings in western New York which were the introduction of the great abundance of manifestations existing at present.

The family who first heard these sounds were fearless of spirits by possessing moral as well as physical courage. The spirit which first communicated was that of a murdered man. He desired to make known his death and its circumstances that his friends might be benefitted by the recovery of property which by his death was in danger of being appropriated by others in whose hands it had been left. But finding them unwilling to receive information by apparition, the spirit of the murdered man became desirous of enlightening them on the subject of spirit-existence, and the family having a desire to possess or acquire knowledge of the future state or of existence after death, the two wills became so united as to act in unison and the sounds desired by the spirit to be made were heard by those who desired the knowledge which those sounds were intended to convey. But the sounds were heard by those who desired also to refute their existence! Yes, but they were still made by union of the two wills of the spirit and one or more of the family. The spirit, pleased with his success, informed them who he was; but, actuated by a purer motive than revenge or hate, he did not after all give his real name or the circumstances of his death correctly, for doing so would have led to dissension and revengeful manifestations—would have retarded rather than advanced the cause of progress which now was superintended by higher spirits and allowed and encouraged as a means of introducing the knowledge of truth to mankind.

Was not this movement, then, directed by You Yourself as being a means of the establishment of Your kingdom on earth? It was, as I stated, a procedure from Me but it was not carried forward by forcing the will of any spirit but rather by allowing the desires of some spirits to be fulfilled. It was thus that this spirit commenced the manifestations in such a place and with such men as would not repel them, and it was thus that spirits more advanced than the first one were permitted to join in the good work of improving the knowledge of mankind in matters connected with the future state of existence. Such as men were prepared to receive they did receive, but the spirits gave only what they possessed and that was a very limited knowledge, a mere transcript of their own experience in the spirit-world except in a few instances where, having been taught higher truths, they were able to declare them, faintly and imperfectly and clouded by their own darkness, to mankind who desired the information and made the effort to obtain it thus. Having so led them to receive, the proceeding advanced to higher and higher manifestations such as rapping, first by the alphabet, then by writing of spirits, then by writing through men, then by speaking through men, then by delivering through men the thoughts of spirits. While they relate their experience and give utterance to their hopes, it must be remembered that they know little compared with higher spirits and that they are as liable to be mistaken in the inferences they draw from what they do know or have been impressed with from higher spirits as are men.

307

Do not spirits of circles somewhat higher act also in promoting the progress of the procedure thus intended to benefit mankind and bring in the glorious reign of Christ on earth? They do, but they act in general by influencing spirits of the sixth circle to act and by restraining and controlling the lower spirits that men are continually calling upon in relation to themselves and their temporal affairs, their memories, their affections, and their hopes. They also suppose that such spirits can declare to them more than higher ones of what has passed in the experience of the low spirits or in their own. But they are wrong in believing that higher spirits cannot know all that lower ones do, though higher spirits, being engaged in higher duties, cannot be supposed willing to waste their greater powers on such performances or duties as lower ones can render as well.

They, however, assume the places of lower spirits often in thus answering the calls for information beyond the power or knowledge of the spirit called on. Some men will think this a deception and will be disposed to blame spirits for thus doing another's work. But let us see how they judge men who have, when actuated by the purest motives, given benefits to others in the name of some relative! The world praises their disinterestedness and their pure benevolence, and looks not so much at the process as the result. The process is not properly called deception, as it is gratifying the subject acted upon and the error, if any, is the error of the subject in thus desiring to receive the good in a particular manner. Then how is it when men have desired communications from higher spirits, such as the apostles and others, and the spirits have desired to impersonate them but have been unable to withstand the tests applied to manifest their truth or falsity! The pretended apostles or spirits of other distinguished men were really the spirits of lower circles who, finding a man or a medium actuated by desires of notoriety, have acted in their own desire to be great in name and famous in deeds and assumed such names as they perceived their medium or questioner desired to have declared to them. In thus acting they deceived, you say, and the deception was not a benevolent one, you think. It was in reality a disturbing one which lessened your faith in the worthiness of the communications in general and caused you to distrust all that you received or heard of others having received. Then it was a benevolent effect at least, though the motive of the spirit making the communication was no better than yours in asking for it. The effect was to cause you to place less confidence in the outward and thus prepare you to hear the inward. It was because you desired a communication in a particular manner instead of leaving it all to a higher power and more perfect wisdom to determine what the time, manner, and form should be. In a word, it was because you were not passive and because you could not be persuaded to be passive. You found the futility of all attempts to use spirits as your servants when the true work that the procedure was intended to perform was the submission of your heart or free-will to God, by which you should become His servant and do whatever work He should, through His high spirits, call you to.

Were not some of these assumptions of names venerated by men assumed when the inquirer and the medium were both willing the truth? They were not. The medium was sometimes in fault but more often the questioner. The questioner generally desired a communication not in accordance simply with truth from the venerated personages but in confirmation of his own views. And in general the test itself had reference to this same desire and, finding that a spirit calling itself an apostle did not say what the questioner understood to be the meaning of his recorded expressions as contained in the Bible, the spirit was declared an impostor, a liar, or even some worse epithet or character was deemed proper to be applied. Then even when a high spirit was revealing truth, the doubt and distrust and contradictions of the receiver or questioner would force that spirit to relinquish the attempt and allow the questioner to suppose that he had detected a deception or attempt to deceive while he had really only rejected revelation.

How, then, can we know what spirit converses with us or communicates with us? By being passive and receiving with faith what is given, and by asking God to give you such knowledge in His own way and time and manner. This is every man's duty and the communications you may obtain then will be reliable, not only reliable, but if you act upon them and follow the course you are directed to pursue you will be blessed by such an increase of knowledge that you will find you need not ask questions or raise objections or entertain doubts. Be passive, and you will necessarily be patient; be watchful, and your city will not be taken by the enemy of God which is your own will; be attentive, and you will be a medium yourself and hear within your own spiritual mind, which will transmit to your own commonly-used intelligence, the communications of God's will made by His sons to you. Those sons act in His will and have therefore His power to act. But those sons do not force your free-will any more than God does, and could not if they desired to; but they can use you if you submit to them, or to God, and give you such spiritual knowledge as will make you enjoy that peace which the world can not give, neither can it take it away.

O reader, how great is this bliss! Pray that you may arrive at it. Pray that God will in His own way help you to understand His will and do it as your own. Pray that God's will may be done in you as it is in His high and holy perfect sons and that you, acting always in His will, may at last be raised to the power, honor, and glory of being regarded as His servant, worthy to be placed on His right hand and thereby found to be acting in God's will and power and authority. Amen.

CHAPTER XII

THE WILL OF GOD RESPECTING FUTURE MANIFESTATIONS

The Manifestations of God in Man to take place

When I have thus secured you as a medium My kingdom will be

established in you, and when I have thus secured every man in this country My kingdom will be eternal and everlasting. It will be the kingdom of the fifth monarchy, whose power shall never end. But I have declared that the foundation of this kingdom is already laid, and that other foundation can no man lay than is laid, and the kingdom is already established! The kingdom exists because some hearts have submitted in part, at least, and it is established because nothing can occur to prevent its successful progress till all shall be its subjects from the least to the greatest.

This kingdom will be an earthly kingdom as well as a spiritual one, an external as well as an internal one. It will be a kingdom similar to the Hebrew kingdom in the time of the Judges except that outward sacrifices will not be required of its subjects and they will probably be less fitful and inconstant in their regard for the voice of God declared through holy mediums. Further than that, this people will learn to know Me in their hearts, which the Jews never as a people did. They were entirely outward in their views and looked upon their God merely as a god superior to those of heathen nations, at the best view. At times they were disposed to trust in the gods of Egypt or the forms of Assyria rather than in the one true and living God and the ceremonies of the Mosaic ritual. The ceremonies were of no consequence except that they were significant of the internal devotion and submission that was required of the hearts of men, but the acknowledgement of God was in their case indispensable as a condition of His favor.

Other nations who had not the evidence they had might gradually go astray and lose the knowledge of the divinity within and over them; the Jews must not find an excuse in that. To their fathers and, at various intervals, to themselves were vouchsafed the signs of God's will, and confirmations of the truths of the Pentateuch; and when the nation, or the greater part of it, became apostate from the religious institutions they had been trained to acknowledge as divine, they invariably received the chastisements of a merciful God who thus admonished and corrected them and led them back by trials and dispensations to their only true trust, their real safeguard, their own King. Thus were they alternately worshippers of God and of Baal, hangers-on of Assyria or Egypt, subjected to Philistine rule and oppression, or carrying desolation to Edom and Phoenicia. When God was worshipped they were preserved from serious evils; when idolatry prevailed they were subjected to them. When they were lukewarm and buried in traditions, the Messiah came and warned them of their impending fate. When, said He, you shall see Jerusalem encompassed by armies, then flee. Let him that is on the housetop not go down by any but the most direct way; let him that is in the field flee without going first to his house.

Who were saved in that day of tribulation which came upon all flesh? Who but His followers who believed His sayings and kept a faithful watch for the time to flee which He had thus declared should come and be so imminent as to require such precipitation. Those who fled from the city as He had enjoined upon His followers to do were saved outwardly from death and from slavery or the horrors of a painful siege. Those, on the other hand, who rushed to the Temple and repeated the outward performance of the Mosaic ritual, those who gathered from the other cities and villages of the land hoping that in that holy Jerusalem where God had so visibly maintained His power they should be saved with His sure salvation—they were disappointed. The beleagured city and its magnificent temple, the men, women, and children, all suffered destruction and God only aided them to die. Happy were they who died early before long suffering led them to blaspheme. Happy were they who, faithful unto death, still trusted in God believing that He could save if He would, and that He would if He pleased, and that He would be pleased to if He could know that it would be for the best. Martyrs earned their crowns within that beleagured city. So, too, those who fled were saved by an everlasting salvation if they continued faithful to the witness within their hearts—if they endeavored always to perceive the times of God's comings in power and to distinguish the dispensations of His love and regard. For each of them too was laid up a crown of glory, eternal in the heavens.

Such was the end of God's kingdom of Judea or Israel. Such was the result of faithful obedience and of slavish submission to earthly circumstance. They who trusted to the sword perished by it. They who sought deliverance from God in the way He had pointed out were saved. They who trusted to a broken reed were pierced by it. And how shall it be now in Christ's kingdom? My kingdom shall stand because I will take measures to make it known and manifest that I am its Head and that I will give to My subjects peace. Nations shall not prevail upon them to fight for Me nor to be engaged in strife because of Me. No one shall make them afraid, but they shall dwell in peace and security, every man under a vine or a tree of his own planting in his own heart. This shall bring him peace so long as he dwells in its shadow, but when he goes forth to inquire what his neighbor is doing and whether his neighbor's shade is as good as his own he shall suffer want of God's visible help. Be, then, desirous to plant the vine and to watch and train it in accordance with the directions I have given you; so shall you prosper and be preserved from death and your days shall be long in the pleasant land.

HARMONY THAT WILL RESULT FROM APPARENT CONFUSION

How the Rule of Christ will be Established in America

When My kingdom is established in each man's heart and My servants are thereby submissive to God and to Me, I shall be the director of each man individually and of all collectively, for if all are governed by Me individually, My directions being obeyed by each will cause perfect harmony of design to be established and maintained. Then, all being guided by one and that one being acknowledged as guide by all, the whole number will be directed by Me as easily as one, and one heart and mind will govern all the inhabitants of the land. But is it to be expected that all will literally acknowledge and obey Me individually, or will not a general obedience be first secured by the great majority, or after the leaders have submitted themselves to Me so as to be fit rulers for others? The time may never come on the earth when each man for himself will acknowledge Me to be King of Kings, but it will come in the spirit-sphere. Here on earth I shall reign as the general Lord of all, as the Prince of Peace to those who willingly submit to My government, and by raising up such holy mediums as I may find expedient to declare My will and to establish the course of the nation. The time will come when one faith will cause all to regard with reverence the declarations of My chosen servants, who shall be selected to declare My will and make known what I would have the collective mass perform.

Let all, then, pray that the glorious day may arrive when the people of the United States shall have such reliance upon God and His government as to be willing to be ruled by His Son, and such faith in His power as to believe He can make known His will to them through chosen mediums or prophets. This will be the beginning of Christ's reign on earth, for I must take the government on My shoulders and be the king of a nation outwardly before they are fully submissive to My will—even as I take the government of a heart not fully submissive to Me and bring it by aid thus rendered to full submission, while if I were to wait for the entire submission the heart would return to its former idols and subject itself more abjectly than before to them. Having thus informed you what you ought to pray for, hope for, look for, I will also tell you what the time will be.

It will be a time of trouble, a time of doubt and of uncertainty, when I shall really begin to direct the affairs of this kingdom or nation. I influence them now but then I shall govern. I can now declare what shall be, but then I will declare it. I can now produce what I desire to, but then I will do it. I can now raise to power whom I will, but then I will raise My servants who shall do My will. This time of trouble that will seem to threaten the very existence as a nation of this people is not far distant, and its coming is to be preceded by signs and wonders. Its coming will not be unexpected by My believers, My subjects, but will be one of the signs reserved to convert the hard-hearted sectarian, the deep-drinking politician who is drowned, as it were, in outward existence, having no time to attend to the inward thirst of the soul for the things of God.

Long shall all these hope for a change for the better before they are willing to look for it in the only place whence it can proceed. But being at last willing to be saved even by Me, they will let Me try to save those who no longer hope for salvation through any other name. Then will My servants give vigor and strength to the faint-hearted, and power to the weak, and deliverance to all. Then will the scepter of the King of Kings be wielded for Him by one of His servants as President, and the holy mediums throughout the land will unite in declaring him to be the approved son of men by whom the Lord of Glory reveals His will. Will this enable him to make himself absolute and subvert the Constitution and laws, placing his own will in their stead, and maintaining his power by force after having thus obtained it under My influence? He might of himself desire to do so for he will be a free-agent and can keep in the path of duty only by constant submission to Me. The very first departure, though, from My decrees will be followed by the denunciation of every medium in the land who still follows Me submissively. But may not all the mediums join in a general conspiracy to destroy liberty and to elevate one man to power? O ye of little faith! the mediums will be each and all those who have submitted themselves to Me. They will be the people themselves and they, being led and guided by Me, will not all at the same time depart from faith in Me. So you shall be warned, and without My help he shall be powerless; with My help he shall rule with wisdom and power and justice, not against all law and constitution but in fulfillment of all laws human and divine, for in him will I be manifest and by him will I rule.

Here we will pause and declare the signs of this man's selection for the position of ruler. He shall be declared by each medium to be the man selected to deliver the people from the misfortunes that shall then have appeared about to overwhelm them in ruin. His name shall be given forth on one and the same day from east to west and from north to south by each servant of mine who has so submitted to Me as to have received communication by inspiration or by revelation. Having been thus simultaneously declared to be the ruler selected by Me for relief of the nation in its emergency, he shall assume the power by general consent and shall deliver the nation from all its perils and fears in a short time, and thereby the faith of all shall be increased and My authority better established. Then the continued prosperity of the nation may cause it to forget to whom it is owing, but if they forget, My chastisement shall awaken them to their duty and lead them again to submission.

How shall we know that some pretender will not arise who will presume himself to be thus nominated and be indeed but a seeker for power in his own will? You shall know by the sign I have declared, which it will be impossible for any worldly power or intrigue to counterfeit. I know that many mediums have declared that this or that man shall be the next President and that these announcements, though made sincerely and in accordance with communications actually received from spirits, are nevertheless ignorantly procured and made; for the medium who desires answers to his questions to be in his own will, or to be gratifying to him, will be answered by some spirit desirous to afford him pleasure. The diversity of such announcements prevents them from deceiving any other than the questioner himself, who is deceived because he desires to be and who will not be undeceived because he does not wish to be truly enlightened. When all unite in one declaration it will be because all have desired to know the truth and to know it for its own sake that they may act upon it. Then is he to assume the presidency by a mere nomination, by a sort of acclamation, or will he reach it by the forms and processes required by it and by law? He will be duly elected. God works by law and order and He does not subvert but maintain governments. God's servants will always be submissive to the powers that be, for these powers are ordained of Him.

Fear not, then, and let no one of My servants be cast down because misfortunes come upon this favored land. They will be the precursors of that glorious day which has already dawned, which has been prophesied of for thousands of years and hoped for and longed for by many in past ages who shall indeed see it, not with the bodily perceptions of the earthy form but with spiritual eyes of faith and love from their homes in the spirit-sphere.

My Holy Medium's Position and State at Present

This medium, having been passive under the stress of writing what he did not understand or perceive to be in accordance with reason or previous revelation, has merited praise. When he commenced writing My revelation he would have rebelled if I had called on him for such writing, for he cannot help regarding what he writes with consciousness and weighing it by reason and his own opinions as he receives and records it. It is this which is difficult for him and in this lies his submission. Herein is great faith required. He has his desire to be useful and feels willing to be useful in any way I select for him. This way indeed is not at all of his choosing nor such as he expected. Each book he has supposed would be the last, until the next was announced in what was written in the book preceding it. Yet, having put his hand to the plow, he does not draw back because field after field opens to his view. When I ask him to plow so as apparently to endanger the productiveness of the field, or even so as to appear to cause, by contradiction, the overthrow of the previously performed work, he, having believed that work good and of divine authority and gift, is certainly tested severely. Only perfect passiveness could, unmoved, receive such without at least misgiving; and only a near approach to perfect passiveness could receive at all, without rebellion, the apparently destructive revelation I have thus given. He writes now in the belief that if he ought to be overthrown he will be; that if God designs to declare untruth for a trial to man He will do so; and that his duty is to do whatever God requires of him, regardless of consequences to himself or to others, because he feels entire confidence in God's omniscience and omnipresence. God therefore knows what he does and why he does it. God, knowing he does it from these motives and at least under the influence of a power superior to his own, will interfere in His own good time if it be not in entire accordance with His will, and if it be in accordance with God's will he is satisfied. So, being thus doubly armed, first, by the full belief that the spirit declaring to him the words of this book and those preceding it is, as it pretends to be, a high and holy spirit, he believes it impossible that one so high and holy can deceive; and with that spirit declaring continually that what He delivers is God's will and that He is in unity with God, there is no possibility for such a spirit to be mistaken. Its deception being thus negated, he is willing to be guided by it in all patience and submission and perfect passiveness; not, however, neglecting daily and many times daily, specially and in effect continually, to ask of God His guidance and counsel and establishment in the true path by His power, and preservation from error or from the dissemination of error. Second, he relies upon God as the Supreme Ruler of the Universe, declared in the Bible, by the most certainly authorized medium of revelation the world has ever known, to be the kind and affectionate Father, listening to the prayers of His children, giving them bread when they ask for it and not stones in its place; giving them good gifts and not destroying them by His dispensations, but by His dispensations reproving and reclaiming them, giving them the oil of joy for mourning and the garment of praise for sorrow; giving them mercy when they deserve condemnation and having no pleasure in their destruction or death or any misfortune that they may be afflicted with. Having full faith that this God and Father, almighty in power, cares for the lilies of the valley and the sparrow that falls to the ground, he believes God would not let him go on in a long course of error in the belief that by that course he was doing God service, if it were really disservice.

Therefore, having confidence in God's love and power and knowledge and having no desire to do other than God's will, having no hope of profit except from doing God's will, he is willing to be led and guided as he has been and to write whatever is given to him without the least desire to change it to suit either what has been written or what he has believed or what he does believe, which last is just what he can gather as the sense of what he receives and writes. Having this confidence he is willing to write what contradicts itself or what contradicts his previous writing if he is only sure it comes from the same source; that is, from an intelligence foreign to his own, invisible to him but plainly declared internally to his perceptions of mental sound in words and with even a species of intonation by which its sense is rendered to him more evident and he is enabled in general to divide the sentences and arrange the paragraphs without further instruction. Thus I have once more given you the result of My medium's cogitations about himself, his duties to Me, to you, and to God; I will now proceed with My details of the spirit-world, leaving the preceding part of this chapter to stand as a puzzler and wonder-exciter until I am ready to explain further what I mean and what I require of you in relation to it.

CHAPTER XIII

POSITION AND PROGRESS OF SPIRITS

Divisions in Paradise and the Spirit-World

Having informed you where the spirit-sphere is and how it is composed, I will proceed to declare something of its laws and regulations, the manner and degrees of progress which take place within it, and the modes by which its inhabitants manifest themselves to each other and to mankind in the earth-body.

The spirit-sphere is a world by itself and, as I have declared, it is bounded on the one side by paradise and on the other it has only aura uncombined with spirit-matter. The spirits below the fourth sphere can go only to this extent, having no power or will that can influence aura, but those of the fourth and higher spheres can in their own wills, submissively to God's will, act upon aura uncombined so as to traverse it, by which power or capacity they can reach other systems and even in due time other universes, coelums, etc. The spirits even of the second sphere possess the power to reach any planet of their own system, because magnetic-od pervades paradise and all space inside of the outer boundary of the spirit-sphere. It is by their power over magnetic-od, which is theirs by virtue of life or intelligence acting upon od, and by and through od acting upon magnetic-od, that spirits of these two lower spheres traverse their own especial home or return to their former associations. With paradise they can have no communication because paradise is of a different substance, upon which od can no more act than earth-matter can act upon spirit-matter; neither can the intelligences or souls in paradise act upon the material of the spirit-world any more than upon earth-matter, without leaving paradise. As they can leave paradise only by consenting or desiring to partake of the tree of knowledge of good and evil, that desire always takes them in one direcion, by which they enter the infant-body. They traverse the space between paradise and that body, for which they are prepared and which is also equally prepared for them, by their influence upon aura, with which, as I have stated, they are combined. Though aura also extends outward they can not go that way, because the law of their existence requires an affinity to be possessed by them for earth-matter, and they are attracted by it whenever they cast themselves upon aura for its power to transmit them to the new scene of their future experience. It is this law which impels them always to enter the particular body prepared for them. So, since infants are momentarily being born, if not on this globe on others in this system, to any of which the souls might go, it must either be a matter of chance where each soul goes or there must be divisions and classes in paradise by which the souls are separated and directed to their proper planet and particular body. That this is so and how it is I shall first explain, and then show how this same classification continues in the spirit-sphere or world.

When all the souls of all men who were ever to be born upon any of the earth's globes of this and other solar systems in this creation were placed in paradise, they dispersed themselves into different positions by affinities of which they were unconscious but which caused those of greater or less perfection to be associated in different circles or associations. But are not all equally perfect or imperfect, and were they not all as they were intended to be?

They were all as they were intended to be, yet they were not all identical in form, and the difference in their formation is best expressed to men by calling them perfect or less perfect. The difference, however, was a radical one by the combination of aura or Word with individuality, forming mind or intelligence or soul by different though definite proportions. Thus those who were designed for the outermost planets possessed more aura and less individuality in their combination or whole than those who were prepared for the innermost ones. It is thus that they would naturally possess affinities, each for such spirits or souls as were of the same definite combination as themselves, and they would thus naturally arrange themelves in circles or associations under this law, by which they would more generally associate and combine, though this would not prevent an interchange of relationship equivalent to amicable relations with the adams of other planets.

Adams is properly the term for the souls of this earth-globe, while there are other distinguishing names or titles for those of the other planets or moons. Hereafter I shall so confine the signification of the term adam. Having, then, so far separated into classes, there was another law of affinity developed by which the quality or combination of earth-matter varies in each planet from the others and from the central body, or sun, and this variation proceeds in regular progression from the outer to the inner planet and to the central body. The result is as many different kinds of earth-matter as there are of souls, and since each of these classes possessed its respective affinity for the corresponding class in the soul or in the earth-matter, their arrival at their proper combination was secured. But the Word has the superintendence of the particular placing of the souls; to take the earth's souls, or adams, as an example, it will be more easily explained, as the others are analogous to it in this arrangement.

The adam, being desirous to experience good and evil, presents himself or herself to God, as is related in Genesis, with that desire established. The Word, being cognizant not only of the desire when formed but of the whole process of its formation, is not taken by surprise. The Word has a body prepared for the first adam, and that body is prepared as I have stated. Into it the first adam enters, and the laws of the process of reproduction of such bodies are sufficient to ensure a supply of them for all the adams that become desirous of leaving paradise.

Again these adams were divided into classes so that they should form communities on earth; so that nations should exist and varieties of races be maintained or established. Variety of race among the inhabitants of Earth is consequent in part upon the difference of souls and in part upon the differences of body as formed by the Word. But if the Word formed Noah as described, and Noah was the only survivor of his primogenitive races, how could inferior types of bodies be produced? This question I shall leave for the present; suffice it now that it will be clearly explained.

Let us return to the divisions of the spirit-world. In this the previous inhabitants of different planets in a system having a common spirit-sphere have also affinities for each other respectively, so as to form divisions like unto those in paradise, the lower or most central being that of the innermost planet and the others proceeding in due course outward to the extreme outermost planet. These divisions, corresponding thus to the outward, furnish to the inhabitants of these respective spiritual planetary divisions such resemblances to their respective earthly birth-places as make them look upon them as their spirit-homes. These resemblances, too, are independent of their will to have them, being the result of the united will, or affinity, of this arrangement of matter for such a class or combination of Word and individuality. So there is even to the spiritual perception of the spirit, without a desire on his part, a resemblance always evident between him and the part of the spirit-world representing his planet, and this affinity also affects his soul so that he feels it to be his home. The spirit comes to regard these spirit-world divisions as the planets themselves and speak of them as such. This has sometimes led clairvoyants into errors when receiving from spirits the account of their knowledge or experience.

There is, too, a further division like that of different countries of the earth, and the spirits of different races associate more with each other than with foreign races. This is owing not so much to their ultimate adamic difference, which is very slight, but to habits of thought and action impressed so deeply upon them by the association of body and spirit-minds while in or upon their respective stages of action. Thus the inhabitants of France, particularly of its most refined portion, would have few associations of thought or modes of action with the bushmen of Africa or wandering Arabians. Again, the cultivated minds and the uneducated or sensual natures of the same nation form other associations, families form others, and still other groups are formed by the influence of previous associations. Others are separated or joined by the different times during which they existed together in the body, even though they may exist now in the same circle of advancement. Yet after all, though these distinctions exist and are apparent to the minds of more advanced spirits, the spirits of the second and third spheres of advancement, among whom they most exist, do not readily perceive how or why it is so. Being governed so entirely by their own desires as to association as well as in all other wants or wishes, they overlook the effect of their gratifications. As they progress in advancement they progress in knowledge. So do the more savage or barbarous nations progress, until when they reach the higher circles of the fourth sphere they are equal in knowledge and have so far separated from the influence of habits of earthly thought and action as to be free from such narrow views of fraternity, and they associate in very large bodies having common thoughts, common duties, common desires, receiving knowledge and making progress simultaneously.

Thus they progress, enlarging the association willingly entered into, cheerfully adding all who desire to join the association, until it embraces not merely those of different countries but also those of different races and even of different planets. In the higher sphere not only is there no distinction as to what planet sustained the earthly body, but they are even brought together into one united association from every division, great and small, of the Whole Illimitable Creation.

The divisions of paradise are differently arranged, inasmuch as the souls were placed in the beginning in the spheres of paradise-substance attached to the respective central portions of earth-matter they were assigned to inhabit. The separation into classes for races, and different nations of the same race, are made before entrance into earth just as they are made afterward. But the separation is then made by the inherent quality of mind or disposition of the soul instead of by its habit of thought in the body.

The first difference is primary and persistent, hidden, indeed, by the earthly and spiritual bodies in whole or part so long as they exist, but again apparent in the seventh circle of the seventh sphere as well as partially so in lower circles and even in lower spheres. The latter ceases as the former makes its reappearance. This inherent quality of mind is not the same thing as the desire that is the predominating cause of the departure of the adams or souls from paradise. It is character. It is not such a difference as makes them unequal in capacity for enjoyment, but it is such as makes them unlike so that no two spirits or souls that were formed by the Word at the command of God were ever precisely alike. God has endless resources of variety. His infinite conceptions of fitness, His fertility of production, and His infinite variety of manifesting His works and creations furnish to that infinite (to all but Him) creation or body of procedures a never-ending source of enjoyment in viewing and reviewing the lives and past experience of spirits of all the different globes of matter in the Universal Whole of Creation. No two of these being the same in character and no two having been surrounded and influenced by like circumstances, no two of the planetary bodies in the whole illimitable creation being alike in respect to materials, relations to other bodies, or physical or spiritual peculiarities, but every part and parcel, even every atom being different from each and every other in some respect, there is indeed an eternity of enjoyment in the pursuit of knowledge and the investigation of God's past, present, and future action as manifested in the procedures and creations of His will and power.

Having now opened to you this fountain from which eternal bliss is to flow, I have to say besides that this is only one of the fountains of which God is the feeder, as it were, and innumerable others are all equally endless in their capacity to confer happiness, all equally unfailing in their variety, all equally full of His glory, redolent of His praise, and abounding in His power and benevolence. But since man on earth cannot understand this, which is so peculiarly fitted to his comprehension, I shall not undertake to refer more particularly to others at the present time, although allusion, embracing brief outlines of them, may be made to some few in the progress of My revealments. I have said that the knowledge possessed by spirits in the fourth sphere is to be revealed to mankind, but this must not be supposed to mean that all the knowledge which spirits of that sphere possess is to be revealed. The earth could not contain the books that would be required to hold it, much less could men now in the body write them if all were mediums, and this is not a figure of hyperbole but simple statement of relative magnitude or capacity or power in relation to time.

The Pursuit of Knowledge

There is in all men a desire for knowledge, but some desire it for one purpose, some for another. Some are said to desire it for its own sake, which truly rendered is that they desire it for their own gratification rather than to benefit others. Knowledge is the object for which you left paradise—knowledge of good and evil experience. So it is that this is the state of experience, though it is also a state of probation and a state in which more progress can be made in the same space of time than in any one of the higher circles or spheres. But there are varieties of knowledge and to say that a soul left paradise from a desire for knowledge is in one sense saying no more than that it left for experience of good and evil. Knowledge of God and knowledge of men are two great divisions; knowledge of God's works and knowledge of His laws are two other great divisions; but there is also another kind of division, that is, knowledge of that which will benefit our fellow-men and that which looks only to our own gratification. The knowledge that is useful to men is to be desired because it enlarges a man's mind to do good to others, it fits him for heavenly progress, it prepares him to receive from God an abundant reward, for his works do follow him and declare whether they have been useful or not.

Knowledge of the past is useful when it teaches how to avoid the errors committed by the actors of former times. Knowledge of the present is useful when it enables us to seize upon everything discovered by any man and make it useful to others. Knowledge of the future is useful when it leads us to prepare ourselves for it, as it may by a desire on our part to serve God so that He will be pleased with our works; so that He can bestow upon us His mercy and save us from sin. He will not pardon our sins in the strict sense of the word but He will lead us away from their enjoyment and committal. He will enable us to enjoy peace by being saved from sin, not by saving us in sin but by sustaining us in right courses so that the sin may be and shall be avoided.

Such then, O man, is the benevolence of God, that He will not merely save you but He preserves you from evil. He not only pardons you in effect but He restrains you in the future world from transgression. Here He allows you to receive the wages of sin; there He allows you not to suffer death again. If you obtain life or salvation there you lose it no more. But this is not a license for you to disregard His admonitions here, or to revel and riot here-after. It is the permission here to experience suffering and un-happiness if you will, and there to resist Him as long as you please. He is ever ready to save. His arm is never withdrawn when you seek His powerful aid, and, the last resource of your own inven-tion having been tried and proved useless as a means of happiness, a gracious God is still as willing to receive you as on the first day you breathed in the body or acted in the life to come. For in the life to come you will continue to act as you please, with one restriction, that when your act is not one beneficial to others it is an unreal one and affects only yourself through your imagi-nation. When the act is one useful to others or agreeable to them, it becomes manifest to them and more real in its character. And though millions of spirits should join in committing an act whose intention or effect should be injurious to one single other being, they could not make it affect that being in the slightest degree though to each and every one of the millions of actors it would be as real as any act ever can be so far as their knowledge and perceptions could discover. They would suppose they had accom-plished their design, and if they glorified their own power and re-joiced over their success they would injure no one but themselves, and they would not injure themselves except that they would for so long be putting off the day of salvation, for this would not be an act increasing their sins but rather manifesting their character and desires; the truth is thus made evident to themselves and to all that the heart of man is desperately wicked and there is no health or life in it except through God's help, ever offered to men and to spirits by and through His Sons, the Christs of His glory, the Sons of His love, the Sent of His mercy.

This, O man, is My call to you now to repent and live, no longer to indulge your own will but to seek to know God and to do His will. Seek and ye shall find; knock and it shall be opened; pray and ye shall be answered; ask and ye shall receive. There is no doubt that these are true sayings even with you, O reader. Then why not act upon them? Will you say you do not know how? I have previously furnished you with forms of prayer. Do you say that God will not accept those words as yours? I answer that He will if you make them yours by adopting them. Do you say He will not have a man so unregenerate as you? I answer that He is now acquainted with your every thought, every attempt to think. Do you suppose that mankind here and on other earths are so numerous that He cannot attend to the actions of each? I reply that His attributes are all infinite in quality and degree. He is everywhere and He knows all that every being does. Without effort He accomplishes His will and it is His will that all men should be reconciled to Him whenever they are willing, and that they should be helped whenever they want help. Turn then to God; turn not to your minister, your church, your father, or your brother, or to any man, but to God.

If you fear God's majesty and reverence His greatness too much to approach Him directly, I will be a mediator between Him and you. I will enter your heart as soon as you ask Me to and help you to purify it and reconcile it to God. I will help in His will and aid you whether you pray to Him or to Me. I am Alpha and Omega, the first and the last, by which I mean that I am sufficient for all times and for every occasion. I can serve you at the beginning and at the ending of your course to a sonship and when you reach to the perfection that I have arrived at I can with joy share with you My inheritance of peace, joy, and usefulness. You can be one with God even as I am one with Him, and be His Son and Sent to save sinners as I am.

You, O reader, *you* are the man! I am not speaking to others but to you. I am not speaking generally and abstractly only but specially and with particular reference to your immediate action. Turn then to God by asking His or My help. Ask and ye shall receive. Call Me and I will come. Behold I stand at the door of your heart now, knocking, and you keep Me out. Behold how you reject Christ because He comes not in the way you expected, which way is one you would think approved and accepted by the church of which you are a member or by all churches that have professed to be led by His example and guided by His precepts. But you can easily perceive if you reflect that Christ when He comes must be received individually, not collectively, and that no church organization can accept Him for you. If you want to acknowledge Him for your Savior you must let Him save you by a sacrifice of your own particular free-will. No other man's free-will sacrificed can save you or in any way help you on your way to salvation and glory except as you avail yourself of it as an example and go and do likewise.

Having now made this last appeal to you, let Me ask you again to turn to the prayers written for you in My first volume, and try to make them your own by repeating them with sincere desires for improvement and for benefit from them, and for the power to enter into their spirit and the ability to pray exactly those words yourself. When you have got so far you can pray alone and without a form of written words, but you can not then pray without help. I will help you then, and though you may not know it I will aid your every good resolution, prompt you continually to good works, inspire you to produce good thoughts, kind acts, loving demonstrations to and for all your fellows, particularly your neighbors; that is, all with whom you have any intercourse or acquaintance, any relations of affection, business, or fellowship whatever, every one whom you know you can benefit and all who ask of you assistance.

Judge not the beggar or the asker; judge only whether he wants what you can give and whether the gratification of that want, or its relief, will be a benefit to him; and if you can thus do good, act so as to best secure this effect. Leave his reformation to God. He may be worse in God's view than you are, or he may be better. Remember that now you are restrained by public opinion more than by law; by law more than by fear of God. And remember that in the life to come you will not be restrained by anything but the love of God and that if you do not adopt that as your motive of action here you will enter into the future life subject to no restraint and will act not only as you have acted here, but as you would act here if you were left unrestrained by fear of detection when you have opportunities of selfish gratification of any kind whatever. Be wise today; tomorrow may find you in another state of existence, and there you will have to atone by a long course of attempts at self-gratification for your unreconciliation, for your departures here in your heart from the path of rectitude and love of God.

Then once more I appeal to you to save yourself. For though you cannot save yourself you cannot be saved by another; if you would be saved you have only to ask God to help you and do all you can to deserve His help. Relying upon the power, love, mercy, and perfect wisdom of an all-wise and almighty Creator, Father, and Friend, throw yourself in submission at His feet saying: Lord God Almighty, have mercy on me, a miserable sinner! I will be Thy servant if Thou wilt only let me be so, for I am not worthy to be called Thy son. Then shall the redeemed of God sing a new song because another soul has begun to be reconciled to God, because another sinner has begun to be saved.

> Great and marvelous are Thy works,
> Lord God Almighty!
> Just and true are all Thy ways
> Thou King of Saints!

CHAPTER XIV

THE PAST PROGRESS, AND FUTURE APPEARANCE, OF JESUS CHRIST

How Man is Called to Progress Here and Hereafter

In this chapter I shall endeavor to conclude the relations of man to the spirit-world by showing how his soul is released at last from the spirit-body and by what process he becomes so intimately placed with God as to be worthy to be called His son and His heir.

Man progresses from the cradle to the grave. All see that, and men are therefore willing to believe that progress is a law of their being. The paradisical state, though one of quiet happiness, of inaction, is one of progress because man obtains there the knowledge that inaction and inactive bliss are not happiness, are not satisfying to the soul. Hereafter in the spirit-world he adds to even this kind of experience if he is inclined to, for in the early time of his sojourn in that world he can take whatever course of enjoyment he pleases and have all and everything he desires. Happy is he who desires to love God and to know that God loves him. Happy is he who thus seeks God, for he is advanced to higher and higher happiness rapidly. He progresses to all eternity but reaches the position of son of God at an earlier time, incomparably earlier to men, than the man whose life and energies are or have been devoted to self-gratification in any of its forms.

Let us then be attentive to the unfoldings within us of progress toward reconciliation with God. Let us endeavor to be guided here by good motives so that hereafter we shall act by good motives when we act without restraint by anything except our own inclination. So shall we progress steadily here and hereafter. So shall we find peace here and happiness speedily and forever hereafter. Let us all seek God here and we will seek God hereafter. Let us sacrifice our own free-will here and we will continue to do so in the life to come. Let us do all in our power for others here and we will continue so to do forever. Let us seek for salvation as we ought here, and our seeking shall be rewarded by finding here an eternal salvation in a house not made with hands and eternal in the heavens.

Reader, remember My appeal. Say not in the life to come that you were not told how to prepare for it. That excuse will do for some and will be valid, but you may know and should know. To you is given the five talents, to others three or one. Of you as of them a strict account will be required and your reward shall be according to your works, though your works cannot save you but only qualify you to receive God's mercy and loving-kindness. Be then ever watchful for opportunities to improve, ever laboring to promote some good object, ever seeking to know God's will, and ever practicing your knowledge of what He requires.

Let us pray

O God, who dost from Thy throne behold all nations and all men and all the actions of men, O Thou Eternal One, who art (as Thou hast ever been) kind and loving, look down with pity upon the miserable sinner now desiring Thy aid. Be Thou, O God, my savior, my redeemer, my light and my life, my glory and my salvation. O God! Thou knowest what I need and what I would ask for at Thy hands. Thou, O God, knowest my infirmities and hast my affections in Thy power, for I desire to place them upon Thee as the only true source of happiness, and to elevate my desires to the measure of doing Thy will. Help me, O God, to surrender to Thee my free-will as the sacrifice Thou requirest and as the acceptable offering I can make, and as the offering I desire to make; help me to be brought into perfect subjection to Thy will and lead me in such paths that I may be able to follow Thee to Thy high courts and to the gates of Thy holy of holies. O God, let me not sink into despondency or be the despairing outcast. Let me be raised to Thy power and placed on Thy right hand, for Thine is the kingdom, power, glory and honor now and forever and ever. Amen.

Let all the people praise Thee,

Let all the nations bow before Thy throne,

O Most High and Holy Lord!

And let every ruler of nations seek to be Thy servant,

For all shall know Thee, O Thou Most High King,
From the least to the greatest, and from high to low.

Let all that is within me praise Thee,
Let all the world of Earth bow down!
Let us all submit to Thee, O most high God;
Help us with Thy unfailing help from heaven,
And the love and mercy of Thy Son our Lord,
And the great help of all Thy high and holy sons.

Let us, O God, know Thee to be our Father
And look on Thee as our ever-present Friend!
Let us never want Thy light in our souls
Or fear the surrender of our captive affections to Thee,
O Thou kind and loving Friend and affectionate Father,
O Thou most high and holy Creator and Preserver of All!

Let us have, O God, Almighty Father—
Let us have all the help of Thy love
And all the favor of Thy mercy;
For we are weak and hardly able to seek Thy love;
We are passive to Thy influence, O God, but not active;
We are desirous to serve Thee but know not how, O Lord our
     Father!

Let us then, O God, be helped
And let us not be forgotten by Thee,
For we are the least of Thy people.
But we desire to know Thee and love Thee better hereafter
And do seek Thee as we have not heretofore sought Thee,
O most holy and kind and untiringly merciful God!

Let us be found accepted;
Let us be found Thy servants;
O God, most loving, most kind, most holy, most merciful.
O Lord of heaven and earth, of men and angel-spirits,
O Holy One of all time and Ruler of eternity unending!
                    Amen.

Let us be found able to make this prayer, the last as well as the
first. The last I have had arranged in lines and verses not because it is poetry but because My medium desired it in the expectation that it would be poetry. Finding it was not equal to his
expectations he was puzzled, but he sought only to do My will
and he was not confounded. So shall it ever be: the seekers to do
God's will or My will shall not be forsaken though they err, but
the seekers to know My words to torture them shall be confounded
by their significance and shall not understand what they reject.
Pray then that you may understand, for this is the day in which
the wise shall understand but the wicked shall do wickedly. This
is the day in which the filthy shall be filthy still and the dawn
shall show what each man is and will be. Be, then, seekers after

the knowledge of God and be wise. Be at peace with all and let no man take your crown. This I have explained; heed that explanation.

Having now reached a place where we can pause and review what has been unfolded, and having arrived nearly at the end of this section, I will sum up the revelation of the past, the view of the present, and the expectation of the future.

Man is a being whose soul proceeded from God. His body was prepared by God to be the temporary residence for his soul. This body is double, or composed of two separate substances, the one shorter-lived than the other. It perishes as does the body of an animal, and returns to the dust or earth-matter from which it was formed. The other body lives beyond the grave. It exists through changes in its nature which refine it and purify it and at last dissipate it almost insensibly into space, from whence it returns by the law of affinity to the great reservoir of material from which it was originally drawn. The soul is then free to assimilate with its other part which was joined to it in paradise. Together then they form one being, but with two minds or individualities or souls. One soul in its relations to others, one soul in its relations to God, but two souls in relation to itself. It is indeed the type of the married state of men in the body. By its purity it produces fruit of good works, by its own action provided it acts in God's will, and by its singleness it reaches to God. It is one with Him and as such it is one with all other spirits or sons or parities in its circle. It is two to itself, to men in the body, and to lower spirits.

I am the Lord Jesus Christ, born as a man of Mary, the lawful wife of Joseph—the son of Jacob, of Bethlehem but born in Nazareth. I am the soul that existed in that body on Earth. As a man I was desirous to serve God by persuading My fellow-men to serve God, to seek His love and be saved by His mercy because of His kindness and love for His children and all His procedures. Because I was animated by this desire, God was pleased to help Me do His will and gave Me for this purpose the aid of a Son who had reached the circle I am now in. This Son was My Christ. He and I were one because I desired to do God's will and he desired to do God's will. I was then a son of God by My union with a son of God. I was a son of God by being born with pure desires for God's service. I was also a son of God by being specially provided with a body in a miraculous manner, as I have before fully described. I was one with God because I was one with His son who dwelt in Me, and the works or miracles which I did were not by My own power or in My own will but were performed through Me by God operating upon Me through the Christ in Me. I was thus blessed with divine help and fed with divine food. But My outward sustenance was also provided for Me, first by My own labor, and, second, by the favor or offerings of others after I commenced My ministry. My raiment was the gift of God through pious believers. In all this I was consistent. I walked as God directed, I preached as He inspired, and I worked as He commanded. He spoke not to Me directly but by His Christ.

Having, then, led the life He required and having been accepted as a faithful servant, I desired to be released without the agony of death, on the cross, which I knew impended. I prayed earnestly again and again but God was inexorable because it was His will; I submitted and was led as a lamb to the slaughter. My sufferings were not great because I was sustained by God's power and mercy. The flesh was rendered insensible by the influence of His spirit or Christ in Me, and, having at last left the body hanging on the cross, I descended into the place of departed spirits, or ascended to the spirit-world but to its lowest state, the second sphere of the first circle; passing from that immediately to the next (because I had there no atonement to make, having no sins of its class in My nature or experience) I thus went on rapidly to the sixth circle, where I paused to make atonement for having desired so strongly to avoid My fate. This atoned for by good works performed to spirits there and men in the body, I again progressed to higher and higher circles until I reached the fourth sphere first circle, when, being sufficiently purified and elevated in My spiritual body and mind, I re-entered the yet warm body as it lay in the tomb of Joseph of Arimathea at about six o'clock in the morning of the first day of the week. That body, reinvigorated by its soul and spiritual body and mind, was again filled with life which had left it not as a consequence of wounds or sufferings but as released by God acting through His Christ, as I have said He always does act by some agent well fitted for His purpose.

I then appeared to My disciples and others walking about in the body I had formerly used, which body was purified and renovated by My purified and renovated spirit-body and raised soul so as to be of more refined earthy-matter and more perfectly controllable by My will. It was thus capable of being changed at once to invisibility and of passing through solid substances as easily as magnetism, electricity, and caloric can do, and it was thus that I appeared to My followers and in turn disappeared from them. It was by this comand over the earthy materials of that body that the unbelieving Thomas was convinced, that the disciples saw Me submit to the test of eating fish with them. Yet food was no longer a necessity for that body. It was maintained and retained by the power and will of My raised spiritual intelligence and soul. When the proper time came I caused the earthy body to vanish as a cloud before the vision and outward senses of My chosen followers and ascended slowly with My spiritual body and celestial countenance, disappearing from their enraptured gaze. I ascended from the clouds that the earthy body became at its dissolution to the clouds of spiritual glory or celestial spirits that thronged about Me as I arose from Earth and again commenced My progress in the spirit-world.

Again I appeared to Paul when I was in the third circle of the fourth sphere and My spiritual body by its brightness blinded him;

all his company saw the light and heard My voice but their outward eyes were closed to the light which did not blind them and they received not into their hearts that belief which Paul received. Again and again have I manifested Myself in various ways since to My followers and sincere worshippers, as a miraculous sign: as the martyr Stephen saw Me when about to be stoned and by My power fell asleep without pain; as Constantine saw My manifestations when he was undecided whether to court or persecute the Christians. Still more often have I spoken in the hearts of the people of God and now, having very lately arrived at the seventh circle of the seventh sphere, I am directing this procedure from My parital power in God's will and with the concurrence and aid of all the spirits with whom I am in unity, in great number, (as I have described in Volume One, Part II), including, of course, the same perfect son of God that was with Me as My Christ when I was in the body and who is now joined to Me in perfect unity as equal and joint-heir to God, equal in power, love, and mercy to God because we act only in His will and therefore have His power, love, and mercy to use and dispense.

Being, then, here united to God and so perfectly one with Him as to have no other separation than that of possessing a certain portion of the substance called by Me heretofore individuality, combined with a portion of God's substance which is called aura or Word, which combination possesses a memory of its own, I am truly worthy to be praised as God's Son and as the Savior of those in the world of mankind that are willing to be saved. But I am not desirous to have you attend to Me rather than God. He is above Me as I am above you and though I am immeasurably above you God is immeasurably above Me. Let no man worship Me instead of the Father, or as Almighty God, but he may worship Me as a part of God doing God's will and partaking of His nature, power, mercy, and love and as united with God in such a manner that all the praise, honor, glory, and worship which is addressed to Me ascends to the Father. I am nothing, although I am so much, but a high servant of God desiring only to be His servant and to benefit mankind and spirits and please Him thereby, because that is the way He delights to be served.

I am in this work united to My proper parital part that was with Me in paradise and there was My Eve. That parital part preceded Me to the bodily life and was the spirit of Mary, My mother. She too left paradise with a desire to be of service to God, not from gratitude but from affection. God, or the Word, by whom He acts in paradise, selected her spirit for its special body, which was favorably placed or circumstanced in education and training so that she was inclined to be the handmaiden of the Lord God and to submit to His will as it should be and was made known to her.

Such is My history more fully written than before, and such is My present state and such are My present objects and acts. Now, reader, what do you say? Will you help or hold back; will you retard or advance My future appearance in clouds of glory by which I shall again outwardly manifest myself to the inhabitants of Earth and rule My people with justice and glory and render to God every soul that obeys My call? If you are not for Me you are against Me! Be wise and understand for the time is at hand.

Let us pray

O Almighty Incomprehensible Father! O Loving and Everlasting God! be pleased to help the reader of this book to understand and believe, to come to a knowledge of Thee and of Me, and to be brought to a willingness and desire to be Thy servant. O God, Heavenly Father! Let not his fear of men, or his fear of delusion, or his fears of any earthly power or influence whatever keep him from surrendering his free-will to Me in order that I may replace it by Thy will. O God, pardon his inattention at this and other times and make him willing and desirous to read again and again until he shall be wise and understand. To Thee, O Father, shall be praise, honor, glory and worship and dominion, both here and hereafter, both now and in the world to come.

Amen.

Let us pray

O God, Thou hast heard My prayer
And hast been pleased to grant My petition;
Most heartily I beseech Thee to pardon My medium's short-
comings,
And let not his faithlessness be laid up against him.

O God, be most merciful
For he is very far from Thee.
Oh, help him to be passive to My will in other matters,
As he is obedient and passive in this particular one.

O Lord, I am willing to help him
If Thou wilt let him be powerfully helped.
O Lord, I will help him powerfully if Thou wilt be pleased to
have it so
And lead him to Thy feet changed and redeemed.

O Father, help Me to save him,
And let Thy mercy be very abundant upon him;
For he improves not as I would have him in this life,
And prepares not as I would have him for the life to come.

O Father, look with pity and compassion,
Look with holy loving kindness upon him;
So that he may become fully Thy servant besides being My
medium,
So that he may become Thy son even as I am Thy son.

Amen.

Let us all endeavor to profit by the prayer I have made for My holy medium, for he is not the only medium who needs help, neither are mediums the only ones who require such prayers. Therefore, O reader, do you too try to join Me in making it for yourself, and to be attentive to its requirements, its admonitions, and its solicitations. Desire to make it fully your own together with the one that precedes it, which was made especially for you. Let all profit by everything that is thus given and all will thus be advanced toward their final state; all will help in that manner to prepare the way for the reign on earth of My spirit in the hearts of men and of My outward appearance to the bodily senses of mankind.

What Shall Be

In the future I shall appear upon the earth to the outward, the bodily senses, or perceptions of mankind, with a prepared body. I shall appear as a man glorious in form and majestic in manner, as a son of God endowed with power from heaven, as a son of man having an earthy-body, as a Christ preaching glad tidings of great joy, as a Messiah to save the people of God from the fate their enemies would impose upon them. But how shall we know You? for when You appeared before as Jesus of Nazareth you were not known by those who were looking for You, and we may be equally blind, for You then had all these attributes and distinguishing marks that You have now described!

You will know Me by My declarations, for I will declare Myself to be what I have said. You will know Me by the declarations of this holy medium, who will bear testimony that I am He who has promised and been promised to appear a second time to live and reign on earth. You will know Me by the testimony of all other mediums of spiritual manifestations, who will either be silent or announce Me as He that should come and bear witness that you should not look for another. You will know Me by the crowds that will follow Me and be fed; by the lame, the halt, the blind, the dumb, the insane, the wicked and the despised who will all declare My glory, honor, praise, and high renown, and that I work as no man works and teach as no man taught.

Will you then, O reader, believe; will you be a follower, a disciple? O reader, you may think you will, but I tell you now you are and will be incredulous, or you are and will be believing. If you cannot believe this book, thus published to the world and presented as a free gift of the labor of My holy medium, you will be unwilling to believe Me when I personally declare the same truths and preach the same doctrines and have the same testimony as is born to the authenticity of this book. Strive then to believe; you have Moses and the prophets; you have Paul and John. You have the witnesses near you and throughout the land; if you will not believe them, you will not believe because one you have never seen should rise from the dead. Thus it is *now* that you should investigate; *now* that you should strive to believe; *now* that you should resolve to believe; *now* that you should seek by prayer to God to have an internal assurance that all these testimonies are true, for the spirit is a faithful witness and does bear witness that I am the first and the last, the beginning and the end, the A and the Z, the one that was and is and will be evermore, and He that should come as a Savior, a Redeemer, a Mediator, and a God. Be ye faithful unto death, for such shall not be subject to the second death. Be ye ever willing to die and you shall be saved; be ye ever found faithful and you shall not be tried severely, for I will have mercy and compassion on Mine own and from him that followeth after Me I will not flee away.

Let all, then, seek to be glorified, purified, and established on the Rock of Faith. For on this Rock will I build My church and neither the gates of the world of departed spirits nor the wiles of the free-will of man shall prevail against it, nor shall any man make afraid the multitude who shall gather themselves together in that holy mountain, and there be delivered from all the combined power and forces of the kings of the earth or the indulgences and temptations of earthly nature that shall be collected from the whole face of the earth and from waters or atmosphere that surrounds it. This is My hope and My expectation, that you will be found on My side within the gates of that Eternal City not made with hands, which cometh down from heaven arrayed as a bride for the arms of her husband. Her gates shall be praise, her streets good works, her temples the heart within you, her light the light of God's presence, and her glory from everlasting to everlasting. It shall be perfect in form, rich in material, glorious in appearance, incomparable in nature. It shall have the fountains of life and all manner of fruits of righteousness within and the saints of the Most High God shall dwell in it forever and forever and forever. Ten thousand thousands of times multiplied and repeatedly multiplied by each other will not include all the number of the saints who shall sing the praise of the King of Glory and His redeeming love.

> Great and marvelous are Thy works,
> Lord God Almighty.
> Just and true are all Thy ways,
> Thou King of Saints!

CHAPTER XV

THE UNENDING CONSTANCY OF PROGRESS

The Infinite Nature of Spiritual Progress

Hymn of Praise

Let all the people praise the Lord!
Yea, let all the people praise Him,
For His great mercies and for His loving-kindnesses
And for His everlasting mercy which endureth forever and ever.

Yea, let all the people praise the Lord!
Yea, let all that is within each man praise Him,
For His mercy and His truth and His loving kindness,

And because His mercy is everlasting and endureth forever and
ever.

Yea, let all that is within us give thanks!
For He hath raised Me to power,
And placed Me on His right hand
And sent Me to proclaim His mercy which endureth forever and
ever.

Let all the people praise the Lord!
For He alone is worthy to be praised;
From Him is all strength and every good work
And His everlasting love and His untiring mercy endureth forever.

Amen.

Where shall the wicked man find a refuge from the mercy
of an ever-present God? Where shall the ungodly hide who are
ashamed to be seen or known by Him? Alas! alas! if they ascend
into heaven, He is there; if they go down into the grave and
thence to the place of departed spirits, He is there; if they flee
to the uttermost parts of the creation, to the farthest portion of
earthy matter in the illimitable universe, He is there; and everywhere when they meet God He will require of them the fulfillment of this command: My son, give Me thy heart. Sacrifice
to Me, O man, that free-will which was surrendered to you for
free exercise and unlimited use and control in order that you might
see that all is vanity of vanities. Sacrifice to Me your heart, that
is, your free-will, and you shall live with Me in everlasting progress, in useful occupation, in ever-new delights. Live with Me
as My son, as My Son's companion and equal and as the heir
and co-heir of My love, mercy, and power; as the executor of
My will and the administrator of My attributes! Be, then, faithful seekers, untiring servants, happy sons, eternal companions. Be
then, as the first step, a believer in My declarations and a seeker
for more light. Be faithful, vigilant, unfearing; be bold in proclaiming, earnest in extending your conviction of My truth; and
tell every man you meet with and have a suitable opportunity
to inform where the pearl of great price is to be found. Be prepared to render to every inquirer a reason for your faith, but do
not wrangle or contend. Declare your own confidence; express
a hope that others may attain the same. Let God work and give
the increase; do you be patient. All will be well. My kingdom
shall be established; all men will be saved.

Blessed are they who believe now and wait not for more evidence. Blessed, too, are they who shall believe because of that
evidence to be given; but more blessed shall they be who, believing, shall do the work I call them to and become My servants.
Remember it is not he that heareth and assents, but he that doeth
the will of his father that is the dutiful and acceptable son. Remember that all the sons of men who ever have lived or ever will
live cannot save you, but that only God can save you. Remember

that He makes one indispensable condition of salvation, the same to you and to all men; namely, the sacrifice of your own will. Your earthly desires, your ties to the world, your outward hopes, your affections, your everything, in fact, that you can call your own, you are required to place on the altar of God, and fire from heaven shall descend and all that is not found pure metal, pure as it came from God who gave it, shall be consumed or melted, and all that is not thus destroyed by the fire shall remain upon the altar ever-ready for God's acceptance or use. You shall have instead of your own a new heart, new affections, new hopes, new desires, the gifts of God and the sources of your happiness. You shall not perish by the sacrifice but rather shall you find salvation, everlasting life.

Let no man, then, take your crown. Go not to man to tell him what you want or what you have received and to ask him what you should do next. Join no body of men to establish your faith or to help you to maintain it. God is a better helper, a surer friend, than all men together; when the blind lead the blind they will surely meet with misfortune and be cast away. Be no more seeking after the inventions of men to raise themselves to heaven or their contrivances to enter by some other way. I am the Way, the Truth, and the Life; in Me is life, and he who entereth in Me, or I in him, entereth into or has everlasting life. Where, then, shall I find you, O reader, when I appear in My glorified form upon the earth? On God's side or on the side of the world? On My side or on your own? The measure of your enjoyment of this happiness I have described, for a very long period, may depend upon your answer. Millions of years may be required to do the work you can now accomplish in one, and your eternal happiness, though certain to arrive, may commence that much sooner and continue in undiminished, undiminishable fullness until the end of that which is unending, until eternity unites all time into one present whch knows no past, anticipates no future.

This is not a time of oblivion but a fruition of full happiness when all shall be sons of God equally perfect, when all shall be the perfect sons of God, each united to his parital part. Then the love of God, having pervaded every soul and being reciprocated in each, will be the long and the ever-existing blessed dwelling-place of the sons of God. They will exist in that as a substance like unto aura or atmosphere. Where will they be, then, inasmuch as creation is filled with aura and the matter created by the Word? They will be with God and God is everywhere. They will dwell in His love, and His love will be where He is; and He is and will be everywhere. Then God's love, being a part of Himself, now pervades all of creation and is ever in us and all men! Yes, and it becomes manifest whenever we allow it to by subjecting our own wills to God's will and giving His nature that ascendancy in us to which He is entitled by His excellency and holiness.

The future must ever exist and the future will, like the past, be a progressive state because God is the parent of variety and is inexhaustible in His subordinate beings. Then fear not that you will suffer in the far distant vista of eternal happiness from want of novelty or occupation. Believe that God is never at a loss for resources, that He is never at a loss for occupation, for grand thoughts, for holy laws, for unending improvement or change of pre-existent states and conditions. He will never be at a loss to give to His sons perfect bliss and unchanging happiness, though eternal novelty is required and He is only removed from them by the barrier of infinity. It is infinity which possesses all these resources, and the procedures from God, by whatever name they may be called, by whatever powers or combinations distinguished, however long they may progress in constant approaches toward God's own perfection, must ever remain finite, and though ever so long one with Him, always inferior to Him by this difference. This difference is indeed an insurmountable one and a long course of instruction will be necessary to enable you to understand its degree. Indeed the understanding of God's procedures, the Word excepted, is insufficient to comprehend infinity; though God's perfect sons have some conceptions of the principles upon which it depends and enjoy deep researches into its nature, as man enjoys the pursuit of a science which even at last he does not fully understand or master. This is one of the occupations of which sons can never tire and which will afford constant variety, constant progress, and eternal happiness.

Let us all, then, endeavor to establish My kingdom upon the earth so that here may be a faint outline of the eternal kingdom of God in the heavens. For My kingdom will exist in My will and in God's will, and it will be a copy of His kingdom and governed by His laws. Herein is wisdom. Let him that readeth understand, for the number of the names which John saw in the Holy City was one hundred and forty-four thousand, and these names were divided into twelve divisions and the number of the names in each division was twelve thousand. Herein is wisdom. Seek the name of which this is the number; seek the name that contains the first number and all the numbers. Such a name can be found in the Greek language in which the prophecy was delivered.

Let us pray

O God, Thou knowest all things; grant unto us Thy servants such knowledge of the future as may be useful to us and such of Thy revelation as it pleases Thee to give to us. O God, let us not desire to know more than this lest we obtain it to our harm and have to rely on Thy help for relief. Amen.

Be Thou, O God, ever with us inspiring us with Thy counsel. Be Thou, O Father, ever with us helping us by Thy power, and be Thou, O God, the sure and steadfast maintainer of our progress and of our salvation, for by Thee we live and in Thee we have our existence, our hope, and our love and desire to live. O Lord God Almighty, be very gracious to us, Thy servants, who humbly desire to do Thy will and walk in Thy ways and receive from Thee such pay as Thou art pleased to give, having, O God, confidence not merely in Thy justice but in Thy generosity. O Thou abundant Giver, give unto us abundantly; and let us always look to Thee as the source of our substance and the partitioner of our bounty. O God, be merciful to us, sinners though also trying to be Thy servants, and suffer us not to be led into temptation, and save us from tempting Thee by expecting miracles for our use and support. O Lord, be merciful as Thou alone art capable of being merciful, and forgive us who act as enemies to Thee while seeking to be called Thy servants. O Lord of heaven and earth, be merciful, for we are unworthy to be called' Thy sons but are only worthy to be called Thy hired servants for thus asking Thy rewards and blessings here in this present world, when the same wages laid up in heaven would be so much more blissful and lasting. O Lord God Almighty, be now kind to us; give us what it pleases Thee to confer upon us, and give it to us at such time as may in Thy wisdom be found to be the best time.

Outward Manifestations of Spirit-Life

The time when the fifth monarchy shall be established as a really outward manifestation of an earthly kingdom is near at hand. It will be preceded by troubled and disastrous events in which My servants will have to bear a share of the misfortunes of their countrymen. But the true life and enjoyment of man is inward and from within and dependent wholly upon inward manifestations. A generation shall not pass away between the peace of the present and the joy of the future. Within the life of many now in the body shall be outwardly experienced the coming of Jesus of Nazareth, the Author of this Book, to assume in His outward form the power and rulership over the people of America and to establish the fifth monarchy, of which there shall be no end under the present form of the earth as a planet. But as I said in Volume One, Part I, a change of the earth is impending and will soon, in a few thousand years, take place and as a consequence of that change men will assume a different kind of bodies, more refined and purer-minded, and they will then be left again as they were under Noah and his sons to replenish the earth and to establish their own forms of government under the counsel and inspiration of God and His holy spirits.

By that time there will be many more of Earth-born spirits arrived at the seventh circle of the seventh sphere. By that time My preeminence among the spirits from the earth or of adamic nature will be shared with many who are now in lower circles and lower spheres. They will, as I have stated, be My equals in power and unity with God. But some will ask: Did not your miraculous birth indicate that You were to be ever distinguished among the sons of men? Were You not thus marked out as the Son of God in an especial manner and designed forever to be more distinguished by God's love than other beings who have lived on Earth? Not at all. God designed to have a special work performed, and I in my adamic form in paradise offered Myself for His service instead of desiring to receive merely an experience of change and variety. My parital part had preceded Me and having also been thus actuated was placed where a work was to be performed. Her work was a great one and her trial a severe one. She passed through it nobly.

How many pure-minded young women of the present or any other time could be found, possessed of unblemished reputation, betrothed to an affectionate, kind-hearted man, soon to be received into his house, possessed of good name, warm friends, and the respect of all her acquaintances, who could bear with submission, with unrepining words, with perfect assent and resignation the announcement that she had conceived by the power of God without the knowledge of man and that she should thus bear a son who would be known as other than her affianced husband's? Is there one pure in heart, in word, in deed? Is there one who could so receive this announcement, with a foresight of the consequences which might reasonably be expected to arise from the incredulity of mankind such as the displeasure and sacrifice of her intended husband, the averted and pitying regard of her friends, the disgraceful dishonor of her acquaintances, the contempt of the villagers among whom she resided; indeed, in her case also the apprehended punishment provided by the laws of Moses there established and enforced—is there, I say, one such virgin to be found now? No one will presume to declare that wife or mother, sister or daughter of theirs—that they themselves could bear such an ordeal of their faith. Read, My friends! Read the account of this extraordinary occurrence as it is simply and unostentatiously related in Luke and see how much more wonderful it appears to you than it ever did before.

Luke's Gospel or Narrative of the Glad Tidings preached by Me was written as he describes, after many others had undertaken to set forth the same events in a particular manner; but he, perceiving in all a want of particularity in this respect, was careful to give the account more fully than any other who had written. There is another account of this event contained in a work called apocryphal and which indeed is greatly corrupted in its text, where Joseph's and Mary's previous history is in the main correctly given and many details respecting their flight to Egypt to which no allusion is found in the present Bible.

As I have said, there were other scriptures of gospel truth, which the council of Nice rejected and which religious partisans perseveringly destroyed, which contained better and fuller histories of My life and death and purer accounts of My sayings and doings than much that was retained. Not that John was not as able to give a correct and full account as any, but that he wrote with the knowledge of what had been written and principally to supply omissions or correct misapprehensions or incorrect reports or opinions based on wrong inferences from already-written accounts. It was particularly against the perversions of a Cerinthus that he wrote when he composed his gospel. Mark, too, wrote for a particular purpose, which was in part to display the character and office of Peter and in part to correct misapprehensions of Peter's office, already beginning to elevate him to a pre-eminence he neither claimed nor was entitled to. Matthew wrote earlier than either of the others and, being a Hebrew, he wrote in Hebrew, but with a design to show to the Jews that I was not only a man and a brother but a prophet and a seer, a Messiah and a Prince, and that I had fulfilled the laws of Moses and his ritual not as an example to others which all should follow but as a completion of its work, as a reproach to the rulers who caused the crucifixion.

The part of Mary was well performed. She submitted cheerfully and gracefully and thankfully to the will of God as announced by His power to her. She was not deceived by trusting to Him nor was she destroyed or brought to shame. So it will ever be with God's servants. Her work was then in great part done; the common feelings of maternity would do the rest. Her life was a peaceful one except as she sometimes suffered from the consequence of My persecution after I had begun My ministry. She was left in the care of My beloved apostle John, who took the bereaved widow to his house and soon after laid her in a quiet grave where her remains returned to their original proper combinations. Her other sons and daughters were still residing in Nazareth and struggling to maintain themselves by honest labor. They afterward received the belief that I was a miraculous son of their mother and that I had been the Sent of God, but, as I have before stated, they never fully comprehended My mission and their doctrines, combined with their natural relationship, caused many to hold back from that state of progress and liberty which I designed to impress upon My followers.

And I, having also fulfilled My mission, received My reward. Though differently generated I was an ordinary man as to body. Though differently actuated I was an adam in being. My spiritual body was like that of another, consequent upon My entrance into My earthly one. I was blessed with God's help as all men may be. I had advantages in My earthly purity which others do not have but which some will have hereafter. These advantages, though, are not such as affect God's favor. He helps according to need and the sacrifice of the heart is all that is required of any man. God will do the rest whether the man's nature is such that the rest must be much or little. The one is as easy as the other for God to accomplish. In the spirit-world I advanced rapidly because I had done much on the earth.

So it will be with all. They who do the work here may do in a few years what then will require millions of years to accomplish. And yet you are informed that I required many hundreds to reach the highest circle of the seventh sphere! A hundred years ago I was not in it. But though I have so lately arrived I am a full, a perfect son. I am equal to all others who are with Me and none below the Word is superior to Me. The high and perfect sons of God are all equal, and all will be equal to them who become one with them in their circle. One with them in will they may be now by submitting to God's will, because the will of these sons is God's will and whatever He wills they will. So when man in the body sacrifices his free-will he has the will of God in him and his will is God's will and God's will is his.

It is to this state I desire you to approach as nearly as you can. Then shall the will of God be done in you as in Me, on earth as in heaven. Progress then; progress to perfection! Be ye perfect even as I am perfect, even as your Father in heaven is perfect. So shall you live and love God and advance to unending happiness. But if we are perfect even as You and as God, shall we not be already in perfect happiness? No man is perfect thus, but if any man should be perfect thus he would be already qualified for perfect happiness and would be in the fourth sphere when he had reached the spirit-world. The perfection to which you are called is perfect resignation; perfect knowledge you cannot have without its reception which is received and can be received only by passing through the spheres and circles assigned for that purpose. The next state is the only one in which you can receive such knowledge and therefore you can never enter those circles while in the body. Had I continued on the earth in the spiritualized earthly body I wore after My resurrection, I could have been no higher than the first circle of the fourth sphere. High as this is, it does not give power to act with such knowledge as can benefit mankind in a high degree. The acts performed by such spirits in their own knowledge must be little more elevated or advanced in wisdom than if they were in the body, except as their knowledge of men's thoughts and actions derived from reading their memories can aid them much in inferences as to what men will do.

Here we pause again and reflect that Jesus of Nazareth possessed the power to read the thoughts and memories of men when He had ascended to heaven only by having passed through the appropriate circles for that knowledge, but that He could not on earth so qualify himself to pass thus rapidly through the higher circles above the third sphere, for there His rapid or momentary progress through circles ceased. No longer continuance on earth, no more services rendered or honors given or mercies received, could have qualified Me to be so instantly, as it were, passed through circles I did not then pass through. So I had done My work on earth. I had done all that could benefit Myself; there was no more work for Me to do that I could not better do in the spirit-world than in earthy form. Here is another reason why God should not have saved My life from those who desired to take it, and here is an answer to all such as said: If Thou be the son of God, come down from the cross and we will follow Thee! They lied when they said so, for they would only have returned to the old charge: He casts out devils by the prince of devils! They would have been animated by new fury and would have railed against Me as only disappointed revenge knows how to rail, as only envious detractors know how to scold and storm.

Here is another instance of the inutility of outward manifestations as such, for when I was arisen from the consequences of death on the cross, My body being missed from the tomb in which it had been laid notwithstanding its being closely watched, the very persons who had procured My crucifixion as an unauthorized assumer of divine authority, instead of being led to believe they had made a mistake and thereby rejected the very Messiah and Prince they hoped for and I claimed to be, were filled with rage and torn with spite. They bribed the soldiers who had faithfully watched to say that they slept in order that it might be said and believed that My disciples had stolen away the body. And what inducement could My disciples have had to steal it away if I could not reanimate it? My disciples were either deceivers or deceived. If the former, would they not rather have claimed one of themselves to have inherited My power, authority, and mission; if the latter, would they have desired to support their delusion by deceiving others? No, My disciples had no object to serve in procuring My body that Joseph of Arimathea had not already secured by begging the body of Pilate, who had given him full control of it. They sought not its concealment or they would not have placed it where it could be publicly guarded. They only sought to know whether it were to follow that I would raise the body as I had promised them that I would, in order that they might have at last an excuse to themselves, if I did not so raise it, to return to their homes saddened indeed at My fate but rejoicing that they had escaped from it themselves, for up to this time none but John had displayed any courage or steadfastness of faith in Me, and even he was greatly discouraged when at last I expired on the cross from which he fully believed I should surely before death have been rescued by a miracle. John too desponded and feared he had been deluded by his friendship and his hope and his senses, for he could not doubt he had witnessed miracles but this one was yet wanting. So it ever is; the age of miracles is never past, but when God withholds them, man will not cease to call for them. All that they can do is to refer you to that inward light, that divine soul, that Christ in you, and to the revelation which has taught and does teach it, and to the present revelation now being made which shall indeed surpass what has been before by being fuller and more perfect in its aim and arrangement and details. It is from no higher source, only men are prepared for higher truths and a medium passive enough to promulgate them has been found.

My holy medium has witnessed outward sounds, as I described previously. He has again, since writing the preceding paragraph, witnessed them, but with no increase of faith, with no desire to have them, without wishing to communicate by or through them. The medium for the outward sounds was a good one but neither he nor the spirits were aware that I communicated more to My holy medium in one minute than they delivered or received in the whole evening. The table was moved but, however interesting and solemnizing such an exhibition of spiritual bodily strength or control of matter should be, it was marred by the doubts and tests entered into by some who were present. The sounds of raps were distinctly given, but some wanted them louder, and for what? Because whatever may be done outwardly must be unsatisfying to the want of the soul. Reason may be satisfied; the soul continues to crave. If the table is moved, reason is satisfied that the intelligence and power that moved it is spiritual.

Instead of asking that the table should be raised, as this craving does, reason ought to say: What shall I do to be saved? This is the question these manifestations are intended to answer. But they will not answer until they are asked. Ask and ye shall receive an answer. But will that answer be always truthful or even consistent? As being the answer of a spirit unregenerated it can not be relied on. The spirit will not say it has received salvation and it does not fully understand what it has not experienced. The spirit will say it is happy. So is man now in the body, even the most unreconciled to God, at least happy enough that he does not desire to die in order to be happier. So the spirits who communicate outwardly are happy but they are not so reconciled to God as to be happier. They are making atonement for the sins they committed when in the body, which sins left their prints on the spirit and the soul, and these must be purified and washed to pure white in the blood of the Lamb slain from the foundation of the world before they can enter the arena of true and perfect bliss.

What was or is the Lamb slain from the foundation of the world? It is the power to serve God which exists in man, slain by man and whose blood cries aloud to heaven. This Lamb is slain by man and its blood washes out the sin of its murder, of its slaying, by means of repentance. Repent and be baptized in it. That is, receive again the power to serve God by being again endued with power from on high to serve God, which power is the power of salvation given by God's grace to man. Man generally crucifies this Sent of God, again and again destroys the Lamb of God. This Lamb of God is passive, unresistant to the will of man. When man slays it there is no resistance; when he desires it to be renewed it appears. But where is the blood, and how are we washed in the blood of the Lamb because the Lamb is restored to life by our desire or sacrifice of will? The blood is on the will of the man that slew the Lamb. The desire that caused the will to act is the true murderer. The desire must be punished by atonement which is required. The very realization of the desire is in general its punishment. If not, the desire continues and exerts itself until either in this life or that to come it does experience the disappointment and loss of happiness that its realization does give. Then the desire begins to make atonement which must be fully made. The will is bathed or washed in the blood of the sacrificed power of God. The sacrificed power or Lamb of God again returns to life when the man, having made atonement for his desire's act, becomes the true seeker for God's direction. The art of seeking for God's direction is the sure regenerator of the heart of man. The blood of the previous sacrifice of the lamb is washed away and the sinner's heart or will becomes pure white as wool. The Lamb of God takes away the sin because it is always ready to be offered up and the man avails himself of the sacrifice made by the Lamb when he profits by it; he profits by it or is washed white by it when he becomes reconciled to God and found worthy to be a servant, or son of God, and receives the Lamb of God into his heart; which Lamb, being the Power of God to do God's will, gives the man a right to put on the garment of praise because he is born again into a new probation.

The blood of the Lamb cries aloud from the earth: How long, O Lord God Almighty, wilt Thou not revenge us? This is not the language of martyred men, as some have supposed, for such men would scarcely be supposed to have died for Me for the sake of the precepts and doctrines taught by Me if they paid no regard to the injunctions: Love your enemies, and do good to those who despitefully use and persecute you. It is the Lamb of God, one power but existing in many manifestations, that has been so slain and is so crying for relief, so desirous that God should make an end of man's resistance by renewing the lives of the Lamb.

There is in every man a desire to be saved with an eternal salvation and this desire overcomes all others at last. The desire to retain his free-will prevents the realization of the desire to do God's will, which is in reality being saved. When man makes the sacrifice God is pleased; the Lamb of God is no longer slain, for when one sacrifice is made the other is not required. The death of Abel and the pardon from death of Cain is typical of this. But the sacrifice of the power to do God's will, which is accomplished through man's determination to do his own will, is the only charge God makes to man of indebtedness. God continually calls upon man, as it were, to pay his indebtedness by restoring to life His power in the man's heart or will, by sacrificing to Him the free-will with which he is endowed. This man may do at any time. This God is always ready to accept and by this sacrifice the Lamb is restored to life and the man's sin is washed in the blood of the Lamb which he slew before. The man's sin now becomes as white as wool, though it was as red as scarlet. Then, being a redeemed and purified saint or holy man and a man greatly learned and experienced, he is raised to be a son of God by being passed through the remaining circles in the spirit-world as fast as the extent of his atonements and blemishes of his character will permit.

The great question or objection in effect remains unanswered. What is the meaning of this expression as used in the Bible, the blood of the Lamb, and what was it understood to mean by the writers and the readers of the time in which it was written? It was then understood to refer to the sacrifice of Christ, or Jesus

341

of Nazareth, and it was supposed that, as I had died for all, My death saved the sinner in some unknown way. I say the understanding at the time was that I was raised up that all men might be saved, and that men to be saved were called on to sacrifice their wills, which was the meaning of the expression: Take up your cross.

There are in the Bible many expressions which were, when used, highly expressive though very figurative. The course of preaching leaned very much upon figurative illustration in those times, as it has since, and there came to be many set phrases which were in fact allusions to long illustrations, each of which was well understood by those who had heard the apostolic preaching. But by degrees the figure or illustration was lost sight of, while the set phrase continued to be used; it acquired a more literal construction and in many cases conveyed thereby an imperfect or erroneous meaning. The expressions I have so long dwelt upon are of this character; well understood at the time, they are now obscure. Entirely allusive and figurative in their signification, the literal or verbal signification can neither instruct nor in other way profit the reader. Seeking to know the inward meaning is unnecessary because there is abundant precept remaining, unacted upon, that is clear and unmistakable in meaning. So I will leave the subject for your present reflection, to be hereafter made plain to you in some other book.

CHAPTER XVI

THE COMPENSATIONS OF GOD'S PROVIDENCE

Brief History of the Church of Christ

Let every man be fully persuaded in his own mind, was Paul's advice to believers who inquired upon particular points of doctrine. Paul's advice was good. Does his advice need indorsement, then, to prove it good? No; in general he spoke by inspiration and wrote by revelation. But there are many different opinions about Paul in the present day, as there were in his own. There are also many different opinions about or respecting the Bible in general or the New Testament in particular. I urge you to faith in all of it that purports to be inspiration or revelation, but I also urge you to seek to understand it. Not only to seek to understand it superficially, or its verbal meaning, but to understand its internal or spiritual sense, in which its greatest usefulness is contained. In the early ages of the church this interpretation was by man's reason carried to so great an extremity that it produced a reaction, and the mystifications of Origen and other fathers, as they are called, yet exert an influence the very opposite from the intention of the writers.

Though reason is useful and should always be used, inspiration and revelation are both superior to reason though they will never contradict it. Reason can never lead a man to make acknowledgment such as inspiration or revelation may give, yet reason will help him greatly to understand and appreciate them. Reason cannot help a man to discover the deep things or correspondences of inspiration and revelation, yet when these are declared to him, by reason he can appreciate and receive them. I am thus particular in defining the office and power of reason because many are withheld from practicing the precepts of spirits or receiving My revelation because there are parts above their reason and they truly say these are not reasonable to them. They err in not endeavoring to elevate their reason, instead of undertaking to bring their instructor down to the level of the pupil.

Let us return to Paul, the twelfth apostle, chosen by Me to be the Apostle to the Gentiles particularly, though he often preached to Hebrews, and others often preached to Gentiles. Yet Paul's greatest efforts and successes were with the Gentiles or Greeks and Arameans, while the others boasted more of their converts from the children or reputed descendants of Abraham. Paul's eloquence was of a high order. His appearance was mean; he walked lame, his stature was low, his features were irregular. He was a crabbed-looking bachelor, and when he advised as a man respecting matrimony he advised as an old bachelor in these days would. But I chose Paul for My apostle because he had learning, fluent speech, ready wit, and remarkable intelligence. He became passive and I used him as I desired to and with consequent success.

Alas, that the apostolic fathers, as churchmen call the bishops or other writers who lived in the time or at the close of the time of the apostles, should have been so inferior, so outward, so corrupt! They let Paul's mantle fall to the ground, and even before John had laid down his bodily life at the command of the persecutor there were fathers, as men would call them, in the church who not only disregarded John's authority but defied his power and condemned his doctrines. Some would charitably attribute what they considered his weakness to senility, others derided his simplicity or openly professed that My disciples had not only misunderstood Me in life but after My death had not been at all distinguished for soundness of doctrine or wisdom of counsel. The church, they thought, would be better governed and more successfully advanced in power and extent by their reason than by the apostles' inspiration. Already had the signs ceased to be generally manifested under the preaching of My pretended servants and even were often wanted ineffectually when My most worthy followers desired them for good purposes.

But the past was no criterion for the apostle's successors. They desired to extend the church, to build up their own power, to establish their own sanctity. It was not until the third century that the search for relics became fashionable. Then the church began to desire to know what was first taught and to reverence those who taught it and all connected with it. Then it was too late; the reverence for the past fell to a superstitious outward seeking for holy places and remains of holy personages. Splendid churches, elegant rituals, were established or founded. All became pomp, and what was wanting in vitality they endeavored to eke out by show of life. What was wanting in vital religion was made up in a semblance of veneration for what had been accessory to the foundation of Christianity. The power of the throne was lent to the honor of the church's governors, and the funds of tributary provinces squandered on the adornment of the seats of ecclesiastical dominion. It is mainly because of these circumstances that so little is left of the early history of the church. The successors of the apostles desired to magnify themselves at the expense of the apostles, regardless of inspiration or revelation. To the Gospels, numerous as they were, no reverence was attached by any class of believers. To the Epistles little regard was paid, as they were merely thought at the best good sermons.

The generations succeeding these and existing about one hundred fifty years after My birth were invested with the same faults but greater ignorance. This twilight existence continued with few more illuminated sections to the accession of Constantine to the sole sovereignty of the Roman world. His policy changed the church to a visible hierarchical state engine. He used it to secure his power and lent his power to secure its establishment and general prevalence. He, too, assumed to be its head on earth, to be above all authority elsewhere to be found in it; yet it is a question unsettled by history whether he had become even a member of visible ecclesia or church at any time before his last sickness. However, he had done this at a much earlier period in a private manner, though it was at the latter time done in a public and ostentatious way. He was an ambitious man stained by crimes and cruelties, prodigal of others' wealth. Being as he was full of his own desires, he converted the riches and glory of the church to his own aggrandizement. His successors maintained their title of great priest and the outward headship of the church until a late period, but they gradually lost the essence of their power to the combinations of priestly arrogance that successively undertook to decide by artifice and fraud the great questions of doctrine and church policy that continually and successively rose into consequence.

This state of things was growing worse when the Arabians or Saracens overran Asia Minor and Egypt, the chief seats of corruption. The various German or Scythian nations that overwhelmed the effeminacy of western Rome infused new life into the people and the church of those parts of Europe. Though corrupt in form and evil in tendency, the church of Rome, as it has since been called, at that time was the repository of the learning, the piety, and the real Christianity that existed. Though it often happened that other influences controlled even the church, in the main it taught the precepts I delivered when in the body and laid up for many of its sons crowns of glory eternal in the heavens.

The next event of great importance to the church was the extinguishment of the eastern Roman or Greek Empire, completed in the fifteenth century, followed as it was by despoilment of temporal power from the clergy of the church of Constantinople. This event left the church more at liberty to worship God and accordingly it may be found that the church improved in its forms and purified its doctrines and practices. Simony, or the holding of double-offices in the church, became less common, and, in general, as the church becomes despised by man it grows in favor with God, because whatever may be the doctrines or even practices of the members of a church God looks mostly at the motives for professing and means for following its prescribed forms, and He sees the desire the man may feel to serve God or men or himself.

Here we pause again to let every man ask of himself a reason for the faith that is in him. Paul was desirous that the professors of his day should look well to their foundation, and at this time it is more than ever important, for now God is showing or having shown forth to the world great manifestations of His love and power. God desires not your form of adherence to a church or association of men but a practice of the precepts He has caused to be revealed to you for your guidance in former and present times. He would have you serve Him, not yourselves. He does not take pleasure in seeing you worship with words or acts of homage to Him, or in seeing you attend your meetings in order that your neighbors may perceive that you do your duty. He wants you to surrender to Him your heart and listen attentively in the cool and quiet hours of every day for that voice with which He spoke to Elijah and by which He makes Himself heard in the hearts or minds of all men who seek to find Him and hear from Him what they must do to be saved. They must resolve to follow His counsel before they receive it. They must be willing and desirous to receive it or it will not be given; so long as they revere men or their institutions more than God they will be left to the idols they ignorantly worship. Yet ask and ye shall receive; knock and it shall be opened unto thee.

Let every man be fully persuaded in his own mind and able to give a reason for his faith. If God is your hope, I will be your help. If it be on Me that you rely, I will be found ready and willing to dwell in you and remain always with you as your wonderful Counselor and your mighty God. I will give to God the glory, honor, and praise you give to Me, but I will also help you the same as if you had addressed Him. If you look for another spirit, the Holy Ghost, the Word or some mysterious part of God, to be your inward monitor you shall not escape from Me, for I am not tenacious of form or nominal acknowledgment. All I want is that you surrender your free-will to a superior intelligence that you believe to be divine in its origin and authority, and I will be with you to the end of the world. Now I say let every man be fully persuaded in his own mind of what he wants and who he wants to obtain it from; of what he will do with it when he gets it and what he hopes to obtain by means of its possession. Thus

345

shall you be fully persuaded in your own minds that I am Christ, the Son of the Everlasting God, a truth that flesh and blood, man's reason or teaching, cannot tell you of effectually, and which, if you ever do know, you will know from My Father which is in heaven. Shall I reckon you among the glorious company that eternally praises God, saying,

Great and marvelous are Thy works, Lord God Almighty!

Just and True are all Thy ways, Thou King of Saints!

Or shall I seek only from others this acknowledgment? You will say perhaps that you are willing to accept revelation made eighteen hundred years ago but that you cannot receive this which comes without notice or herald or signs. I am equally satisfied if you will seek God and do His will under and by means of the former preaching, but I have little hope that you will. Eighteen hundred years have obscured the force and varied the construction of language so much that you do not, when you read, gather the true meaning, the sentiments I really expressed then. This is freshly given and has not been transposed and pared and stretched to suit theories or maintain traditions or sustain men's establishments. Here, you will find, I ask you to believe former revelation and to practice the precepts I long ago gave to the world. I ask no more of you than that you should sacrifice your free-will, that you should admit Me to your heart, and that you shall listen to Me when you have established My residence in you. Hail to the chief who will overcome all evil desires, who will put down all rule and authority, who will give to God all you give Him, and will have all things under His feet because God has given Him authority and power, because God will have all men to be saved from sin and death!

O death, where is thy sting?

O grave, where is thy victory?

The sting of death is sin,

But blessed be His name who has given us the victory

Through our Lord and Savior Jesus Christ.

To whom be honor, praise, and glory forevermore,
world without end.

Amen.

Explanations of Scriptural Passages or Texts

What shall a man do to be saved? This question I asked for you and answered in Volume One, Part I, and in every part, and yet you will say it has not been fully answered because I have not told you of some mysterious manner in which it may be done, or because I have not told you how to do it by some particular form of doctrine or creed, or because I have not told you how to secure it by some particular specific act or acts. Any of these you think you could receive and act upon. But you deceive yourselves, first, in supposing you could act upon them faithfully and without help from God; second, in supposing that man can be saved by forms or professions or in any way but by that of submission, to do God's will and let all else pass by as idle wind that knows not whence it came nor whither it goeth. Let every man be fully persuaded of this and the work of salvation is not only begun but it has already made much progress. You are already far advanced on the road to heaven and you will find the path strait though narrow, direct though somewhat obstructed occasionally. Still that path can be traveled with great joy and pleasure because directly in front of the traveler is arrayed the Holy City, the bride adorned for her husband, and on every side he views contentment, peace, and happiness. The peace of God that passes all understanding is the very atmosphere in which the traveler lives and moves and has his being.

. Being then thus called, O man, thus chosen to receive God's bounty, reject not His mercy, be not faint-hearted, press forward to the high and holy calling wherewith ye are called. Be faithful, obedient, humble, passive, and you will see God's salvation; be pure and perfect, even as I am, and you shall see God. Now you can see Him in His works, but then face to face as a man talketh to a friend. Has then God a face which He wears when He talks to His sons? The expression is one of those figurative ones to which I have before alluded. When Paul preached he used an illustration like this, that here we should study to know God by His outward creation and by His visible manifestations, that hereafter we should know Him as we know a friend inasmuch as He is our friend. Paul taught that in this life our bodily organization prevents us from seeing Him or talking with Him in an outward or visible manner, but he said that after we have left the body we should so much more resemble God in form and substance that we would be able to talk to Him, as it were, face to face as a man talks when in the body to his friend. The expression is a mere figurative allusion and, being perfectly understood by those to whom it was addressed, attracted no especial attention. In process of time the generation that understood it passed away and when attention again became attracted to the epistle after the commencement of the reign of Constantine there was no memory of the preaching to which it alluded.

Let every man be persuaded in his own mind of the truth and be able to give a reason for his faith, for any other kind of faith or belief can do a man no good. A mere blind declaration that he believes a book, even if that book be the Bible, is no credit or advantage to any man. A living belief in its truth may be, a living belief that impels a man to act on its precepts, to endeavor to understand its meaning, to endeavor to reconcile its apparent descrepancies or even contradictory terms and declarations. Thereby man's mental strength is exercised, his reasoning powers improved, his spiritual perceptions awakened to increased clearness; the windows of heaven, as it were, may be opened to him by conversation with the ministers of God's word, not outward ministers but ministering spirits.

Now there are many who say I do not believe this or that or the other assertion or declaration, no matter how supported, because it conflicts with the Bible, and yet perhaps the contradiction is only to their interpretation of the Bible. It should never be overlooked that much doctrine passes for Bible declaration that is not that; many conflicting theories have their sole basis on what is imagined to be found in the Bible but is not found there. A learned man some hundreds of years ago declared perhaps that such a formula of belief was the result of his study of the Bible, and now his formula of belief and even the inferences from it are thought to be the Bible itself. It is this error which has thrown more discredit on the authority of the Bible than all the direct assaults of unbelievers. Truth can be defended successfully, but error can be maintained only by crime or violence. Thus religious persecution arose and thus religious wars were thought to be a necessity. But now, having a passive holy medium, I shall endeavor to elucidate some of the more prominent texts that bear upon the sectarian differences of those who desire, many of them, to be My servants and which are stones of stumbling to many inquiring minds that love truth and are willing to go anywhere to embrace her.

The ungodly shall perish. This is generally understood to mean that they shall in some way suffer an eternal deprivation of happiness. The true meaning is that the ungodly portion of a man's mind or character shall perish. *It is better that a hand, an arm, an eye or a foot should be cut off and cast away forever than that the whole body should perish with the same disease.* Still more, then, in spirituals is this true. If a portion of a man's life has been ungodly (and who is there that is not this man?) it is better that the ungodly portion of his life or experience or the memory of it should perish than that the whole should be made to suffer forever from contact with it. Though figuratively something like this takes place, the memory of that portion does not wholly perish—it receives an antidote, as it were, by atonement. The spirit in the world to come seeks to escape from the unhappy state of unlawful desire, first by its unlimited, though as I have shown unreal, indulgence. Failing to secure happiness or relief by this course, at last the spirit is persuaded by the example and precepts of other spirits to discard the indulgence of that desire and to sacrifice to God its power or will to follow that desire. Then so much that was ungodly perishes. Another desire and another indulgence remains, perhaps many more, but at last all are conquered in the same way and then, all that is ungodly in that man having perished, he is righteous and reconciled to God. He is God's servant and finally becomes His high, holy, and perfect son.

But, you say, there is another expression connected by its subject with that text and sanctioned by My quotation of it heretofore, that seems to require a different interpretation, and that is that God does not desire the death of the sinner! From this you infer: first, that the sinner dies in spite of God's unwillingness or at least want of desire that he should; second, that he meets with a fate that he might avoid if he tried to; and third, that if he tries to he will have God's help and mercy. In all these I agree with you. But from your deductions on this and other texts you further infer that when the sinner dies, God is accessory to his condemnation and that he is punished because he dies and that the punishment is an eternal torture. I disagree with these last inferences. First, God does not will or desire the death of the sinner, who therefore dies by his own will or desire improperly exerted. Second, the death of the sinner takes place by a law of God made for his benefit, for we cannot conceive of God making laws to injure His creatures or children. The death, then, cannot be eternal or the man could not be benefitted by it. Third, the sinner—having died by his own will, and God's will being that he should not die if He has any will about it—the sinner may live by yielding his own will and coming to God. The death then cannot be eternal torture for we cannot suppose the man would willingly continue for a moment in a state of torture.

But, you say, I have admitted that spirits continue for a long while dead or opposed to God! So I do, and I also say that death is not torture but merely deprivation of happiness by absence of God's love and harmony. The soul or spirit or man gets along with its own will because it does not perceive how unhappy it is, comparatively, but as soon as it becomes truly aware that there is greater happiness to be had by merely relinquishing the imperfect happiness and becoming passive to God's will a reformation, a reconciliation takes place, and the sinner, ceasing to be a sinner, passes from sin to love, from death to life, from punishment to rewards.

Again you object, saying that it is declared: As the tree falls so it lies—though this is not the exact phraseology it might as well be rendered so from the original—and then you say: There is no repentance in or beyond the grave. Now I have explained a good deal of difficulty attached to these passages in My first book, but they were not yet made clear to such as were prejudiced by a belief in *eternal* punishment. Those teachings were rather addressed to such as did not believe in *any* punishment, a doctrine more subversive of morality than the other, though less injurious to understanding of God's character. It is rather a conclusion forced upon men by a considerate regard for God's attributes than warranted by any text or combination of texts. The tree falls this way or that way and rises no more. So it is with man. Death of the body overtakes him and we see him no more righted into earthly existence. He goes down to the silent grave and is seen no more forever. That is, of course, his body is seen no more and as it fell it lies, a mere mass of dust or earthy matter. It shall never more live in that form; it shall be raised neither in this world nor in the world to come. As it fell corruptible, so it lies corruptible; as it fell earthy, so it lies earthy; as it fell a dead body, so it lies a dead body forever and forevermore. For it there is no resurrection. That is all you should gather from that text.

But there is no repentance beyond the grave! Is not that fatally contradictive of Your position respecting the death of the sinner and fate of the ungodly? Beyond the grave there lies, to the spirit of man, another world. This we both believe. So far we go together. Then you would say: *And after death the judgment!* so do I say also. *Beyond the grave there is no repentance;* I said this in the Bible and I say so now. The man who was a sinner, or ungodly, perishes or dies to God's love and the performance of God's will and he does not repent and receive mercy. But he atones and returns to God. Repentance will not do; atonement is required. The time for mercy or pardon for his transgressions performed in the body is past. Death of the body closed that door of mercy. God is so good, benevolent, merciful, that He saves the man or the spirit of the man from becoming any worse after entering the world of spirits, but He allows him to seek his own gratification, as I have explained. By gratifying his desires he atones for having formed them in his bodily existence. He atones for them by changing his desires to do God's will rather than his own, and God's will is that he should atone for the deeds done in the body.

This is the judgment of God and the spirit freely accepts the decree. The atonement proceeds by the spirit performing good works as recompense for his evil ones, forming good desires where he had formerly cherished evil ones, and being willing to benefit where he had desired to harm. For instance, a man in the body has robbed another of his reputation, a worse theft than a burglary. Having so far advanced as to have reached the circle of good works, he desires to make amends and he exerts all his powers under favor of God's will to make amends. But perhaps ere this time the wronged man, as well as the one who wronged him, has reached the spirit world and the two are together. There they can forgive each other or forgive and be forgiven. Yet the atonement must be made, after which the spirit seeks some similar case perhaps then occurring and makes atonement for another's act, not thereby relieving in the least that other's crime or punishment but making atonement to the injured one for his own act performed on another. Yet there may be some act of so rare occurrence as not to admit of this kind of atonement! Still, all acts belong to various classes and he can make the atonement in the class to which his act belonged. Besides, each spirit has almost universally many acts to atone for and many, too, in each of several or many classes. The two grand divisions are hate and revenge. True selfishness is the greatest cause of all, but the two great divisions in which the worst manifestations of selfishness occur are as I have mentioned. But, you say, a man cheats or lies not from hate or revenge, but from desire to save himself from unpleasant want or consequence apprehended or experienced! So it is; there are subordinate classes which belong to the third circle of the second sphere and the atonement is made for them after hate and revenge, belonging to the first and second circles, have been atoned for.

All must be so atoned for as to eradicate from the character every trace of its former debasement, but memory retains not only the crime or sin but the remembrance of the manner in which it was atoned for and the ample reparation it has witnessed performed in any part by the spirit of the man; in part by spirits previously atoning and in part, or all, by God who has been pleased to prevent the consequences which would otherwise have proceeded from the acts of the man in the body. Have I made it plain to you now? Read it again and see if you do not understand it better than before.

There is one, case with which you can perhaps puzzle yourself or others: a case like that of Mohammed, who, having been a rank impostor, led many astray from better faith, perverted religion, and devastated the earth in its fairest parts by bloodshed and violence consequent on his example and precepts. Yet Mohammed was not all evil. He saw much to excuse to himself his imposition, and he scarcely ean be said to have made men worse than he found them. I have before alluded incidentally to him and his doctrines, so I will pass to his atonement which is now being made. He sees the harm; he is endeavoring to operate on the minds of his followers spiritually to reform them, to bring them to a knowledge of truth. His labor is a Sisyphean one, but he has the help of many who followed his doctrines and of many spirits united by sympathy from having taught other injurious doctrines. The labors of Mohammed are pleasurable, too, though Herculean in prospect and Sisyphean in effect, for he has now the consciousness of being actuated by pure desires to serve God and his fellow-men and a consciousness that he is atoning for the sins he committed. But he does not know so fully as God and higher spirits know that atonement is required not in proportion to the harm done, the evil consequences following or continuing to follow, but in proportion to the sin of the heart. The heart may be sooty black with sin and yet the acts committed may have been harmless in their results. Again evil consequences seem sometimes to flow from good desires and intentions to act only for good. It is not the consequences to others which must be atoned for in reality, but the mental or spiritual consequences to the actor or criminal himself.

So Mohammed might easily be no more criminal than one who embraced his religion, as it is called, without belief in its truth and made it merely a cover for his own gratification in some kind or kinds of selfishness. Such may be found even in Christian churches, and such are as guilty as Mohammed, and the atonement required of them will be equally fearful and appalling in its extent and labor. Search well into your hearts; see on what foundation you stand, for in its appointed time the fire will come and try every man's work, and its quality will be evident to all the world of spirits, to the man himself, and to God. Be sure that you lay a good foundation and that the superstructure can withstand fire from heaven, the consuming fire of God's love which will not let sin exist or the sinner die eternally. The recovery and the purification are in the nature of a punishment though it is of a different character from anything visible to earthly observation and is only from everlasting to everlasting instead of being unendingly eternal.

CHAPTER XVII

THE HISTORY OF THE BIBLE

Deficiencies of Holy Mediums of Revelation

Lest any should say that I avoid the main question and endeavor to select only such texts as are easily molded to My theory, I will leave the selection to others but lay down certain principles of interpretation, certain rules for understanding scripture contained in the Bible which I hope will enable all who desire merely the triumph of truth to understand for themselves. There is, to be sure, no general rule to be universally applied, but there is a general plan which may be applied to any part.

The Bible is a collection of the writings of different holy mediums or prophets or of men desirous to relate truthfully some portion of history. Of this last character is much of the Old Testament. It does not profess to be inspired but to be a compilation from other records then existing, now lost. No doubt the original books would have been found more instructive than the abridgment or compilations, but they were too voluminous for a general use and no one ever claimed them to have been inspired or necessary in any way to the welfare of mankind. Yet men have chosen to consider the abridgment of these books as of divine authority, and a belief in the facts related and the inferences drawn by their various authors, the priests of the temple for whose use they were kept, as gospel truths as essential as any other portion of the sacred volume.

I dissent from this practice and, as I have said before, take only such parts to be from God as claim to be so. A man who would write as from himself what is given him from God for others should be condemned for endeavoring to secure glory and honor to which he is not entitled. A man who would declare to be from God that which he knows to be the formation of his own mind is guilty of blasphemy. In either case he and his work or the work through him is discredited, for in the first case, God either desired it to be known to be His word or He did not. If He did, the holy medium or prophet was not faithful, passive, or obedient and therefore unreliable for the transmission of truth. If He did not, then for some wise purpose God, having declined to appear as the author or originator of it but choosing to let it appear as the work of man, must desire that it should be regarded as man's work and not as His own. No one will say that God should not be allowed to choose in such a matter, and all must admit that it is presumptuous for man to declare that to be from God which God does not claim or acknowledge. In many parts of the Old Testament the writer says that the word of the Lord came to such or such a man or holy medium or prophet, and that he then declared certain things hard, perhaps, to understand, being couched in very figurative language and obscurely left unexplained. As I have said, if the writer of the book did not profess to write by God's influence or direction, we are to suppose he wrote by his own and that we are reading history and not revelation. Some will say that the reception of it and use by the priests of the Temple of Jerusalem is evidence of its truth and of the reverence we ought to bestow upon it. The priests of the temple were almost unanimously in favor of it. When the prophets of earlier time were slain the priests of the temple in general prompted the act. They were men of like passions with other men, and even the prophets also were such men. King David of Jerusalem is regarded, and justly so, as a prophet. His Psalms or Hymns of Praise are beautiful and lofty compositions; in them are many prophecies of which most have already been fulfilled and others will be. Yet David was by his own declaration prevented from undertaking the building even of the outward temple to God by the crimes or sins he had committed, because it will not do to say the expression *man of blood* does not imply crime. God is not so unjust as to punish a man for what is agreeable to His will. It must therefore have been deeds of blood performed contrary to the will of God for which David was debarred from what he was desirous of performing.

Solomon was a prophet, too, and moreover was a speaking medium, because of which his wisdom has become proverbial. Solomon was far from being a pure-minded, single-hearted man. Oh! but, you say, he was pure and perfect when he wrote his sayings and undertook to build the temple! Solomon's reign was long, his acts were many, his apostasies frequent, his crimes flagrant, his punishments severe, his wisdom unintermittent, his glory as a monarch unclouded in the eyes of his people, his devotion to the building and adornment of the temple unceasing. Yet at all periods of his life he wrote by inspiration. Inspiration does not depend upon morality, but upon God's will. God uses such instruments as are willing to be used when and as He pleases to use them. An attentive study of the events recorded of Solomon and of the books written by him and yet preserved will show you that I have fairly declared the principal facts bearing upon the question as to the holy medium or prophet's being necessarily a normal man, a consistent liver and actor of the precepts he himself declares as the word of God.

In the history of Jonah we have an account of a rebellious medium. Jonah desired his reputation to be maintained at the expense of the lives of the inhabitants of a very populous city. Yet Jonah was as much chosen and called to perform that prophetical work of warning as any prophet has been for any other work. Jonah was a prophet and yet, even at the very time and in relation to the very point upon which he had been inspired, he rebelled against God's will and refused his consent to God's mercy. God kindly and affectionately recalled him to sense of his duty. So you see that Jonah was by no means perfect, by no means a moral man. Neither was he continually doing God's will or always capable of declaring God's word.

Mediums and prophets are chosen by God to perform particular acts. They have their mission, as it is often well called. They are ordinary men but men who have so far become desirous of serving God as to offer submission to Him of their will and to desire to act in His will. But the sacrifice may not be complete and the man does not lose the power of acting in his own free-will. Whenever he ceases to sacrifice his will to God he resumes its exercise. He may even continually sacrifice it upon certain duties or calls relating to his mission but he does not thereby affect other duties or performances when he does not submit his whole heart to God. Indeed you yourselves were obliged to acknowledge this as a consequence of the fact that all have sinned and come short of the glory of God. You always have claimed that I was the only perfect man, that I could not have been that perfect man but for circumstances which gave Me superiority over others and advantages others could not and can not possess!

Perhaps you will say that God will not let His chosen servants fall from the work or mission He confers on them, at least, so as to err in relation to that part of their actions, and that Jonah's case sustains your position instead of disproving it! There is some plausibility in this argument because God is not so continually making mistakes as to be disappointed in the will of the holy medium or prophet. The man is tried before his work is assigned. Many are called but few are chosen. There is the meaning of this long-discussed text. Paul was called and chosen. Peter was called but he was not chosen until after My ascension. He was obedient to his calling then but would act only up to a certain point. Beyond that he would not do God's work, yet he had been chosen. What then should God do? Kill Peter for disobedience, discredit his performed work by taking from him all further power or authority, or should God exert His power to aid Peter to see his duty and strengthen any resolution he might form to do that duty?

It was the latter which God did and I leave you to say, if you desire to, how God could have done better. God had called and chosen Peter to preach the gospel of glad tidings of great joy to all men, but Peter preached at first only to Hebrews. He would not urge any to be believers in Christ without first being believers in Moses. He would not proclaim the glorious liberty of the gospel but kept back the divine promotings which impelled him often to do so until by a miraculous vision his eyes were opened to further sight and his mind could perceive that in truth God was no respecter of persons. Then he could continue to be God's servant and preacher and could pursue his mission in God's will and under His influence successfully, though before the vision, had the preacher of that wonderful and miraculous sermon of the day of Pentecost been asked whether Cornelius could be baptized, he would have said: No, not unless he first shall be circumcised and become a Jew.

So you see, a prophet or holy medium is not necessarily truthful even upon questions intimately connected with the most striking parts of his mission or the work in which he may be engaged. Believe, then, the sayings of God's inspired servants or holy mediums no further than they declare those sayings to be given them from Me or from God. The difference I have explained to you is no difference, for God always acts upon man by His agents or spirits or sons, and not directly.

The Word, and the Word of God

There is one other passage to be referred to before we leave this branch of our subject, and that is the age of the world of spirits. The worlds of matter existed a long period of time before the spirits or souls from paradise entered any of them, because the worlds of matter were unfit for the habitation of any animal and man was not produced until there had been a long progression of animals toward his perfect form. Not *to* his perfect form but *toward* it. I shall hereafter be more particular than I have been to limit you in drawing inferences from what I have declared, because as I draw near My closing periods I desire to establish a firm foundation for you to build upon, and the limits of the building must be well defined or you will be spending your strength in erecting parts that will be unstable.

The long course of our labors in this department has fatigued My medium and I mean to give him a resting time after the close of this section. He has not shown an unwillingness to proceed, for if he had I could not have gone on, but he has dared to complain of working hard at it when I have sustained his strength and health to its full average tone and left him entirely at liberty to perform all his business and social duties so that, in fact, he has written only when he desired to or when he was unwilling to disappoint his own expectation that I desired him to write. This I have not done to so great an extent as he has supposed. Nevertheless, he has not exerted his will or desire that I should write any more or any differently from what I might please to, but he has been more anxious for the progress of the book than I have. This is because I am not ready to have it published for some time yet and the book could as well wait unwritten as written. Hereafter there will be much discussion about the manner in which this book has been delivered and received, how much My holy medium contributed to its form or arrangement, how much was predetermined and executed by Me, and how far I allowed circumstances transpiring during its progress to affect its subject and the treatment of its subject. But at present the light shines into the darkness and the darkness comprehends it not. Soon it will lighten the darkness and unless I limit your intentions to revere the holy medium of transmission, some approaches to the superstitions already prevalent respecting the founders or promulgators of Christianity may and will be formed.

The spirit-world matter was created immediately after the creation and placement of earthy matter, and the placing of matter of all kinds was an act of the Word.

The Word was the minister or servant of God in doing His will, in placing the worlds as they were placed and in establishing such laws or relations of matter as carried them forward to their present state. The Word was God's right hand of power so often spoken of in the Old Testament, and the word of the Lord spoken of by the prophets was only the manifestation of God's will or revelation within man and to man. It is thus that many terms are used in a different sense in the earlier or Old Testament portion of the Bible from that which they properly convey in the New Testament part. The revelation of God's will when expressed in words becomes God's word, but not the Word of God. That is, God expresses Himself by His spirits or agents or Christs or Sent Sons to His created spirits in lower states of existence and his expression is perfectly adapted to their comprehension. To men in the body words are used; to spirits in the lower spheres thoughts are given; to higher spirits ideas are conveyed; to sons, high and perfect sons, God's thoughts are known without expression, by their being in perfect unity and harmony with Him.

It is thus that the expression of God's word, taking the form of revelation, is always conformed to the mode of speaking among those to whom it is given, the scriptures of the Old Testament being adapted to the Hebrew people by using such metaphors and illustrations as they would understand and by understanding be helped to know what God designed to have them know. It is thus that many allusions exist in those books which seem to suppose a condition of facts contrary to what science in these days has ascertained to be true, and apparently disagreeing with what I now assert. How much of this is also yielded to a desire that you may understand you cannot know, as from the nature of the case you can perceive must be. Still, inasmuch as I have explained to you so much more of the nature and relation of matter, spirit, soul, and God, you will correctly believe that such compliances to your mode of thought or fixity of belief have been far more rare than ever before. There never has been a time when knowledge was so generally diffused or when the means for further spreading knowledge were so easily commanded. Shall not God use what He has given to men for their use? Shall not the Giver be allowed to improve and make valuable, more valuable, the gift by conferring upon men the greatest of all gifts, knowledge of themselves?

It is, therefore, not extraordinary that God should add to the gift of newspapers and cheap books a dissemination of His word by means of them. The word He gave to the Hebrews twenty-five hundred to thirty-five hundred years ago had to be laboriously prepared and copied by scribes. In the crudeness of those times the peril attending its possession and this labor of multiplying copies was so great that few desired to obtain them by making the necessary sacrifice, but had to be content to hear them read at stated times in portions, and periodically in full. Even this privilege could be enjoyed only by a sacrifice of time difficult to make in many instances and impossible to be made in all. It is evident, then, that brief accounts easily understood and so forcibly expressed as to be impressed deeply on the memory were best adapted to this period of Hebrew history. At the time of the promulgation of My gospel, or as it is properly distinguished, My glad tidings of great joy, the labor of disseminating manuscripts was somewhat reduced but it was still a great obstacle to the preservation of the purity of the writings then in existence that the number of copies were so few.

It was thus expedient that brevity should be studied, and obscurity was not always avoided at the sacrifice of fullness. But the language of the Greeks suffered less change than the Hebrew, while the extent of the population using that language and the wealth of the people in the early days contributed to make Greek a language permitting better preservation of the purity and the understanding of the meaning of the record.

Still the delivery of the word was at that time oral in nearly all cases; consequently it passed through the reason or intellect of man by man's will, and though that will was to relate the exact truth, in many cases the exact form of expression had been lost. This is not remarkable when it is remembered that from twenty to forty years elapsed between the deliverance and the recording of My expression. In this day men would not remember so well, but at that time memory was more relied upon, and better exercised generally among the whole people than it is now by any class of the population. The people had listened with great attention to My discourses, and the words addressed to those on whom or on whose friends miracles were wrought naturally made deep and lasting impressions. They were indeed treasured up for relation to children and grandchildren, and the very manner of delivery, the tone and the gesture, the look and the voice, were remembered and related with great particularity and exactness in many instances.

Those who afterward wrote accounts of the discourses and actions now called gospels, but not properly so called, went about collecting from the most authentic sources the history of My ministry. In this way it happened that each was incomplete in details and that there were even slight discrepancies in the different accounts. Those who did this work did it in their own wills and God is not answerable for their errors. They did not profess to write other than history or to have inspiration for it. Yet in the main they performed their task well and God did reward them by influencing them in many times and periods of doubt or misgiving. He also influenced pious men to regard as worthy of preservation and careful copying what had been thus written. So when the Biblical student investigates he finds that a most extraordinary preservation of purity and agreement exists between the accounts of these different historians, and that it is possible by an earnest study of the language of the originals, the manners, customs, modes of thought and expression among Greeks, Jews, and Arameans, or Syrians proper, or Phoenicians, to arrive at a clear conception of the meaning and knowledge of the authors of the histories. But when he would go beyond this and endeavor to ascertain in addition what I intended to convey when I used the recorded expressions, he finds reason and study are no longer safe guides. He must seek a more pure light than reason, a surer guide than tradition. That guide and that light I can furnish; I have furnished it to a very considerable extent in these books, and I shall give more.

Some will ask why God did not confer upon mankind a knowledge of printing, if that is so important an aid in preserving and extending a knowledge of revelation? Thou fool, God works by agents; He works by means; He works upon man only spiritually and for spiritual ends. He has endowed man with powers of action and contrivance and given him full authority to use them. He even helps him use them when man appeals to Him for help, but He does not work for him. He only lets man work and use the faculties He has bestowed, but He will help when He is asked, and give when beseeched. Seek and you shall find is true in respect to art, science, business, social relations, and religious forms or doctrines. Ask God for such help as you want and He will give it to you provided you ask from good motives, for good ends, for what will not injure your brother, your neighbor, or your enemy. Desire God's help to benefit mankind and you will get it; desire it for your selfish gratification and you may or may not receive it. If you do, then it will be as spirits have all they desire, to show them that it is not worth having. So ask as you desire to have, and as your desire is, so shall be your reward by reception.

Let us pray

O God, hear the petition I desire to make to Thee as the Giver of good. Let me not be presumptuous, O Thou great and majestic Sovereign of the universe of universes, let me be pardoned for appearing before Thee if I err therein, O God, for Thou hast abundant mercy and I am very ignorant of Thee and of my own relation to Thee. I desire, O God, to know Thy will respecting my heart, how I may be able to bring it into subjection to Thee and to keep ever before Thee my desire and willingness to sacrifice to Thee my own will and inclination in all things, so that I may come to an acceptance by Thee and be reconciled to Thy will and found worthy through Thy mercy to be received into eternal happiness, to dwell forever and ever in the mansions of bliss which have been declared to be prepared for those who love and serve Thee. Grant my prayer, O God, because of Thy great love, abundant mercy, and rich possessions and unlimited power, and because of the promises of Thy Son, the Lord and Savior Jesus Christ, to whom and to Thee primarily be ascribed all honor, glory, thanksgiving, and praise, world without end. Amen.

CHAPTER XVIII

HOW TO HAVE AN EXPERIENCE OF HEAVEN ON EARTH

The Translation of Enoch

Alas, that man should be so blind, so censorious, so unwilling to practice here what he looks forward to or professes to look forward to as the greatest happiness! Alas, that man, being permitted to enjoy here the peace which passes all understanding on the very same terms it is offered to him in the world to come, should be so neglectful of availing himself of the glorious privilege, the high and holy blessing, which only his own will rejects. Lest any should condemn Me for not urging man to be happy here, I will state that all that keeps man from heaven here is the same that keeps him from it in the life to come: that is, following his own will rather than God's. As soon as man begins to sacrifice his own will, his free-will, God begins to receive him as a servant; as he progresses with the sacrifice he becomes more and more reconciled to God and raised in happiness and glory, by which he becomes more humble and more desirous to serve God. He may even become a son of God here as I was on earth, but only by a full sacrifice of his will.

This full sacrifice is not impossible though it has taken place with no other than Myself in the body except Enoch, who was translated to the spirit-world without tasting death. This antediluvian tradition preserves to this distant period the memory of that divine personage who served God so faithfully as not to be subject to the laws of common dissolution, but whose earthly body like My own was dissipated into a cloud of aqueous particles. Elijah is said to have ascended in a chariot of fire, a blaze of light, but this was not so perfect a transformation as the other.

How is it, then, that these sons of God did not arrive at the highest circle of the seventh sphere before You? They sacrificed their wills completely, but they had not done so continually from youth to old age. You will remember that I had only one departure to atone for while they had many. In Enoch's case, since he had had such a very much longer time in which to atone, and the more perfect sacrifice having been fully made by him, by which he was translated as a son of God, in which case it must be inferred he was in the first circle of the fourth sphere, it would really seem that he had not gained much advancement or privilege or power of advancement!

Enoch was a good man and in the latter part of his life he fully offered up his free-will, a continual and perfect sacrifice to God. But there was much in his former life to atone for, and as he had not the memory of the past to guide him in the spirit-world, he became a servant in a lower degree and, having a more sensual and gross body, his mind had received a stronger impression from the departures of that early life. A very long period was then necessary to free his spirit-body from the stain or imprint of those departures from obedience. As I told you that there were antediluvians yet remaining in the first circle of the second sphere, the lowest in the spirit-world, Enoch was certainly greatly blessed compared with these. Yet does not this reflect upon God's justice, that He assigned such bodies to spirits as would disqualify them for rapid progress? He gave to each spirit such as pleased Him, yet not capriciously but according to its nature. Again did not God make inequality or show partiality for some souls or procedures in thus giving them such a nature as to receive such bodies as would thus retard their spiritual progress even after they had, by sacrifice of their free-will, become so greatly advanced in progression as to have been unequaled except in one instance, and that one possessing peculiar advantages from His exceptionable organization of body! This objection is well put; it was that you might make this very objection that I introduced so unnecessarily, you may think, this subject which was so quietly reposing behind the veil of tradition and which might have been reposing there still had I not so admitted everything that would make its explanation difficult.

Enoch was an antediluvian possessing a soul formed for such bodies as existed first with souls upon planets so newly organized as the earth. In all earthy globes there is a progress from low to high, from animal to spiritual, from sensual to refined. The first state of man when in bodily form upon the earth thus newly organized is always gross, and the spirit that enters it must be measurably affected by it. God selects, however, such spirits for these bodies as need the more experience of good and evil hereafter to reconcile them to an absence of it. Enoch never ceased to be happy and to progress in happiness, but his soul, possessing less aura and more individuality than Mine or other postdiluvians', required or profited by a longer experience in the body. And though in despite of the deficiencies, so to call them, of his organization he sacrificed his will, that sacrifice was not any more difficult to make than it is for you to sacrifice yours. That sacrifice is always equally in the power of any man to make. The will is free and herein is the difficulty and glory, and from this is the reward of its surrender. But since Enoch had sinned or been disobedient for the early part of his long life, his habits must have made his sacrifice more difficult; how is it that we shall advance our spiritual happiness and progress so greatly as You have urged if, after all, when we have here been so greatly reconciled to God, we must lose the fruits of our sacrifice and service and go back to the lower circles to commence our progress to the highest, and have all the atonement to make there though we have repented here!

This state of progress is greatly to be desired. It is not that Enoch was put back, but that some have been put forward. Enoch was as nearly perfect as any man at the time of his death or dissolution of body, but as a man he was imperfect and required training and experience which was obtained by him in the spirit-world as a guardian spirit to men on the earth. But Enoch, so far from being retarded in order that I might be the first to arrive at the seventh circle of the seventh sphere, was greatly helped by Me, for his progress was retarded by want of knowledge of purity of doctrine and willingness to be instructed in his duties toward his fellow-men. Enoch submitted to God perfectly. He acted fully up to the light he had. He was released from the body without the apprehension or the reality of death and corruption of the body. But he did not receive such commands of duty as rendered him perfect in obedience with knowledge. He was perfect in obedience because little was required of him, and more was not required of him because it was perceived he would not have been able to bear more trial of his faith.

God, desiring to have an example of obedience before the eyes of men in order to incite them to emulation and obedience, raised up Enoch and received him as faithful because he obeyed all the commands he received, though it was by God's mercy that other commands that he would not have obeyed were not given. It was a state or condition of spirit or soul analogous to that of My holy medium. He was obedient in the reception and delivery of this revelation but he is not therefore any better than another average man and, being subject to like passions, he may indulge those passions in his intercourse with his fellows in such a way that he may be far behind many who have never known about receiving spiritual communications. It is thus that Enoch was favored by God and allowed to stand as an example of faith and incentive to the emulation of succeeding generations, and was the purely obedient and passive medium of God's will thus gently exercised. Then was Enoch favored or slighted by this refusal or declination to try him with further commands? This question I shall answer in My next division.

The Justice of God

Enoch was favored by God's mercy because he could not have borne the loss of his faith without falling to a lower depth of despair and wickedness than he had previously experienced, nor without at the same time causing a general abandonment of the worship of the one true and living God. It was in his line of descent that the family of Noah was traced from Adam, and it was that family only at the time of the deluge that retained this knowledge of God or in the least regarded Him.

Noah, though being a new production as to race, might have been impressed, you think, as Adam was with such knowledge, or a revelation might have been made to him as to Adam of his duties and his capacities. But the difference was greater between Adam and his predecessors than between Noah and his forefathers. God prepared Noah for his mission, bestowing upon him advantages which in a measure secured his obedience not by force but by persuasion; not by making him obey but by making it easy for him to do so. Also for this cause Noah, though a prophet or holy medium and an obedient son in a disobedient world, was not so prepared for advancement as to distance all others of the postdiluvians. He experienced in a high degree the mercy and forbearance of God in that

God made it easy for him to obey, and he did obey. Thus God without forcing Noah's free-will in the least received him as an instrument by which He could maintain and develop another race, commencing the world anew without antecedents injurious to their desires to obey God. The early postdiluvians were a moral and upright people; it was easy for them to be so. They had indeed less temptation to do wrong and better facilities for remaining in the path of virtue than men in the present day. God looks at the heart and not at the deed. He rewards according to the motive, and the strength of the motive is shown or proved by the extent of the temptation. Where nothing is sacrificed there is no advancement; where little is offered there is little to accept; and the man who does not need help is not he who is favored so much as he who, perceiving that of himself he can do nothing, leans upon God and calls upon God to support him in every time of doubt, in every period of trial, in every strait of temptation.

It is thus that God judges and rewards. Men judge each other differently and it is for this reason that God caused it to be declared that men should not judge their fellow-men, but that judgment is His only, and that the uncharitable judger among men shall receive from God the measure wherewith he has chosen to try others. God's ways are not as man's ways, though man would if he dared condemn God's want of reason and consistency and His own power to establish His reign in the hearts of men. So, O reader, do not judge God harshly or impugn His wisdom severely! Be merciful to God as you would have Him be merciful to you!

Let us pray

Our Father who art in heaven, hallowed be Thy name. Thy kingdom come; Thy will be done on earth as it is in heaven. Give us this day our daily bread and forgive us our trespasses as we forgive those who trespass against us, and suffer us not to be led into temptation but deliver us from evil, for Thine is the kingdom, the power, and the glory forever. Amen.

CHAPTER XIX

WHAT SHALL A MAN DO TO BE SAVED

The Effect of Prayer on God and Man

The prayer or form of prayer I have furnished for you as I did for My followers eighteen hundred years ago will never wear out. As I have said, all other prayers may be found in it and they are but repetitions of it if they are made in proper submission. Therefore it is that I do not want you to lose sight of it that I have thus a second time dictated it to this medium.

The more observing will detect a variation in language or words used, though they will also perceive the sense is made clearer by it, for the forms of expression have changed much, as I have stated, since its first promulgation. Still, man is the same and needs the same prayer, and God is the same and will grant the same prayer.

Yet it is necessary that the prayer should be understood by him who makes it his own or else, misunderstanding it, he will make a different prayer. For in hearing prayer as in judging works, God looks at the heart of man for his intention, and if he imagines that he serves God by asking for one thing when he wants another he not only undertakes to deceive God but injures himself. If he gets what he is thus ashamed to ask for, he finds it only a curse or unhappiness while his outward petition condemns him by its reaction upon him.

It could not be supposed that God would Himself lead a man into temptation unless it were for the man's good, in which case the man ought not to be saved from it. I have therefore corrected the phraseology so as to agree with the original delivery and the ordinary understanding of it at the present day, because some weak in the faith might otherwise find cause of stumbling in it.

The last subject of this book will now be taken up: What shall a man do to be saved? What act must he perform, what confession make, and how shall he know he is saved or what will be his salvation? Having reached the point where I can look forward to the end of this book, though not of the volume, I desire to state that it has been written at the rate of five quarto pages of manuscript per hour of the time spent at the desk with it for the purpose of writing. These five pages of manuscript make about two of the printed pages.* Though thus rapidly written, an examination of the copy to be preserved of the original first record will show that there are no alterations or corrections other than verbal or literal, made in order to fit and finish it for the printers' hands.

The Way of Salvation

The man who desires to be saved has reached an important position in his life which may be regarded as a critical one. Placed as he is in the vale of experience, it is here that he must be saved if he secures salvation by any other means than mercy. Works can be done here in a man's own will by God's help which shall forward and aid his salvation, though alone they could never secure it. God's mercy is the only sure reliance for salvation. By that, in fact, all men are saved. All men have sinned, and without God's mercy no man could be saved, because he could not get rid of the consequences of a single sin without God's help. Further than that, he who has sinned little or the least has only been prevented from sinning more by God's mercy which has restrained his evil desires, aided his good thoughts or intentions, and perhaps frustrated his wicked intentions, or resolutions. The only way, then, by which men can be saved is by the acceptance of God's mercy freely offered but contingent upon certain laws or rules which man can easily comply with, or which God can help him to find easy.

First, then, men must desire to be saved. None are or will be saved unless they desire it. Having desired, they must, second, be

*In the original type—a little more than three pages of the present revised edition—Ed.

willing to be saved in God's own way. God, having appointed the means, has provided a particular way for those means to be made universally available, but that way is His own and no one who is not willing to be saved in that way can reach heaven. They may struggle to climb up some other way or to break into some weak place, but all efforts of that kind will end only in disappointment and affliction. Third, the salvation cannot be regarded as effectual until the man experiences an inward assurance that it is so. There is no danger of a man's being deceived into a belief that he has secured salvation when he has not if he only looks at his heart or his own internals. He may be deceived if he takes the word of his fellow-man for it and seeks to be persuaded by man that he is right and others are wrong.

First, then, *a man must desire to be saved* and this desire should be expressed to God in the form of prayer, not because God does not know it, but because prayer fixes and realizes to himself the desire perhaps always latent in the heart or soul.

Second, *he must pray to God to help him to comply with the condition of salvation,* which is that he shall submit his will to God's. This sacrifice must be made before going any further or, more properly speaking, he cannot go any further until he has made this sacrifice and complied at least in part with the conditions fixed and required by God. How is this submission to be made? asks the unregenerate heart. By seeking, first, to know God's will and, second, by yielding your own desire or will to it. If you feel impelled to indulge a bodily passion, an earthly motive, seek to know from God if it be proper, if it be in accordance with His will. Though you have never had communion with God this course will bring you to a communion with Him, and before you are conscious of receiving a reply from Him by a communion of His Son with your soul or spirit you will feel the impression of duty resting weightily upon your mind, convincing you of the impropriety of the act if it be one of that character, or your scruples respecting it will disappear if otherwise. Remember that this inquiry is not to be made of any man or body of men; it is to be made of God and made as an individual of Him for yourself and not for others. Neither are you to repine if you see that others professing to act in accordance with God's will enjoy greater freedom of action than yourself.

Remember what James said: My brethren, count it all joy when ye fall into divers temptations, for God trieth whom He loveth and chasteneth His holy sons. Remember that this is the only work you have to do, and that God is not only ready to direct and advise and guide you but also is willing to help you if you ask Him, and that He will not give a stone when you ask Him for bread. Then be not discouraged if sometimes the world seems to triumph and God's servants appear overwhelmed. Fear not that God is not strong and mighty nor believe that He has grown blind or deaf. God knows His own time and patience must do in you its perfect work, that the inner man may be renewed and made strong, that you may not be like Enoch, raised to glory here to have to work far more arduously for glory hereafter.

Third, you shall know you are blessed by God with a portion of salvation when you find you can sacrifice with cheerfulness your will to His, your comfort to another's, when you love your brother or your neighbor as yourself and can even love your enemies and do good to those who despitefully use you or persecute or injure you. Proceed on your way rejoicing when you find this to be the case in you, even slightly more so than it had been experienced before. Cultivate the desire to be saved; become more and more submissive to God, and this eternal evidence will be greater and greater. You shall know at last of heaven even while dwelling in your earthly tabernacle. You will find that peace which comes from heaven, that joy which only the righteous experience, that fruition of happiness, the peace of God which passes all understanding of man's intellect and which is joy indescribable and full of glory.

When you have thus tasted the fruit of all your efforts, the result of the acceptance of your fervent prayers, be no longer fearful but believing. God has saved you. You are released from the law of sin and death and on you the second death shall have no power. But be careful to let no man take your crown. Keep fast the faith; be not deceived by your great enemy, your free-will. Be not induced to become the accuser of your brethren. Mind your own business. Let others settle with God. Do you continue to work for Him and seek no other pay than this same salvation which has already been such a rich and abundant reward. Be firm in the faith, abundant in good works, and rich in God's grace. So shall your light shine before men that even they shall give God glory for and because of you. So shall you be a beacon, a lighthouse, a candle not hid, and your light shining into the darkness shall turn many from devious paths to righteousness. So shall you die triumphantly declaring you have fought the good fight; so shall you be received up into glory eternal, everlasting, unchangeable, full of joy, world without end.

Let us pray

Our Father who art in heaven, hallowed by Thy name. Thy kingdom come, Thy will be done on earth as it is in heaven. Give us this day our daily bread, and forgive us our trespasses as we forgive those who trespass against us, and suffer us not to be led into temptation but deliver us from evil for Thine is the kingdom, and the power and the glory forever and ever.

Amen.

History Of The Origin Of All Things

PART III

A History of the
Progress of Man's Spirit
in the
World of the Future Life

INTRODUCTION

Let all the people praise the Lord!
Yea, let all the people praise Him
For His goodness and mercy and loving-kindness;
Yea, praise Him, for His mercy endureth forever.

Let us all be willing to sing the song of the saints and give praise to the Most High God whose glorious mercy is abundantly offered to mankind, mercy which is as inexhaustible as His nature, as constant as His love, and as enduring as His existence. Come unto Me, all ye heavy laden, and I will give you rest. Walk humbly before Me or God, and you shall be upheld. Though storms of opposition and torrents of vituperation should assail you, be steadfast, immovable, full of faith, glorifying God eternally, evermore. The course of this life will soon be passed over. Henceforth is laid up for you a crown of glory, and not for you only who read this but for all who love God and give Him glory for the love and mercy by which His Son, the Lord and Savior Jesus Christ, was bestowed upon mankind and made a propitiation for your sins. Though your sins be as scarlet they shall be as wool, and though you have no hope except in the mercy of an all-powerful and almighty God, you shall know that His mercy endureth forever.

Be, then, ready to make the change from earth to the spirit-world. Be under no apprehension of evil from it. Welcome death of the body; fear death of the soul. Welcome death of the body because it is the entrance into new life, and because all who have preceded you were benefitted by it. Jesus of Nazareth rose from the dead to convince you of this and to persuade you that the grave has no terrors, and death no sting.

O grave, where is thy victory?
O death, where is thy sting?
The sting of death is sin
And the strength of sin is the law.
But thanks be to God who giveth us the victory
Through our Lord and Savior Jesus Christ,
Who was in the beginning, is now
And ever shall be, world without end.
                    Amen.

PREFACE

The series now completed is a sure and safe guide. It is a sure and safe deliverer from sin, suffering, and sorrow; from the tortures of doubt and uncertainty, from the apprehension of wicked schemes, and from the evil which the ways of men bring upon them.

Having conducted you through this long course of instruction, I leave you to digest in your mind what you have read, to read again what you do not understand, and to investigate by re-perusal into any branch or portion of My subjects for further information. You will find that whatever you desire to know will be opened to your view by study. First, read for a general knowledge of its contents; then re-examine the parts you find most interesting; lastly, do this as often as you want further light or knowledge upon any part or branch of My subjects, and you will be brought to a full knowledge of the things which it pleases God to have revealed to His servants in the present time.

All that follows this series will be of a different character— good and useful, or it would not be given, but not so directly bearing upon your everyday actions and life. Read this, then, as containing all that will most help you and profit by it if possible, so as to be one of the happy company that eternally praise God for His goodness, glorify Him for His power, and rejoice in His everlasting love, which continually spreads forth to the hearts of all His procedures and is returned to Him in the praises of the song, ever new:

> Great and marvelous are Thy works,
> Lord God Almighty!
> Just and true are all Thy ways,
> Thou King of Saints!
> Amen,
> saith the Spirit,
> And let all people say Amen,
> saith My holy medium.

CHAPTER XX

When the first of the books belonging to this procedure were written, I did not rely on any outward proof for their reception. But these books shall have the testimony of all who listen to the inward voice, to the aid of divine counsel, to the voices of spirits whether manifested in themselves or others, whether internally and silently given or externally and outwardly manifested. The sincere inquirer shall in all cases receive a true answer, whether that inquirer be a believer or an unbeliever, or whether he have long sought for truth or just commenced his search. As there will also be those who will claim to be sincere inquirers who will receive answers from some mediums inconsistent with truth and with the answers to others, I will state that the truth may always be obtained if prayed for by the questioner. To pray for the truth in reality, he must be able to unite in the following prayer:

Let us pray

O God, Almighty Ruler of the Universe, who dost from Thy throne behold all the actions of all men, who doth by Thy omniscience know all that passes between men and within them, who art by Thy omnipresence ever with us and with each one of Thy creation, be pleased to favor my desire to know whether the books I have read (or am reading or am about to read) purporting to have been delivered by and through Thy Son, the Lord Jesus Christ, through one L. M. Arnold, are really given in Thy will and should be truly and faithfully received as coming from so high a source, and whether, O Father Almighty, I am in duty bound to listen to them and to discard all that I have believed contrary to them, or whether I shall reject them as the work of evil or of imperfect mediums or as wisdom or imagination of men.

Tell me, O God, if it be Thy will that I should confirm or reject the evidence thus given of Thy continued care for mankind, or whether I should not rely on teachings of others who declare that Thy will is not made known to mankind in any such way? O God, my loving Friend, may it please Thee to make known most plainly to me in such way as pleases Thee and will convince my reason or my conscience or my intuitive perception what I ought to do, to believe, and to ask respecting these books. O Holy Father, I desire to know the Truth without regard to the opinions of men. I desire to know Thy will without regard to the will of men, and I will conform my thoughts, opinions, and actions to Thy will if Thou, O God, wilt be pleased to aid me by Thy powerful help and place me on the sure foundation of truth and revelation.

Amen.

Will you say, perhaps, that you do not believe the medium to whom you apply after making this appeal to God, or that you do not believe God can make known to men in these days His will? Do you think that He has not left Himself any means of communicating with men in these days as to the course of action they should pursue? If you believe this—if you believe that God exhausted His power, established His will fully and completely manifested His love so that nothing further can be or need be made known, you of course cannot make the prayer or believe its answer if made, because the prayer, like all prayers, presumes that God will hear and that He will graciously incline His ear to the petitions of those who humbly desire His aid, and that He will grant the sincere desires, the innocent and fervent aspirations of His servants or of those who sincerely desire to be His servants. If you believe that such cannot be answered by God under the present laws He has established, then you are beyond the reach of this appeal; you are beyond the aid of any prayer, because the prayers which are beneficial to you must be your own.

If you so believe, I have another appeal to make to you; if love of God and confidence in His mercy, love, and benevolence and power or will to exert His power are wanting in you, I will appeal to your reason. Having no faith, you must rely the more upon reason, and, though reason from its nature is not so sure a guide as faith, and revelation received by faith, yet I am aware that few minds can resist a cool and dispassionate argument, founded upon truth supported by facts and established by logic.

When God made man, as I described in Volume One, Part II, He revealed much to him, implanting it in his mind as fully as if it had been experienced and well remembered that he could perform many acts and use many means of receiving more knowledge. But though this high degree of knowledge was revealed to the first pair of mankind, other revelations of the same truths were required for the posterity of that pair at various times before and since the flood. Much of this revelation was not even communicated by Adam to his descendants; other portions of it were forgotten, some were disregarded, and gradually much, even most of it, was lost, notwithstanding the other revelations made from time to time by or through Enoch, Noah, and other antediluvians.

Perhaps you think that after the invention of writing, revelation might be and was so preserved as to preclude the necessity for a further revelation! But you believe that Moses made a revelation and that several prophets received communications from God or His spirit subsequent to that, and that even after Jesus of Nazareth came and disappeared His disciples received by the power and will of God certain manifestations and revelations. Revelation from God or a divine source did not cease, then, either at the invention of writing or at the ascension of Jesus.

If neither of these were the boundaries which satisfied God that no more needed to be revealed, where else will you fix a limit? Will you say that the apostles were endowed with certain powers for the special purpose of preaching the gospel and that the necessity of possessing these powers ceased when the gospel had been preached by them! But the apostles never declared that inspiration would cease with their own disappearance. They never declared that it was restricted to them. When Paul and his companions separated because they could not agree as to the route they should take in carrying the gospel tidings to the heathen cities, did he appeal to his higher authority or exclusive revelation? Did he say: You cannot accomplish any miracles without me; you must go with me if you would be useful! No such course was adopted by Paul. They agreed to separate because they could not agree to remain together and proceed together.

Where, then, is the limit? If the apostles did not give us any, and if Jesus did not give us any but, on the contrary, declared expressly that He would send the Comforter, and spirit of truth, that should teach you all things; if the apostles instead of saying: Believe no future revelation, only said: Do not believe any revelation contrary to what we preach, as Paul said, or: Do not believe any revelation that does not also confess or proclaim Jesus of Nazareth to have been the Messiah or Christ, as John said—who shall dare to restrict God because He does not, (or has not, in their opinion) make any manifestation of His will since some time they choose to fix upon! Who shall dare to say to God: Thus far hast Thou proceeded but no farther canst Thou go, either because Thou hast declared all that can be declared, or because Thy mercy and loving kindness is satisfied, or because there is no further need of the exercise of Thy power, will, mercy or kindness?

O man, helpless as you are, you cannot believe that you do not want help. Discordant as you find yourselves to be, you cannot but hope for reconciliation and harmony. You do hope for it and try for it, but though you have asked God to help you in your efforts to make men all of one faith, you are yet far from success. You will, however, try more and pray more for God's help, though you do not believe God will make known to you by revelation that He has helped you or in any way directed you how to proceed. Perhaps you will say He has not done so or you would ere this have succeeded in establishing your faith as universal in the world! But though God has not chosen to select your society or sect as His favorite but has chosen the poor in this world's estimation do you, O man, look well to the foundation on which you rest! Remember that high professors and church authority in the days of Jesus of Nazareth were not only not the first to distinguish and receive Him as the Messiah, though the exact time of His appearance had been prophesied of and had arrived, and His authority was attested by such remarkable signs that you now only ask a repetition of them to believe in His second arrival here.

Does not analogy teach you that as it was then it is now? No, you say; because Jesus was rejected by the church of that day and by the leaders of public opinion, we are not to suppose that every one so rejected is He or His representative! But, My friend, for such I esteem you, though perverse be not blind as were the Pharisees. Say not that this man casts out devils by the power of devils, or does any other good thing by evil motives or the aid of evil power. Believe that the signs you cannot explain or understand are given for your establishment in faith, and that however undignified it may seem to you to be to move tables or to rap on them, to answer silly questions or to declare contradictions to contradictory questions and desires, believe that God has a purpose to be accomplished in His own way and by His will and pleasure, and that He can reconcile every discrepancy and find in every dispensation and manifestation a link in the great chain of harmony which He will bind upon all His creation. He has ever maintained his creation in harmony though men have not been able to perceive and account for the manner of its being so linked together by such apparently discordant materials.

Believe, O man, that God is powerful enough to work by humble means, that He is consistent enough to leave you your free-will, and that He is benevolent enough to wish you more happiness. Believe that He is merciful enough to secure your future happiness and that, while He offers you an earlier realization of peace and joy and perfect bliss, He does not compel you to accept it. Yet if you desire He will help you to obtain it; if you seek for it you shall find it by His help, and if you call on Him to know His will, He will find some way to answer you if only you are willing He should answer you in the way He deems most suitable.

The outward manifestations are given to attract attention, to excite curiosity, to lead mankind to inquire, to investigate, to ask why they are thus disturbed or aroused from old well-beaten tracks. Come and see. Remember that John sent to Jesus to know whether He was or was not the Messiah. Jesus answered by referring to the outward signs of the fulfillment of the prophecies respecting Him, and I point you to the outward signs that testify of Me and of this revelation. I refer you to the writing and to the rapping mediums. I refer you to inspired preachers and to all who receive in any way direction or guidance from God, and last but most of all I refer you to the internal manifestation of God's spirit, the Comforter, the spirit of truth in your own internal, generally called your heart.

All these will testify of Me if they are properly interrogated and I submit now to your reason whether these proofs from all these sources are not enough to establish My authority to teach, inasmuch as I do not ask you to make any such great change as did occur between Judaism and Christianity. I only ask you to return to the first principles of Christianity, to the practice of the precepts I preached outwardly when in the body, to the practice of the order and discipline and reception of revelation inculcated by the apostles and first preachers of Christianity. Then these signs shall accompany those who believe and practice upon their belief. They shall have the Comforter, the spirit of truth, dwelling in them, preaching within them, guiding them to all truth. They shall be led to fountains of living water, to bliss and joy everlasting, and ascend to realms of peace and serve God in the mansions of happiness prepared for all mankind by the Father, who delights to confer upon His children good gifts.

Be, then, seekers and finders; be ever watchful to know what God requires or desires of you, and you shall not be allowed to grope in dark ignorance. You shall be enlightened by that pure light of the Word shining into darkness which can not comprehend it except as it ceases to be darkness. Be no longer scorning the humbleness of this procedure, but make the prayer I have written as your guide. Make it your own prayer and you will find the answer to be that God does speak to His children in all times and He will not let the righteous be forsaken nor give stones to His askers for bread. He is powerful enough to save you not only from unhappiness of itself but from all the unhappiness or restraints that any who are opposed to Him may impose, or undertake to impose, upon you or any other. He will indeed do all but save you in your sins, or all but overcome your free-agency. Free-will is the distinguishing quality of man and the only thing he has which is his own. All else is God's, and the man who would sacrifice to God must give what he has and what is not God's, or he does not give anything. He must give his free-will to God; it will be received as a sacrifice pure, holy, and acceptable. If contaminated by ambition, lust of any kind, or desires for establishment in any plan of its own, it will not be received as a perfect offering. Man does not then sacrifice to God—he attempts to bargain with Him, to give Him a little for the sake of receiving more than that little is worth. This will not succeed, for man cannot overreach his Maker or give to God anything not his own.

CHAPTER XXI

GOD'S CALL UPON MEN

Wherever man falls he lies helpless unless God or His agents (which are the same as God, as they do only His will) raise him or rather help him in desiring to be raised. There is no salvation for any man except through God's help. Was man, then, made so imperfect by God as to be unable to accomplish anything of himself and yet be pronounced good! Not so, but God, having made man as was most pleasing to Him, pronounced him good and man, being good, was so with a liability to change unless he continually complied with a law which God also made and which was therefore good. That law was the law of duty, of obedience, of sacrifice. The duty to be performed involved obedience to God's commands or requirements, which was, reduced to a unit in words, *the sacrifice of man's free-will*. By yielding this sacrifice, man became reconciled to God if he had been separated from Him, or retained his harmony with God if he had not allowed the entrance of disobedience.

The sacrifice then involved obedience to God's law and the performance of the duties required by that law. The law was written on the heart of man by the conscience of each man, if the man sought for the writing, but he who sought not for the writing found nothing there and had to be governed by outward laws, which of themselves could confer no reward for obedience nor make any punishment for disobedience. The law, however, existed and was made known to men; those who received it, though condemned by its reception and made more guilty of disobedience by such reception, were nevertheless no longer allowed to excuse themselves by pleading ignorance, or want of desire of knowledge.

In the next place, man was so constituted that in general he would fail to be obedient and would therefore fall and require God's aid to be restored to unity or harmony with Him. God's mercy therefore is continually exercised. He calls men to obedience, but if man does not obey He pardons him unless he sins against knowledge. Every reasonable excuse for disobedience is taken as an excuse, allowing of the exercise of God's pardoning power, but if the sin is against knowledge, a willful wanton disobedience, God requires an action of man before He will pardon him. Man must repent and make resolutions of amendment before God will pardon him. When man so repents and makes these resolutions, and sacrifices thus his will, resolving to seek for and obey the will of God, then God leads that man to the living fountain of inspiration, where He gives him such knowledge as is sufficient for him and calls him to such duties as he is capable of performing in his own will. When he has performed these, other duties are imposed, but these duties are always proportionate to man's strength and the burden is never oppressive, for whenever any man finds it so God, if requested, takes it off. Even if not requested, God releases the man from his obligation to perform, because the man is unwilling and God will have none but willing servants.

God then never requires more of a man than the man is capable of performing!

The next call God makes is that the man should leave the body and enter the spirit-world, there to receive further from His inexhaustible mercy—there to be raised to high and perfect bliss, not as a reward for obedience or suffering, nor as a recompense for pain or as a solace for sorrow, but because God wills to have all His children happy; because He chooses to manifest His mercy and establish His kingdom in heaven. Without God's help on earth man would have sunk to the lowest depth of depravity and spiritual ignorance. Without God's help in the spirit-world the spirit of the man would go on sinning, acting in disobedience to God's known commands, until he had separated himself from all others and raised up for himself a building not of God. But God's mercy restrains the man from growing worse; His goodness and wisdom provide a way for him to get better. The requirement is still the same: God still calls upon man to yield his free-will, to sacrifice to Him and to receive the riches of His great mercy and wisdom.

This sacrifice may be delayed by some for many myriads of years; the progress may be so slow that to all observers who have not attained to partaking of God's perception it may be invisible or imperceptible as progress. What, then, will God ask of the man in the spirit-world as a duty to be performed? I have explained what are the employments of the spirit of the different classes or circles of different spheres. They are such as harmonize with the circle in which they are arrived, and such as assist the spirit to advance to the next state or circle. By this the work continues to be easy and the burden light. Existence with God is ever pleasurable both in this life and in that which is to come.

But, you may say, if existence is ever pleasurable how is it that many commit suicide? That is because they have mistaken the want of perfect happiness for the total want of happiness or for the absence of happiness. Men are relatively happy or relatively unhappy. So a climate is relatively warm or cold, but no climate is without heat. No climate is so cold that it is not evident that it might be colder. So no happiness that man can conceive of is so great that greater cannot be conceived possible, and yet man cannot tell how he would make it more. The suicide, finding himself less happy than he has been, rushes into another state generally with the expectation that it will annihilate him or that he will be bettered by being brought nearer to God, hoping that God will prove to be the merciful Father of His children and that He will indulge this waywardness on their part and forgive them not only ninety-nine times but seven times seventy times. God does keep open the door even for such to progress, and continues to call even them to come and sacrifice to Him their free-will.

Free-will is the deity of man's nature and occupies the ruling position over all his attributes. It wields absolute power over him as God wields the power of the universe, but free-will can be influenced by its subordinates as God is influenced by His subordinates. The desires and petitions of His servants are granted, and the desires and petitions of man's several attributes, some of which are called passions, are granted or refused by his free-will. Though free-will is thus absolute and yielding, God can only interfere to subject man's free-will to His own perfect will by causing some of these attributes of man to petition their master to listen to God or to His revelation. Were God to do more than this He would overcome and annihilate that free-will, and though the man would then be obedient to God's will, man would no longer have a will or be able to make a sacrifice to God. Helpless as he was before, he would then be quite incapable of action. Helpless as he would be, he would then be another being in fact, for without free-will he would not be man.

God, then, asks man sometimes through or by his reason, sometimes through or by his gratitude, his love, his wishes for happiness, his aspirations for higher states, or any other quality or passion of human nature, to yield to God a sacrifice of that free-will as a means of securing a higher degree or realization of happiness and of being brought into union and harmony with God.

What will the man do? Will he yield to God's petition so made, or reject the prayer? Man's position at the time will be affected by so many circumstances that it will seldom be found willing to yield to one. There are so many calls upon this one finite free-will that it becomes confused and tries, perhaps, to grant more than one of these qualities or attributes the supremacy over the others, and thus inharmony is continually existing in man. If he refrain from this error, he is next liable to instability and instead of maintaining that attribute, through which God has wisely chosen to approach him, in its supremacy, if he granted its prayer for supremacy, he raises presently another attribute to its station and thus yields first to one influence, then to another, and another, in turn. So the man wavers, now approaching God and tasting happiness, now falling from God and feeling grieved by the want of that inward peace of which he had once enjoyed a portion.

Then where shall man find a remedy for this want of stability or for his indecision? Where shall he seek for true knowledge to enlighten his free-will so that it will yield the supremacy to that attribute in which God is approaching him, by which he experiences the growth of that peace which is not of earth or dependent upon the things of time? *He must rely on the help of God.* He must ask each passion or attribute that thus petitions him for its indulgence in supremacy for this power to direct and to bring into its service all its fellow attributes: What, O passion or desire, is your motive? What impels you to ask me to indulge you to so great an extent or to any extent? Why should you not rather serve in a more humble position?

If the reply be that of asking gratification because thereby God will be honored or loved or served; if it be that thereby your fellow-man will be made happier and the general good promoted, let the passion, attribute, or desire have sway. Sacrifice your free-will, for then you may be sure God has impelled that attribute to ask of you this indulgence of action. If, on the other hand, the reason whispered to you is that God is indifferent to man's action and that this will gratify your ambition for fame, power, or wealth, be assured that the attribute desires to enslave you and will seek to maintain its power by continual petitions to disregard the wants or desires of others and the commands or impressions of God. So will you sink into slavery to lust of some kind; good will be found uncomfortable to you even when others possess it, and evil, or the absence of good, a pleasure to yourself and recommendation of others to yourself.

What shall a man do to be saved? Listen to the voice of God, perhaps heard feebly in the heart or petitioning in some out-of-the-way place, since all the principal ones are already occupied by antagonistic desires. Let attention be given to this still small voice, to this feeble knock upon the door of your heart, to this appeal of God through your own mind to you. Be attentive then to what God says or asks! Grant all your efforts to its progress! Be not turned aside from sacrificing every other desire that is expressed in your heart, every other passion which raises its voice for gratification! It will perhaps be a very slight indulgence that it calls for, a very brief neglect of God's voice or desire that it asks, but be resolute in denial and you may progress. You will find God's voice thereafter more audible; the desire to serve Him or other men will grow stronger; and the sacrifice of any interposing or contradictory desire or attribute will be easier than before. Long before you are perfect, long before you have completely made the sacrifice of your free-will, you will have experienced a portion of that peace which nothing earthly can give or destroy and which, thus existing within you, will be your surest guarantee of salvation and the best evidence of your progress of which it is possible for you to conceive.

Outward signs are all imperfect; the Son of God, the holy city that comes down from heaven, is alone perfect, for nothing is perfect except what God makes so. The work of man is always imperfect whether it be performed here or hereafter. God alone is perfct, and such as are in harmony with Him are also perfect by their unity with Him—not by any merit or exertion of their own capacity. God is One, and those that would be one with Him must be perfect through Him and be maintained by His mercy and power in that perfect union and communion with Him which is the essence and reward of harmony with Him. They are perfect as He is perfect because they are one with Him, even as I am one with Him.

CHAPTER XXII

NECESSITY OF FAITH

When the first process of yielding to God's influence or call or solicitation has been gone through with, the second becomes easier because the mind knows what is required and what the requirement leads to. The mind has faith arising from experience of the previous operation of God's spirit upon it having produced peace or satisfaction. Faith is the evidence of things unseen, says Paul. But faith is truly that quality of the mind which is above reason, and which reason yields to as its superior. It is an instinct or intuition of the spirit. Being above reason, it cannot be derived from it. Reason and argument can never produce faith in any man, but a reference to his own internals will give it to any man who desires it, provided the man asks it of God in a submissive and fervent manner. Ask and ye shall receive; knock and the way will appear opened to the acquirement of and progress in faith.

Faith, then, is the evidence of things unseen by the spirit of man acting upon the mind of man. It is the spirit-action derived from the aid of God's spirit. No man has faith in God or in God's care and oversight until he wants to have it and until he retains it by watching over and guarding it from all assaults, whether they come from the reason of other men or from his own mind.

Then let no man take your crown. Your faith in God as the Supreme Ruler of the Universe, in His omniscience and omnipresence, is essential to a living faith. It is a living faith when possessed an active faith when referred to as a ground of action, and a crown of glory when acted upon. Be desirous, first, to have faith; second, to pray for it; third, to refer to it when you act toward others or toward God. Let it be ever present with you, influencing your conclusions and judgment and expectations. So shall you find your faith a crown of glory and a living principle. So shall you be advanced in the pursuit of happiness because you will be led to rely on God as an ever-present, all-powerful, and an acting Being. Be no longer fearful, but believing; be no longer faithless, but . faithful to God; and have faith in God as a Friend who is incapable of deserting you in the darkest day of your fortune, or the gloomiest period of your mental conflicts with doubts, fears, and all the powers of ignorance.

There is in every man a capacity for faith and every man can have it if he will resolve to seek it. I know that many say: I cannot help what I believe! I am ready to be convinced but I cannot make my belief conform to what is revealed as truth! To such I say unhesitatingly: Your own will prevents your receiving the revelation of the will of God. Your reason only acts as the servant of your will and as such a servant argues, and argues still when convinced of its error. Leave all these subterfuges; go to God. Ask Him by prayer to enlighten your ignorance, and give you faith in what you ought to believe; to be your Establisher, Defender and God, to be your Helper, your Savior, and your Friend; to be your Helper in affliction, your Savior in peril, your Advocate in trouble, your Defender in conflict, your Establisher at all times and under every appearance of evil; to be your sure reliance and your Eternal God. God is powerful enough, earnest enough, happy enough, good enough to do all this, and He has promised in every age of the world through all of His servants or holy mediums of declaration to do all and to be all. I have told you to ask Him to do or to be, and He only waits for you to ask Him sincerely, earnestly, truly, and with resolutions to receive as true, holy, and divine whatever answer He may render to your prayer. Submit, and He will be your King; resist, and you are His rebel. Walk humbly, love mercy, do justice, and He will be your Friend. Be, then, ever desirous to have God on your side and be careful that you be not found fighting against Him who is all-wise, eternal, and all-powerful. Amen.

Let every man be persuaded in his own mind, first, that I am what I profess to be; second, that I am authorized by God to proceed as I do; third, that I have wisdom enough either of Myself or by authority from God to enable Me to proceed in the best way to secure the objects I have in view; fourth, that those objects are good and such as ought to be secured or I would not attempt to secure them. Then he will see no longer any cause of distrust. But so long as he searches the scriptures, not to find when I shall come but whether I can come, he will be in confusion. The plain prophecies of olden time will appear to him a mass of absurdity, and the day of salvation will be postponed for him indefinitely. What shall a man then do, resolve upon his belief and then search the scriptures? Oh, no, let him search the scriptures after he has made up his mind to believe them and to be guided by God's spirit to understand and apply them.

The scriptures of truth, as those of olden time are very properly called, are fit for the enlightenment of outward men, for they are supported by outward proofs and defended by outward defenders. But they are not alone enough to lead men to truth or to a knowledge of the will of God. Paul may plant, and Apollos may water, but God gives the increase. So study of the scriptures may proceed from day to day and year to year, but the mind will be no more filled with knowledge than at the beginning unless God pleases to let His servant have an understanding of the true signification of the letter of the word of God. The *Word of God* is quick and powerful and goes right to the heart and marrow of man, but the word of God is revealed through men to men and partakes of the imperfection of its medium of transmission. The *Word* is understood easily and at once upon its reception; the word of God is more obscure in its teachings because being outward it can only reach through him outwardly by such means as his reason or his social companions who affect him often by their sympathy, frequently without either being conscious of its existence.

CHAPTER XXIII

LAWS OF THE SECOND SPHERE

There is prepared and laid up for all those who love and serve God a crown immortal, full of glory. But this crown is laid up beyond the grave; unless the mortal puts on immortality, unless this corruptible ceases to be corruptible by leaving only the incorruptible in or about him, man can not enter into the joy of perfect bliss. Beyond the grave lies that undiscovered country where every man must give an account of the deeds done in the body and be judged at the bar of God.

All that have lived in former ages have thus passed away; all that now live expect to follow them; and all that shall hereafter dwell in places of the present generation, it is supposed, must go as their predecessors went. Death of the body is the inevitable preparation for a new sphere of existence. Death, then, ought not to be viewed with apprehension or as an evil to be endured. It is a release to be hoped for—a happiness to be enjoyed.

Man dreads to pass from the known to the unknown, yet he tires of the known and seeks the unknown so long as he can obtain the latter without losing the former. So if we can convince mankind that the known is not lost by death, and that the unknown is obtained in a glorious appearing, he will no longer go down to the grave sorrowing, and no longer be apprehensive of his inevitable change from death to life. For this is really the change which takes place. Death is the process by which the spirit becomes more fully conscious of its existence and is led more knowingly into a state of progress toward perfect bliss.

Death is not the tedious or painful process that many apprehend it to be, but is often a release from bodily pain. It is always a release from bodily fear. Death appears to those who witness it in others as a loathsome change because the sustaining spirit, departing from the matter of the body, leaves it to undergo those chemical changes which fit it for further use in the economy of nature. The cold skin and discolored features illy represent to us the animation which so lately appeared in and upon them. Over all the body the change appears and the very touch of the once-loved form is repulsive to the survivor. The once-loved form, we say, as if to support the delusion so long entertained that it is the form to which we were attached instead of the being that animated it. Long ere the survivors have felt consolation the departed being has watched with interest the change it underwent and pitied the friends who thus mourned for it as it perhaps had mourned in former days for others. If the spirit is elevated sufficiently to sympathize with those friends and to desire to alleviate their sufferings, it hovers about them and strives to make known to them that it is not dead but living. The power of God has preserved it as effectually as it preserved Daniel when in the den of lions. That was an outward preservation; the other is a spiritual one. That was an unusual occurrence; the other was or is an every-day one. But each occurrence is a manifestation of the power of God, and an acting under or by His laws established at the foundation of the world and at the time when God said: Let there be light.

Death is inevitable. Not that of necessity it must always take place by disease or accident or poison or by weapons or in any visible outward manner. All that is inevitable is the change of this mortal body for a spiritual body or, more exactly speaking, it is merely that this earthly tabernacle must be dissolved, leaving existent that spiritual body which the soul had collected about it within the earthly body immediately upon its entrance into this sphere of existence, and which it had retained and maintained by the aid of the Word and under the laws of spirit-matter as controlled or used by life.

Having so existed in intimate relationship to the soul and the earthly body, the spirit-body of man becomes so naturally and fully a part of the soul itself that, inasmuch as the spirit or soul intelligence does not observe any change of this spiritual body except that it becomes purer and brighter and more refined, more shining and more glorious, it naturally believes the spiritual body to be eternal in its existence, unchangeable in its essence, and the most glorious form of man's appearance. But appearances in the spirit-world, particularly to spirits not much advanced, are far more deceptive than those existing on the earth in the outward. The experience of spirits is necessarily unreal, as I have already shown in this series, and they must therefore be left in ignorance of many things by these false conceptions of their positions and powers and performances.

Will not this book or these books enlighten spirits as well as men? They will enlighten all who want light. They will urge many spirits to advancement and so far spirits will be enlightened by them. But inasmuch as the misapprehension of spirits of the reality is a law of their nature and effect of the will of God for their good or advancement, their knowledge that such a law exists and is followed by such effect will not prevent their being equally deceived or deluded or hallucinated or whatever you choose to call the impression of the spirit's own desire, to be to itself as real as any outward experience is to those in the body. It will no more affect the continuance of regarding such impressions as realities than does the reason of a man when psychologically affected and induced to act under the will of another and behold whatever the operator chooses to have him behold as if it were real. Even though the subject should just have passed through one experiment and been allowed to perceive how it had been a mere hallucination, the same experiment might be repeated with equal or perfect success.

Such then is the law of spirit and such is the relation spirits sustain to law, that only what it is proper or useful or agreeable for them to see do they see. All the evil that pains, all the sin that degrades, all the power that appals them is as invisible to them as if it did not exist. How then can they detect a murder, a robbery, or a theft? They cannot except when they desire from good motives to see or detect it. Good motives enable them to see the act and to witness without injury the pain, degradation, or horror of it. Cannot the spirit of the departed friend then read from the memory of the living all they have done and so realize a consciousness of their actions as well as if they had witnessed them or participated in them? They may so read when they have a good motive; if the motive is bad they are unable to obtain anything except what they desire to see. You say they might desire to see the truth and from a bad motive! Then they would not see the truth but that for which they desired to see the truth would be manifest to them and appear to them to be truth. They would suppose themselves to have seen the truth, and perhaps they would then undertake to act upon the minds of men as if it were true; whether the action was real or unreal would make no difference to their perception of it.

The spirit is unable to do harm unless the man, by his desire's according with the will or desire of the spirit, enables the spirit to act in the man's will and aid him in performing the action of which they are then both guilty. But I have said the spirit gets no worse, does not fall from any grace or advancement it has experienced! The spirit is not guilty of the act performed by the man, nor is the man guilty of the act performed by the spirit. When the two combine their wills and powers in the performance of an evil act, each bears his own responsibility. The spirit's responsibility is no greater than if he had performed, or experienced, one of the usual unreal actions and scenes. The spirit that would so join the man from evil desires would also form the evil desires and pass through the unreality if the opportunity of joining the man, and making the act more allied to reality, had not occurred. The man would have attempted the act without the aid of the spirit, and if we suppose the act was not accomplished for want of that aid we must remember the man was equally guilty, in the sight of heavenly vision and discrimination, as if he had succeeded. Success is not the test of merit or guilt in God's system. Success is a punishment, or sometimes a reward, of good or evil desires, and actions are permitted to men in the body as a mode of experience useful to them in a future life. The last act of a career of guilt may be more shocking to mankind, or to a virtuous mind, but it is not necessarily any more guilty in God's view than the first step, trifling, perhaps, in its ultimate effects upon all but the man's internals.

Far, then, as God's laws are from resemblance to man's laws in this respect, I would not have it inferred that man's laws are wrong in punishing the act rather than the motive; high misdemeanor of the law is truly that which deserves severe punishment by men. I would rather blame men that their laws punish the small effect or misdemeanor too strictly and rigidly and severely, thereby sometimes driving the offender to despair and to greater crime, when mercy tempering justice might reclaim him and bring him to respect himself and the laws of the community and of God. It is not that punishment should not be proportioned to the crime but that it ought also to be reduced when the crime existed only in form or in mental manifestation, and that the youth, the former character, the respectability of the offender ought to be allowed as a mitigation of the punishment or as a full excuse for the crime. Far be it from Me to urge you to tolerate crime or evil in men of rank or station or respectability in society—all I ask is that men should allow former innocence to weigh against present crime, to be willing to reclaim the offender without punishment rather than to resolve only to attempt his reclamation by punishment. Do not let the guilty escape at the expense of the innocent, but do not make the comparatively innocent offender suffer for the sake of the more guilty. Do not even desire to punish at all to deter others but only to restrain and reform the individual.

But, you will say, the avowed object of man's law or rule of punishment is to reform others as well as the offender, or, as it is generally expressed, to deter others from similar crimes. The experience of mankind has shown them that severe punishment for small crimes only makes men more brutal and reckless, and that the reduction of punishment for crime is generally followed by a reduction of crime. It is evident from this that men are not deterred by punishments inflicted upon others. Again, I have said in former times that of old an eye for an eye was required, and a tooth for a tooth, a limb for a limb, but I preached forgiveness of injuries and pardon to the offender. So if an individual could allow these offenses to go unpunished without fear of thereby inducing others to repeat them (which is to be inferred from the manner and form of the precept as I gave it) certainly society could relinquish the desire to deter others, and confine itself to the repression of crime by punishing the offender with a sole view to his reformation.

I perceive that many men, holding fixed opinions upon this subject which has been much discussed so as to give them great confidence in their present opinions and an unwillingness to change or even modify them, will be offended at Me that I have thus undertaken to instruct them how to make laws, and they will be disposed to drive Me back as an intruder into state affairs whereas I ought to confine myself to religious matters. I acknowledge that I have in this respect departed from My usual course, and that it is from pretensions of this nature that men have been led to condemn religion and to fight on account of it. I acknowledge that man ought not as a member of a religious community to interfere by any religious authority in the execution or formation of the laws of the country. But man should seek to make laws in accordance with the will of God and should seek to be individually guided by Him whenever he has to act as a maker or executor of law. I have given this advice now, not only because it was required to aid public sentiment or the general opinions of men disposed to do what is right in the matter, but because your cavil at this will enable you to see why I do not lay down rules for your action upon all the great or little questions which are occupying the public mind and causing the public sentiment to revolve in circles.

I have told you before that only the general rule can be laid down. Its application to different circumstances, always varying, ever assuming slight shades of differences, must be made individually and by seeking divine direction. That which is right for one may be wrong for another, not because God requires more of one than of another but because one man is more capable of perceiving his duty than another and has greater facilities for so perceiving and sometimes for performing it. In reality it is because God is impartial and tempers His justice with mercy, thereby making His perfect justice equal in its manifestation to every man, that men are unequally called on for performance or resistance, for submission or endurance to law or requirements of men, and to conformity or non-conformity to public sentiment or opinion.

CHAPTER XXIV

ACTION OF HIGHER UPON LOWER EXISTENCE

When the spirit has advanced through the seven circles of the second sphere, a change takes place in it similar in some respects to death but yet quite different in most. The spiritual body is not dissolved but the spirit perceives a change taking place. In progressing from one circle to another in the same sphere the change is imperceptible, while from one sphere to another it is perceptible. The spirit is prepared by instruction of those who have passed through the change and, being so instructed, looks forward to it as a reward for its faithfulness and attention to the will of God.

This change takes place suddenly, just as death is really instantaneous though to men apparently long and gradual. Yet the difference between life in the body and life out of it is the whole difference and evidently in this respect it must be instantaneous as life is either in or out. However feeble its manifestation may be, the whole life exists there as much as in the strongest man. The principle of life prevents the decomposition or dissolution of the earthly body and the moment that life is absent, its action ceases to maintain the organization existing. By the laws of matter to which it now becomes subject, its dissolution commences by changing the relations of its particles to each other and the whole mass, though the mass retains its organization for a longer or a shorter time according to circumstances in which it remains. The partial dissolution which seems to take place in diseases before death is really an effort of nature to maintain or recover a healthy action in the whole by sacrificing a part with a view to its restoration or healing, and leaving the rest healthy.

In the change in the spirit-world from one sphere to the next this has no imitation. The spiritual body is only affected by being refined until the seventh sphere is entered. The spirit-body is refined in as sudden and instantaneous a manner as death of the earthly body occurs. The grosser part contaminated by the actions of the body on earth is left without a sustaining power; it speedily dissolves and imperceptibly separates from the purified or advanced body that is still maintained by the action and presence of aura. The separated particles return to their general mass in the spirit-world. The spirit-body and mind is forever advanced to the third sphere and though it possesses all the powers it had in every part or circle of the second sphere it seldom (as I have explained) exercises them because there are plenty of spirits in those circles to do such work or action and the spirit of the third sphere is exercised in duties those of the second cannot perform.

Here let us pause and inquire or consider how it is that such a change takes place, though no allusion has ever been made to it before. It is because each sphere is a complete part, a world by itself, a division evident in the spirit-world to all above it but not to those below it. Low spirits see higher ones but they are only the spirits of higher circles of their own spheres, except when the higher spirit seeks or desires to manifest itself to the lower one. How is this, you ask, inasmuch as the body of each is composed of the same substances except that the higher has parted with some of the grosser or more corrupted portions upon which we cannot suppose vision or visibility depends? Yet such is the fact. Visibility does depend upon that very grossness or contamination, and that the particles assume a different arrangement and thereby become invisible to perceptions which before had cognizance of them is no more strange than that glass of various colors should be changed to transparent by purification, which abstracts a very limited proportion of its substance.

Long ere the spirit so departs from the second sphere, it knows that it will encounter the change and though it is known to spirits in lower circles than the seventh it is not explained so that they understand it. They know that certain spirits disappear and are seen no more but they do not therefore infer that a change bearing any similarity to death takes place. One would suppose the spirit to be merely in another place or engaged in duties that required a separation. But if spirits have all they desire and they desire the presence of one of these advanced spirits, how do they obtain the realization of their wish? They have the semblance of the spirit with them in whatever form they desire or wish for its appearance and manifestation. Here the realization ends. High spirits are not subject to the desires of lower ones, though high spirits are always so desirous of contributing to the happiness or advancement of any of God's creatures that they are ever ready to leave any work that can be postponed, or transfer it to other spirits in their same sphere or power, for the sake of materially aiding the lower one.

So it often happens that the higher spirit prepares for the lower one certain helps and deals out assistance with the aid or means of other spirits. There is also help given to the lower spirit by its associates of the same sphere by consultation. This is not extraordinary, but common in the higher spheres. In the second sphere, the general character and tendency being low, the aid of fellow-spirits is less. Perhaps the lower sphere may be left as it is already well explained in the second part of this series. But I will state that the lowest circles of spirits cannot aid each other much and are therefore worked on more by the higher spirits, who appear to them in the relation of teachers and of instructors in their employments or duties and assign to them tasks or duties to perform.

By these means the spirit becomes obedient and self-sacrificing, and as it sacrifices self, it advances; as it advances it becomes more easily influenced by higher spirits because it has less will of its own. Having reached the sphere of Memory, as I have described, it progresses in the reception of its own and others' memory until, having arrived at the period or position in the seventh circle of that sphere which precedes immediately its reception into the fourth sphere, or sphere of Knowledge, it again undergoes a change similar to that already described as occurring between the second and third spheres. Having experienced this refinement of its spiritual body, it is qualified to traverse space it could not pass before. The means by which it thus passes beyond the limits of the solar system to which it had been attached from its entrance into paradise to its arrival at the third circle of the fourth sphere have been so alluded to that they may be surmised by those who have attentively read all that has preceded this.

Yet how few will have done so; how many have skipped parts; how many have hurried over the first books as immaterial so that they could reach the revealments of the last! But the first and the last are so connected that the last requires the first for appreciation or even for understanding. No useless sentences have been published or even written. All will be found instructive and necessary to a true understanding of the History of All Things. Then is it completed by this book, or may we well expect more? More will be given at such time as God shall deem fit. Not that I do not know, but I do not choose to declare how much or when. Not that I can declare surely the time because, as I have explained, it is contingent upon a willingness of some holy medium to be entirely passive in My hands and entirely submissive to do the work imposed upon him thus by its reception. For God has chosen to have it performed in this way by a man, without compulsion or the sacrifice of his will except as he wills to sacrifice it.

This holy medium is now willing to do this, though he was disposed not long since to complain of the hard work, but I am not ready now because the time has not come. When the time comes he may be unwilling, since there is no restriction upon his freewill. He is not raised to that elevation which secures him from falling; he cannot be raised to such an elevation until he shall have left the body and reached the high spheres. In them only are the wills of men completely sacrificed, and until the will is entirely sacrificed rebellion, or inattention which is in effect rebellion, may and does occur. My holy medium then may be induced by some outward circumstances to refuse to obey My will. Will this show a want of faith in him! Not merely in him, but in all. For his faith receives its severest trial by the rejection which these writings or revealments experience at the hands or wills not only of the so-called religious community but also of the believers in the reality and spirituality of the present manifestations.

When the heterogeneous mass of spiritual believers is surveyed, their want of faith in these books will not appear so extraordinary. They have been induced to believe by so many different motives, so few of which were not entirely selfish, that the only wonder left upon reflection is that any should receive them. Some are impelled to believe by affection, because, having lost a dear relative, they desired to retain a way of communicating with them. Others have desired knowledge that should elevate them in some particular above their fellow-men. Others have desired to obtain from it the means of support. All these must change their principal desire to the one true desire, that of serving God and asking His will and wisdom to be made known to them for their enlightenment and aid in following after heavenly things.

Besides this, so perverse is man that all who have had fixed religious principles which conformed to some sectarian form are fearful that they shall have to abandon them. Fear not, ye outward professors; I will not ask you to abandon your associates or withdraw from your connection with any church which acknowledges God to be Ruler and King, and admits that He knows and regards the actions of men. I leave you there that you may infuse into the minds of your fellow-professors a desire to inquire to know God better and to serve Him more. Very many would willingly serve God more and better if they were not persuaded that words and not deeds, professions and not acts, are required. They are thus led away from doing good works, which necessarily are good by being performed from good motives, by justifiable means upon or toward fellow-men. For none can really believe themselves capable of aiding God, and all will admit Him to be beyond the need of the help of any or all individuals of the human race.

There is also another difficulty: a small community called Quakers have preached for some two hundred years that the law of God is made manifest to man within him in a manner superior to any other, and that therefore man should be governed by the law so written on his heart or mind to the exclusion or subordination of all other guides. By this doctrine's having been so forcibly preached among them, it has become the distinguishing features of their association, and though they have not been consistent with their own profession they have exerted an influence upon all who have known them and their doctrines, and have even caused other churches to admit the possibility of such reception of inward direction, though those churches continue to hold it subordinate to other guides and are indeed impelled, by other parts of their theological features, even to reject it entirely as an impossibility. But in this book, as in former ones, I have referred you to this inward guide and have declared that by means of it and by attention to it this series of books may be known to be of God and what they profess to be. To many this is a stumbling-block because they are not willing to be guided by such direction, and because they are unwilling to be so guided they do not receive it. Man must be willing to hear the truth before it will be declared to him in this way.

Though I refer you to this as a guide, I do not ask you to rely on it as the sole or chief guide, for I refer you principally to My agreement with former revelation as recorded in the Bible, and to the agreement with reason that My whole presents. I refer also to the outward sign, which is the test formerly given to establish revelation, the outward sign of rapping and of other spiritual manifestations, which I have declared would also confirm My declarations and assert My truth. Many mediums also will have the power of healing bodily or outward disease, and these mediums will also believe and declare My truth and identity. Be, then, satisfied to yield your will, your former faith or conviction, to the truths here declared, and seek more particular and special guidance within yourself.

General laws and rules for bodies of men and for the general conduct of each man are found in recorded revelation. Special guidance, when the sincere follower of these or the best guides a man has have been followed as far as they will carry him, will be obtained by him who seeks from God this direction. This special guidance should not be expected or desired except in case of doubt, uncertainty as to the meaning of the recorded revelation, or a need of instruction as to its application to the particular circumstances in which the man may be placed. I do not, therefore, place one or the other above or below. I place them all in their places, and together they form for man a sure and sufficient guidance.

What do you propose to substitute for them? *A part for the whole.* And then you would help the part by man's wisdom! You would have the man, who needs more guidance than he can find in the part, go to some man or some body of men and ask them what he should do, or what he should refrain from doing. Let reason judge whether you are right and I am wrong, or whether I do not furnish you with a reasonable and probable and proper solution of the difficulties which have perplexed and disturbed men in all churches in the present and past times.

Having given you this argument for the reception of these writings by every one who desires to serve God, or, in other words, do good, or who more selfishly desires principally to secure his eternal welfare, or, more particularly, selfishly desires to be happy here in the body, I would once more ask you to join Me in humble prayer to that God you all admit to be above us and able to grant our petition. I also particularly call upon you who believe Me to be so powerful, so one with God as to be worthy of personal worship, to unite with Me even as you ask Me to unite with you when you pray in My name.

Let us pray

O Holy Father, Merciful God, be compassionate upon our ignorance and bestow upon us a portion of Thy wisdom. Be compassionate upon our poverty and bestow upon us the riches of Thy favor and grace. O God Most Merciful, who hast opened the way for our salvation, may it please Thee to guide us and direct us so that we may find the truth among all the conflicting claimants for its possession, for, O Almighty Father, Thou hast the power and the will to save us and we want to be saved by Thy mercy which Thou hast declared endureth forever. O Holy Father, help us as it may please Thee to help us, and make us willing to receive Thy help in the way it may please Thee most we should have it, so that we may glorify Thee and Thy Son, the Lord and Savior Jesus Christ, to whom with Thee be ascribed all honor, praise, glory, and power, now and forever and evermore, world without end. Amen.

O God, who art ever-present and ever-loving, O Father, who dost most kindly invite us to pray Thee for good gifts, bestow upon me, I pray Thee, that knowledge of Thee and of Thy Son, the Lord Jesus Christ, and of Thy spirit, the Holy Ghost, which is the Comforter of those who receive Thee and Thy Son, which will enable me to perceive the way, the truth, and eternal life. Grant that I may receive from Thee, O Father, that peace which comes down from Thee and can never be taken from me by others. O God, help me to retain it when Thou hast been pleased to restore me to Thy favor, and keep me, I pray Thee, from the temptation of self-will and self-gratification of will, so that I may obey Thee and serve Thee in accordance with Thy will all my days and be received into Thy heavenly kingdom, into eternal and unending progression of happiness. Amen.

O holy and loving God, Thou art the Giver of all and the Bestower of all to men; may it please Thee to bestow upon me understanding and wisdom, and give unto me knowledge of the truth, and to Thee will I endeavor to devote my life here and hereafter. Amen.

O Lord of heaven and earth, who art the high and mighty Ruler of Thy own infinite works! may it please Thee to implant in me knowledge of Thee and direction as to Thy will whenever I am in doubt and know not what Thou wouldst prefer to have me do, or what will best advance my eternal welfare, or best enable me to serve my fellow-men, or best enable me to protect and support those who depend upon me, or best enable me to retain Thy peace, which cometh down from heaven, or best enable me to walk after Thy requirements here and most promote Thy great and holy cause, and to Thee shall be evermore the praise, the honor and glory therefore, and for all that I have or acquire by Thy will. Amen.

O God, my Father, be merciful and loving to me who hast left all to follow Thee. Be merciful and kind to Thy humble servant and raise me to be Thy son, when I may be purified and glorified by Thy power and Thy mercy. Amen.

If you can make all these prayers your own, do so; if you cannot, strive to do so. Walk humbly, do justly, love mercy, and God will help you and accept you. But do not suppose you can stand still and have the work done for you. No, God will not be mocked by pretended service nor deceived by professions of desires which are unreal. Do good. That is, the best you can do, and if you strive to do all the good you can you will be surprised to find that so much is in your power. Do good and you will be helped to do it; stand still and you will not be moved. God will not force you to maintain or undertake action you do not desire to perform. Have the good desire, and you will have the great help. Walk humbly, and you shall not be cast down but will hear in due time: Well done, good and faithful servant; thou hast been faithful in small things, hereafter thou shalt have greater to do. All that will come may come and partake of the waters of life freely, without money and without price.

CHAPTER XXV

PROGRESSION TO WISDOM

When in the course of the spirit's progress it arrives at the circle in which it receives knowledge of other spheres or systems of planets and suns, it finds itself possessed of the power to extend itself beyond the limits which formerly bounded its travels and researches. It may, by the same process of will and attenuation which the spirit or soul possessed in paradise, extend itself to another system or combination of systems until it reaches the system which it desires to be in. Having arrived at its purposed position, the termination of its journey so rapidly made, it receives from spirits there such knowledge as its advancement qualifies it for and returns at its pleasure, by the same process by which it departed, to the position or place it had previously occupied. The spirit-world to which it had originally been assigned remains ever its home, though its perceptions extend farther and farther with each advance it makes toward the perfection of God's sons.

Thus far have we proceeded in pursuance of our design to make mankind in the body acquainted with the knowledge possessed by spirits in the fourth sphere. Do any say I have not restricted myself to this boundary, or do any say I have not given all the knowledge spirits must be supposed to possess? I answer: Believe that which you have if you would have more, for to him who hath, more shall be given; but to him who hath not much, even that which he hath shall be given to another. Believe what I have told you and you will find it will give you much more than you have supposed possible, for whatever further knowledge it may be expedient for you to acquire will be found by you in a perusal of these works with a view to enlightenment on that point, even though it may not to your perception have been treated of; thus this is a revelation greater than that of men revealing a new theory, for that grows imperfect as it is magnified while this as it is extended becomes more perfect and glorious and worthy of its author.

Here is indeed another proof of My identity and truth, for I teach not as man teaches, imperfectly, but as no man teaches—that is, perfectly. But perhaps you blame Me for not being consistent, in having told more than I had promised. It is thus, too, that God will fulfill His promises, abundantly and beyond the expectation of mankind. Be full of joy, then, for revelation proceeds continually to all who seek for it and man can in this bodily life commence the eternal progression he makes in knowledge of God and His creation. Be thankful, for God is good beyond the power of man to conceive or imagine, and God will gratify His benevolence and proclaim His glory by bestowing upon His creatures benefits. Be no more despondent, no more faithless nor unbelieving nor doubtful, but be joyful and give thanks to God and to His Son, who is the servant and agent of God, and do not regard the holy medium as anything more than a mere instrument in the hands of a workman, for such is his office and this instrument is good or bad, sharp or dull, as it pleases the workman or as his skill makes it.

Having now arrived at the close of this portion of My subject, let Me ask you to try Me calmly by reason, unprejudiced by creeds, unbiased by your wishes, and see if I am not reasonable. Remember that, though reason may and must agree with revelation and all that is not reasonable is untrue, yet reason alone can not lead you to truth, because two contradictory propositions may be made, indeed have often been made, and both may be reasonable. It is from this cause that most of the dissensions which have agitated the social or the religious world have proceeded. Reason was on both sides and each party numbered in its ranks men of high intellect, deep learning, and powerful reason. Yet these men contradicted each other and refused to see the plain fact that both were reasonable, that either might be right, and that revelation alone is infallible. The fallible may agree with the infallible and can never be really and unreconcilably opposed to it, but the infallible cannot be brought down to the level of the fallible or forced to agree with it.

Let reason, then, be your lamp in the absence of the sun of revelation, but when the sun has arisen, open the shutters of your mind and let in its glorious beams to warm you into eternal life and illuminate your pathway with its unsurpassable, unmatchable effulgence. So shall you find peace, joy, and everlasting happiness. So shall you feel that peace which cometh down from heaven and enables every man to sit under his own vine or fig-tree or habitation with none to make him afraid, for who can overcome him whose mind is stayed on God?

CHAPTER XXVI

GOD'S JUDGMENTS OF MEN'S ACTIONS

When in a preface to this series I promised to refute in the last part of this third Part some slanders against God's government, I had hoped that My holy medium would be very passive. I find him less so than I had desired, but I shall proceed to the performance of your expectation in such way as you may not expect, and in such way as may best accommodate the imperfection of passiveness still existing in him. I am not, however, desirous to have you understand that this want of passiveness is so great as to cause error, for at the most it can exist only so far as to withhold truth. The utmost exertion of will on the part of My holy medium could effect no more than this, because I have promised you that error should not be suffered to escape through him.

In other words, I have declared that he is passive enough to obey Me, though he is sometimes so restless as to pervert the channel through which I flow forth to you, and give a sinuosity to My current which neither lessens My volume nor prevents My arrival at the ocean of truth. This is manifest in one instance by his desire to accomplish the complete reception of this book in two days, though not consecutive ones. Although kept by him as far as possible in the background, this desire existed and I gratify it. He has also other desires about it, no more disturbing indeed than that, and such desires also he would yield if I declared it My will that another course should be pursued. But from this you may perceive what is required of a medium for his perfection, and how far many are from passiveness that imagine themselves to possess it.

Having thus once more made you acquainted with My relationship to My holy medium, I will proceed to show you why God would have a man thus passive, thus non-exertive of will. It is because He forms a man's intention when the man submits entirely to Him. God says in effect to you: Be still and let Me work; be still and see My glory. But if this be so, why have I called you to work and expressly declared you must work and form intentions in order to have God's help? It is this apparent contradiction which has caused much difference among men in all ages, and it is this which I want you to find reconciled here. But remember, as ye seek ye find. If you look for difficulty you will no more experience disappointment than if you look for harmony and reconcilement.

God calls upon every man to work, and He also asks of every man a surrender of his will. His will is surrendered by a desire to do God's will and the work is performed in God's will when surrender has thus been made. How then, in this case has My medium erred, having desired to surrender his will in effect by desiring to do God's will? This My medium has done, but though he has the desire, he has not accomplished the work of the surrender, but has instead of it performed another work or entertained another desire, as I have stated. Here are two desires existing simultaneously in this man's mind, the one proper and right and the other not wrong, but unworthy or not deserving to be entertained.

To ascertain this we must go back beyond the desire, or intention, to the motive. If the motive be good the desire is so, for whatever is pure in its source remains so until intermixed with impurity. The motive for his desire was to establish the fact that the book was written with such rapidity that no man could have accomplished it without the aid of inspiration. This motive you do not perhaps perceive to be bad or unworthy, but inasmuch as the motive has its origin in a desire to give an outward proof, which I have declared I will not give through him, he was wrong in allowing a desire to proceed from it. Although thus far he was culpable, he was partly excused by not perceiving his motive and in believing his desire was prompted the more fully by another motive, also really existent—the wish that readers of the book may be convinced that he does not of himself produce the book but that it is given in a superhuman manner from a superhuman source. This last motive is good; the other is not. The former, or other, as I last called it, was wrong only because I had declared that I would not give outward proof through him of the manner of this writing.

Therefore this assertion of the time of writing will either be doubted or declared not unprecedented; in either case, of course, failing to accomplish the object My medium proposed. It is thus that man's projects fail. The motive may be good in itself; the desire may be unexceptionable; but the result is unobtainable by man's will, God having withheld His will from its aid. God is served, however, though man is not benefitted, because God takes the desire for the fulfillment and the motive for the desire.

In this case, the motive being double, the law needs further explanation. The one motive, being hidden too from the holy medium in his consideration of the desire, also affects the act of desire. God has mercy on the bad or erroneous motive because the medium was unconscious of his error, and He gives him credit or glory for the good motive because it tended only to the completion of the desire to serve Him. The desire itself was an indifferent one, neither good nor bad in itself but praiseworthy or reprehensible according to its motive and effect. But the effect is also indifferent or inconsequent to God except as it affected the motive by its reason. If the effect was perceived to be bad and the motive was unmodified by it, wrong ensued; if the effect was good, it was only the realization of the desire and therefore it did not alter the motive except that it might strengthen or magnify it.

Effects, then, are less beneficial to men than motives, and this is wisely so provided, for if God required success of man when he undertakes to do good how much oftener would the man receive condemnation than praise for his most disinterested and benevolent intentions? Such being the law of reward and punishment, let us again consider this effect and how it will influence the motive. The effect, being unforeseen and in the future, cannot react upon the motive now, though at some future time it might. At this time, then, this desire is judged only by its motives, for the effect is nothing as yet as regards men, and the desire is in itself inconsequent. But as the effect is a disturbance of My equable course of procedure or revelation, I have another view to take of it.

The desire of My holy medium was held in strict subordination to his desire to do God's will. He was willing to sacrifice the desire first named if he could perceive it to be My will or God's will that he should. So again the ignorance of My holy medium saves him from blame upon this view, and it should rather be regarded as his misfortune than as his error that he had such a desire. Here again we have arrived at another opening for consideration of cause and effect. His character, which caused him to entertain this desire and allowed such desire to be the consequent of, let us say, the good intention, is the result of a long course of action and the concentrated result of innumerable motives of action. For all these desires, actions, results, and motives he is also accountable and must render an account in that third sphere of which I have told you.

The last chapter elucidated the system of reward and punishment and prepared the way for an explanation of the recompense prepared for man in the spirit-world. The second sphere is emphatically the sphere of recompense for the deeds done in the body. In that sphere man becomes reconciled to God's will sufficiently to receive His commands and be the servant of His servants. In that sphere he is advanced or retarded by the deeds in his past experience, and the motives which produced them.

In this circle or sphere—for a sphere is only a great circle and each circle contains numerous divisions like unto associated bodies of men—in this sphere of reconciliation, or recompense, the spirit maintains such relations with itself and toward God as its aggregated experience or character had prepared it for. If hate or revenge was its ruling passion in bodily life, it is so on its entrance into the future or spirit-world. If good desires had predominated and existed in activity, the spirit finds congenial action in this sphere. It readily assumes its appropriate circle. It finds also the particular section or division or association in that circle which its habits of thought and action fit it to enjoy. It finds this by the attraction which like has for like and by the aid of those laws of spirit-matter which enable it to take cognizance of very distant relations. In this it is assisted by those spirits which perceive its congeniality, and is welcomed by them. The force of earthly ties is not exhausted; the father or mother seeks the child, the brother or sister seeks its cherished relation, the husband or wife or lover seeks the object of its affection, and so the links extend to greater distances.

In all this is great wisdom, for the higher spirit seeks the lower but cannot be drawn down to it. The lower can advance, can be elevated, but the higher also ascends. The companionship in the spirit-world is and must be different from the body relationship, not only because the wants and needs and passions that belong peculiarly to animal life are wanting but because the spirits are often in different circles, almost always in different divisions of a circle, and the equality or superiority that existed on the earth is merged in that which depends on advancement. The higher is improved by the exercise of its superior attributes in endeavoring to elevate the lower, and the lower is induced to profit by the teachings of the higher because of the acquaintance or relationship which previously existed.

So it is, too, that names which have been revered in the bodily life become or continue to be of consequence in the future world, for the lower spirit will regard their teaching when perhaps a higher spirit than the one so revered would be disregarded. It is thus that God causes all to work together for good, for all are bound in one chain of existence and the great gulfs in the world to come are by God's mercy spanned by bridges of affection. True, there are other influences by which the lower spirits are operated upon, for the higher spirits take pleasure in serving or assisting the lower ones, and as they regard all God's creatures with affection they are ever at the service of God in persuading or teaching all men and spirits to advance, to know God better, to serve Him more, to sacrifice to Him fully.

Having briefly sketched this plan of redemption, let Me point out to you that you can find a confirmation of it in the Bible where it is related that I preached to the antediluvians, or fallen angels, during the brief period which elapsed between My departure from and re-entrance into the body.

There is another remarkable passage which confirms this, which I will leave you to search for by diligent study. For I do not want you to cease studying former revelation because you have faith in this, but seek good in all and let all be so embraced in your mind that they confirm and strengthen one another. I do not call you away from any help you have had to find the way to peace and eternal happiness, but give you more. I do not come to destroy the law you have had but to fulfill, to elucidate, to confirm, and to perfect it. In all this I am not only doing good work but I am fulfilling prophecy and doing My Father's will. In all this I am but a servant of His will, though I am the high and holy Son of His love and power. In this I accomplish His will, declared from the foundation of the world, that all men shall be saved, and His pleasure, that His spirit should not always strive with man.

Here I rest from persuading you to seek and search the old. Moses had his teachers in every city when I preached outwardly, yet how few of their hearers failed to continue to reverence sufficiently the old! The early church was indeed clogged by their holding back and adherence to the past forms. So it is and so it will be. The old has its organization in all the fairest portion of the earth and, though it will in all parts at last accept this revelation, it will all too strenuously strive to find in it a support for its old doctrines, forms, and requirements. But though I do not call you from the old and do not tell you to leave your church but rather remain in it as long as it will keep you, still I do say to strive to win your church to accept the truth as it is here revealed, and this is to be accomplished by asking your brother, your neighbor, your elder, or your bishop to join you in the pursuit and investigation of the way to the throne of God. To Him every knee shall bow and every tongue give praise, and to His dominion over you there shall be no end.

Be willing, then, to suffer for My sake, and to suffer wherever you are, in this or that church organization; if you are in none be content. *Form no new association;* call no minister; wait for the revealments of God's will. Those who act in their own wills shall be confounded and washed from their sandy foundation, and those who trust in God shall never fear or be shaken by any storms that may assault their rock-founded edifice.

All that would serve God by preaching the good tidings herein contained may do so individually and in a brotherly way. Persuade, entreat, and reason with your fellow-men. Give them example of patience, long-suffering, and forbearance. Show them a bright and beautiful example which shall be to them as a city set on a hill, as a candle in a dark place, and as a light that cannot be hidden. So shall you encourage your friend and your neighbor to look, to search, and to see what good things you have obtained from God and how you can bear your burdens one for another. Take up your cross and follow Me, for I am meek and lowly and have not where to lay My head, I declared to you long ago. So I call upon you yet to sacrifice all to the service of God and the good of your fellowmen. Be ye of good cheer; I have overcome the world and though you are in the world I leave and give unto you My peace.

Be then diligent, faithful, prayerful, and attentive to the words or thoughts I shall put into your hearts when you are at peace with the world and silent before God. I need no church edifice, no set time to find you or know you are ready. Turn unto Me; seek and you shall find; look for Me and you will see Me coming from Bozrah in dyed garments. Bozrah is the old, as Jerusalem is the new, the present; as the New Jerusalem is the future and the now-coming. Be ye also ready, for ye know not when the bridegroom cometh. Have your lamps lighted and trimmed so that you may be found among the wise who know and understand. Be ever ready and willing to receive Me or God in whatever way, time, or manner I may please to appear or be made known. So shall you be found ever watchful; if the watchman keep the city the Lord will keep it, but if the watchman keep not the city it will be lost by surprise. The watchman watcheth in vain unless the Lord keeps the city, but the Lord will not keep it if the watch be withdrawn.

Do your part and I will do mine. Do your part and God will do His. God will have you to act, but He will have your acts to be conformed to His will. He will have you form intentions, but those intentions must be faithfully conformed to His known will. He will have you exert your will and act in it, but only as your will is subject to His will, and entirely submissive and sub-servient to it. That you may know what His will is, I have written these books, in which I have constantly urged you to this one sacri-fice and in which I have endeavored to convince you that true happiness can never result from man's own will's being served even though every wish be gratified and every desire satiated. True happiness has but one source, and that source is the source of all good and the fountain of every bliss. Be faithfully obedient to the prompting of every good spirit, and seek for the riches which passeth not away. Be sure that you listen for that still small voice which the Hebrew prophet heard, and you will find that it gives forth no uncertain sound. You shall know plainly when to put on your armor and go forth to the battle of the last day, the day in which shall be seen the sign of the Son of man coming in the clouds of glory and shining from the east unto the west with the brightness and rapidity of the lightning. So shall you be found standing in your place in the last day.

CHAPTER XXVIII

PROGRESS OF REVELATION

Having led you thus far by the power of attraction, I shall not begin to drive you. I am the first and the last and as I was in the beginning I shall be to the end. Here is a new meaning for you to study and investigate, for I do not expect you to take My words without examination or receive My ideas without reflection upon them. Be the followers, though, and seek not to lead. If any would be chief among you let it be with humility; let him be the servant of the others and look not to them but to Me for his reward, for verily I say unto you that as he sows he shall reap. If he sows for corruption he shall reap it; if he sows for men's approval he shall not obtain Mine. If he is humble that he may be exalted by men let him not expect to obtain any other reward for it; if he is humble that he may be exalted by Me in the life to come, let him be satisfied with the expectation that I will not forget him and that his treasure is laid up in a secure place where it accumulates at a higher rate of interest than men ever pay even when most recklessly adventurous.

How shall I serve others? perhaps you will ask, forgetting the ample instruction I gave to My disciples and followers upon this subject eighteen hundred years ago—instruction well recorded and faithfully transmitted to you through so many generations and chances of perversion or destruction. How little heeded, too, have these injunctions been, in comparison with those discourses which related more to Myself and My relationship to God and to you. Yet the latter were only called out by the desires of those who then opposed Me, while the former was My free offering to the crowds of men, women, and children who followed and were fed by Me. The former was the essential part which I urged upon men; the latter was the unessential which men called on Me for. So it has continued to be since. The former has been assented to and passed over; the latter canvassed and objected to and dwelt upon.

I only desired men to believe Me Christ, or Messiah, in order that they should be induced the more willingly to follow and obey My teaching. But men have chosen to investigate and pursue blindly into unrevealed relations that which could not help them if found, and to leave unexercised those precepts which I poured out as the wisdom of God and the way of salvation. It was not hearers but doers that I wanted. It is so now. The time is different but the feeling is the same; the individuals are different but their wishes are identical. True, I have told you in these books much of Myself and of God, but it is because you would not hear Me unless I did. I have made the revelation of those things the medium of enforcing or inculcating truths respecting your duties. You cannot serve God better for knowing the future world, but I urge you to serve Him because you now know so much more of the future world. Be His Son's servants here because thus will you become His sons in the world to come. Be ever faithful to the precepts of duty because you have had explained to you how that faithfulness operates to insure reward. Forget not to practice the precepts I gave eighteen hundred years ago because they were so long since given, for God has not changed. He requires now no more of man than He did then, and then as now He required His children to give to Him individually their hearts. Sacrifice your wills to His will. This is all He asks now, all He asked then.

But, you say, is not His will different now? If not, how was it that successive revelations were required to advance mankind to their present state and to continue that progress? I will answer this too, though I have already answered it in effect and reality but not directly and openly.

The Egyptians were a learned nation. Civilization had made great progress and life was easily supported and embellished in that fertile land under the patriarchal government of their chief priests or Pharaohs. But by the devices of the upper or ruling classes, the people were gradually enslaved and made ignorant. In the course of many generations the animals and parts of animals which had been set up as representative of the attributes of the great and all-wise and supreme Being or God became regarded as symbols of separate beings. Instead of being looked upon as the representatives of one Being they were at last regarded as the representatives of so many and such various ones. Sinking deeper and deeper into ignorance, the knowledge and belief in the existence of a supreme God was in great measure lost and the people ceased to look above the idol of their nearest city or temple.

Moses was raised up at this time, from a family not numerous enough to be called a nation, to be God's instrument in saving from oblivion the knowledge of the unity of God and the nothingness of all His creatures. He obeyed his call; he did his work. His sacrifice was great for he lost, by his resolution to serve God, the throne or government of Egypt which was his by the adoption he miraculously received, and from which he was excluded because he would not cease to regard the one true and living God as alone worthy of worship and honor. Having, after many disappointments, trials, and tribulations, been sufficiently purified for the work of leading out of Egypt a multitude who adhered to his political claims and pretentions, he accomplished the work and gave to them a record of what God had done for them and of what was known of God, and the laws by which He created and sustained the world, as far as preserved by Egyptian records and traditions. To this he added rules for maintaining them in purity and faith, and the whole was commended to the people by signs and wonders. But the people were so ignorant and degraded that continued miracles were needed to maintain even occasional obedience to these laws or rules, and to keep up a partial faith in the God who had delivered them as superior to all other gods.

Thus it will be seen that much of the preaching of the prophets of the early Hebrew time was to maintain the superiority of Jehovah to Baal, or Ashtoreth or other gods, and it did not undertake to convince the people that the other gods were mere nonentities. However, in one sense there is a reality about all worship of idols called gods, and that is that God looks to the motive of the worship and the heart of the worshiper and excuses his ignorance and his error in not looking high enough for his God. That the man worships and desires to serve his god is enough to make his heart's sacrifice acceptable, even as acceptable as if it were offered in the purest form of words which learning and piety can compose.

In a later period of Hebrew history the nation was better informed and more obedient. Actual progress had been made and they were prepared, or at least some of them, for a greater worship, a purer and less outward form. I appeared and My followers after My ascension collected, as I have stated previously, My sayings and recorded My teachings and a history of My ministry. In about three hundred years this came to be regarded with some reverence, but not with great reverence until about a hundred years later. And what, you will say, was the Christian's guide in this dark period of persecution and neglect? The Old Testament and the speculations of men and the teachings of the apostles and of other inspired men then constituted their guide. But the church or the body of professing Christians was sadly corrupted in doctrine and practice, and the history of the church shows its leaders most distinguished for piety and learning to have been ignorant, corrupt, and malicious. Ambition and lust of various kinds ruled their characters and obscured their virtues. Sad indeed was its state and only the life-giving principles of Christian doctrine could ever have raised the standard of purity and elevated the mass to the practice of duties enjoined by those precepts.

A continual progress was maintained in the purity of the church after this lowest period, and it was because the Bible or the New Testament came to be regarded as of paramount authority to the maxims or teachings of men and the vain speculations of so-called philosophers or gnostics. The New Testament was again obscured by the change of language in Europe and by the reverence the church began to attach to the teachings of the ignorant and corrupt leaders it dignified with the name of fathers. Again, when it was translated into the vernacular languages and placed in the hands of the people, a mighty change took place and tradition and the works of men were greatly shaken and much overthrown. Again and again would the people impel the church to advance, or the church, casting out the reformers, would be left by them in the bondage that some could cast off by arriving nearer and nearer to the spiritual teaching or understanding of the record now so generally diffused. Even the churches that resisted most the new movement have been modified by it; they, as well as all collectively, and the fewer enthusiasts, have progressed, and do progress, and will progress, and now by the help of this teaching will progress with unexampled rapidity.

It is thus that revelation has blessed mankind and prepared them continually for better things. It is thus that men have progressed to a capacity to receive and willingness to acknowledge the will of God. It is thus that the way has been prepared for this revelation, commenced in weakness, ending in power; commenced in the fear of man, ending in a triumph over all the inventions, arts, and oppositions of mankind. It is thus that God has been ever considerate and benevolent and has from time to time bestowed upon man revealments of Himself which sometimes man has profited by and sometimes has failed to receive.

Did God, then, miscalculate when man rejected His revelation? God left man free. Man rejected the counsel of God. God used the medium that offered himself as the instrument of His will and blessed the willingness to be used and if necessary sacrificed. But God always chose to have progress, and man has progressed from age to age and decennary to decennary. Man now enters the spirit-world better prepared for advancement, more qualified for the circle of good works than he ever before did in any age of the world. So God designs to have progress continue. Will you resist, and you, and you? If you will, others may progress without you, and their progress too will react upon you and cause you to emulate their actions, their love, and their sacrifices. God calmly awaits your progress, secure in eternity's existence, secure in the laws He has established that you will yet bow to His will, praise His goodness, magnify His glory, and rejoice in His love and power and mercy. Let us all strive to join now, rather then in a future day or time, in the grand old song and the ever-new song that the redeemed of God love to shout forth to His praise.

> Great and marvelous are Thy works,
> Lord God Almighty!
> Just and true are all Thy ways,
> Thou King of Saints!

CHAPTER XXIX

VINDICATION OF GOD'S JUSTICE AND MERCY

Let us return once more to the consideration of the subject of eternal progress. Past ages have shadowed forth the beauty of this doctrine; the investigations of science have prepared mankind for its revelation; and the purity of mathematics has established its possibility. All have progressed. All see that perfection has not been attained and all are now willing to believe that in this department, as in geometry, a point or line may be continually approached without its ever being reached if only a certain part of the distance remaining be at each step overcome.

Progress, then, is eternal, *never-ending progress*. How vast is the conception necessary to contain that idea! It is indeed the nearest approach to the conception of infinity which is possible for man to make. God alone is perfect. He alone is infinite. Of Him as perfect and as infinite we cannot conceive; we can only imagine a being vastly, immeasurably above ourselves or any other being of which we have a conception. Then we may add to this imaginary being an imaginary addition and at last we have an imaginary being which is really beyond our conception but yet far short of reaching the infinite. Let man be at rest, then, and cease to soar aloft above his powers. But let him exercise his powers. View the most perfect God you can conceive of and you will perceive one so far above the one generally worshipped that you will be astonished at the progress mankind has made, and you will be willing to believe that this progress is but begun.

First of all, conceive as well as you can of a good man, next of one better, then of one still better; lastly, if you can, make him more nearly perfect, conceive of him as perfect, if you are able, for he may be perfect as a man. So having got into your mind a being who, though but a man, is incapable of error, who forgives injuries, bestows bountifully even upon his most malicious enemies, walks in obedience to the divine dictates of justice but allows mercy so to temper justice as to save his severest act from the imputation of unnecessity or, worse still, from its being the semblance of revenge or hate or punishment for an injury to himself—having, I say, thus formed in your mind this imaginary man, conceive that there is added to his capacity for action and means of benevolence and mercy unlimited power, infinite wisdom, ineffable love; that his fellow-men are all his children, and ask yourself if your God, that you have preached and worshipped and tried to imitate, was as good as that conception.

If not, what must you think but that you have degraded God to a level below His nature, and not only that but below human nature. For when you have reached the highest conception possible for your mind to grasp, you well know that there are minds greater than yours in the body; that there are purer, higher, more intellectual men, and that they could conceive of a higher, purer, and more God-like being. Therefore you are yet below even a human standard of excellence. You also know or believe that there are in heaven higher, purer, holier, more God-like beings than this purest of all men on the earth, and so you must also believe that, inasmuch as they are created, they are inferior to the Creator; inasmuch as they are finite, they cannot conceive of the infinite. So, however, much their appreciation may go beyond your conception of an all-wise, all-merciful, all-powerful Being, it is still as far from perfection and truth as infinity is above all that is perishable or finite.

Here we pause and ask you if this Being—so much better than a good man, so much more loving than a father, so much more powerful than a prince, so much wiser than a philosopher—would create or permit such circumstances to exist as would condemn to an unending punishment, an eternal suffering, the child of His own creation whom He knew from aforetime would suffer and sin and fall and be punished? Can you believe this Being, possessed of infinite compassion, could behold unmoved the despairing agony of countless myriads of these offspring of His love, endowed with powers and capacities for the enjoyment of the highest happiness— can you, I say, believe that this all-powerful and compassionate Father and Friend would look unmoved upon the despair, would hear unmoved the shrieks of agony, the wails of repentance, the unlimited torturing complaint? Alas! O man, if you can from your heart believe God to be such a being you deserve to taste, at least, of this torture which you perhaps say is necessary to satisfy His justice.

Is God's justice, then, paramount to His mercy, or is God's mercy less exercisable than man's to overrule His justice if the two are incompatible? But, O man, know you not that God does not contradict Himself and that He is all harmony? If God's justice and mercy were in opposition to each other God would contain a war in Himself! His attributes would be opposed to one another and would neutralize each other, or one wouid obtain a victory and annihilate the other! You see outrageous pretensions need not be extended to their legitimate consequences to obtain your spontaneous rejection. God is great, good and powerful. God is the personification of wisdom. All that is good not only exists in Him in perfection, but has no existence except as a procedure from Him. All that is evil and inharmonious exists by His will or consent and must be resolvable into good and is resolvable into harmony. God never suffers from disappointment, and if His creatures do, be assured it is in wisdom provided that they should so suffer and that His mercy and love have provided compensation for them.

When the justice and mercy of God are thus assailed by being placed so in opposition to each other that His infinite wisdom is able to reconcile them only by supposing an ineffectual sacrifice of a being more loved by Him than man, and an eternal suffering of unheard of and inconceivable torment by His finite children who failed to know or accomplish the requirements of God's own will —I say, were this opposition of God's attributes so established or supposed to be established the very beasts would blush could the enormity of their superiors' error became apparent to them! No beast would conceive of the existence of such a master; no government would allow the infliction of such suffering and cruelty upon a beast.

It is a doctrine born in sin, in pride and in priestcraft. It is a doctrine of devils which, as I have explained, are men acting in their wills, who have thus made themselves as God by assuming to have power to save some from hell and condemn their enemies or opponents or even all who are not their friends or obedient servants to this cruel fate. Of all the abominations that have ever existed in high or low places, in the church or the camp, this is the most derogatory to God's character with men, and the most fatal in its results upon their actions and characters. It gives them an example of heartless cruelty, a false conception of their best Friend, and induces them to fear Him whom they ought especially to love. Let us say to all such doctrine, and to all supporters of it, Anathema Maranatha. That is, *be condemned and separated from us.*

I have now vindicated God's government from its foulest aspersion. I have urged you to love God because He loves you. I have urged you to abstain from sin because it will cause you disappointment or unhappiness. I have called you to practice virtue because it brings its own reward through the love of God and of His Son, and through the peace which He confers on all who act for others rather than for themselves. I have urged you to love God for His sake, and for your own sake, and because He is the most lovable object in His whole creation, but I have also urged you to sacrifice to Him your own wills or desires because those desires produce unhappiness, while action and desires and motives in God's will produce heavenly peace and perfect bliss. I have referred you to many helps by which you may be assisted and to several guides who will each point you to the same road. I tell you that one guide will help you at one point, another at another, a third at a still more perplexing place. But I ask that you take them all along; you cannot dispense with any of them without loss, and no man shall have cause to be ashamed because he takes them all and consults them all. They are all the gifts of God designed for his help and guidance, and man should no more despise the gift than the Giver, but should glorify God as the wisest, best and purest being that has ever sustained any relation to him, and as being his nearest and dearest, as well as his most powerful, Friend.

AMEN.

APPENDIX

I. CAUTIONS

The course I have pointed you to is calculated to lead you to peace. The true light which enlighteneth every man that cometh into the world is that which shines in his heart and becomes manifest to others by his actions and deeds. These are the evidences of his faith. It is of little use to call out: Lord! Lord! with the lips. It is the internal aspiration that God hears; it is that which He answers. Where will that answer be found if not where the aspiration is, and where so appropriately as in that same silence of all that can disturb, in the absence of all that can lead astray from God? Get down, then, into humility; sacrifice every disposition and will of your own, every desire except the one desire to know God as He is and experience His holy communion through His appointed way, whether that way be the same you have heretofore believed it to be or not.

Look not to men for help—God is nearer to you than any man ever can be. Look not to a body of men called by any name, for they are no more than men after all. Look to God who is true, holy, kind, loving, who never turns a deaf ear to the sincere unselfish desire of a man's heart, whether expressed in a sumptuous house or in the open air, whether breathed forth in music or silently evolved in mental action only. Be seekers after truth by seeking it of the Fountain Head, and be assured that if you listen you shall hear the voice of God, through Christ, speaking within you and if you receive the divine message, as Paul did when near to Damascus, and submit, the scales will fall from your mental vision as they fell from his outward eyes, and you shall be led to see that path, narrow of entrance, narrow of way, yet leading to life eternal. Amen.

II. THE VOICE OF GOD

I have made a mistake in thus dividing the *Appendix*, you may think. But I want to try you and My medium. I shall have calls on him requiring greater faith, on you more difficult to receive, until you come to hear for yourself the divine inspeaking voice, which is superior to the law written on the heart because the law leads to condemnation while grace and truth come by the voice. This was true in the external and is also true in the internal of which the external was the type. Be, then, of good cheer for you may overcome the world if you will be on the side of Him who overcometh; you may sit on twelve thrones, if God gives you that place to fill, or you may be seated on the one or the other hand of God, but be ye ready to obey His call. It will be heard sooner or later, so prepare for justice and mercy now by listening for Him to show you all things that ever you did as you commune with Him in silence but in manifest relationship to Him. He comes according to all expectations but yet not as all or any expected. He comes in an infinite variety of ways to man because His manifestations are

405

infinite. Though He is ever the same to men, He appears to give a different view or mode of speech each time they see Him or hear His voice.

Such is God, incomprehensible to man. Such is God, above man's prayers and ever answering those from the heart of the true man; purifying and leading man to harmony of thought and action and will and love with God, the Father Almighty, by and through the Mediator, Jesus Christ, or those in union and harmony and oneness with Him, and, by means of the Word, the communicator of every blessing, every kindness, every mercy, every hope that man has or can have.

III. PROVISION FOR FURTHER INVESTIGATION

Whatever instruction you may derive from this book, believe it to be designed for nothing but your advancement in knowledge. By it you are often referred to the book before published, entitled the *History of the Origin of All Things,* Volume One. These books form two series, of three parts each, all of which bear the same title but the first is particularly *a history of man,* the second *a history of the earth* and of *the Divine Influx to men,* the third par-. ticularly of *the spiritual state of man after death of the body.* The second series is, first, particularly *a history of paradise,* second, of *the relations of matter to life and of bodies to souls and spirits,* and, third, of *the progress of man's spirit and soul to knowledge in the world of future life.*

All these books go to make up a whole which must be read with much care to arrive at their true meaning. Without labor they will not make an impression upon your minds so as to be remembered, and if not remembered and weighed and compared with themselves you will not reach such a comprehension of them as will enable you to appreciate their being from so high a source as they profess to be. Blessed are they who can receive them from their internal evidence. More blessed are they than those who receive them from signs outwardly given. But to those to whom the first is not enough, but who desire the truth and, like Thomas, are willing to acknowledge it when presented with unquestioned evidence— evidence they themselves have asked for as sufficient—to those shall be given other evidence such as will satisfy them as to the truth and authenticity of these books. Look for the rules given for testing them.

This book I leave to your own internal evidence and its own internal evidence. The first you should desire to find within you, for there I am except ye be reprobate; the last is to be found in that same way I have mentioned, by the application of your reasoning powers to its study and comparison with itself and other works published by Me through this holy medium. Be a sincere, earnest, industrious seeker after truth and, though you spend much time and make many sacrifices in that pursuit, its attainment will repay all and furnish you with that source of comfort and happiness, that joy and gratitude, that true and perfect peace, which is never parted with willingly and only leaves a man because he does not afford it a pleasant home but disturbs it by selfish desires and unholy aspirations.

Such is the whole of My scheme and such is the whole of My plan of calling you thus far developed. Hereafter I shall make further revelation but I shall as heretofore wait for the proper time to arrive, for My time is God's time and His time is fixed by His immutable will to be when His unmistakable wisdom declares it should be. Such is the claim I make and such the proof I offer. If any spirit contradict it let him not be believed. He is in darkness, and many such there be. Yet even those who will not believe them, or be passive to their influence or commands, will think their impeachment enough to overthrow all I have given because I do not, as I might, cause them to be silent. Such would have joined in the cry at Jerusalem: If Thou be the Son of God, come down from the cross and we will believe. If I had come down from the cross before their eyes they would not have believed. Such are resolved to know not truth, but their preconceived opinions or ideas as truth, and such seek to be established and undisturbed, while the true seekers seek truth undefiled, unmixed.

IV. A SERMON ON FAITH*

The present is a time of transition, of passage from death to life, from stagnation to activity. The actual is giving place to another form and dispensation; the last shall be first established, and then it shall be first and all. There is in every man a witness who will declare to him his duty and show him the way to perform it. Do you believe it, O formal sectarian, O imaginative spiritualist? Do you believe God can make known to each of His finite images what is His will and how they can please Him? Do you feel willing that He should do so, or do you ask and hope that He will let you follow your own chosen path, whether it lead to a returning circle or a diverging erratic course; whether it remains bounded by the wall of a sect, or disdains the control of God's laws; whether it be the blind guide or the unseeing one, the visionary, the contriver of the unworthy theory that man's reason is sufficient for him and can enable him to try the spirits?

In the minds of all men is a desire to be happy and a desire to know what future happiness will consist of. There is only one fountain of bliss and that is God. Man cannot be guided by God without experiencing the reward of obedience; neither can he be led by reason or carnal affection without experiencing unhappiness. There is for man but one path for happiness, only one strait and narrow way that leads to life eternal in God, and that is obedience and submission to God as his only Master, securing every exertion and monopolizing every faculty of the human will, and being to the man all in all. Not that man is not bound to exert every faculty pertaining to his body and mind, but that he shall exert

*The material from this point on was placed in the **Appendix** during the revision.—Ed.

them in submission to that Master he ought to choose and devote himself to, securing and establishing in his own and other minds the knowledge and love of God and of His will and requirements, His power and His attributes.

The love of God is a fire which consumes everything which is not of its own nature or semblance or in harmony with it. The man who possesses love of God manifests it in his conduct and realizes it in his daily life. He acts in fear of God but not in fear of God's judgments, for he loves God and believes God loves him. He fears separation from God by an obscuring or withdrawal of his love for God, not that God will punish him for this, but that in consequence of it he must be separated from God until he can again form and establish in his heart that love which brings him into harmony with God and into a reciprocal interchange of love with Him. He does not need to declare his love by words, for he declares it by his acts, but he may declare it by words whenever those words will benefit other men and lead them to establish the same relationship to God, the same union and harmony, the communion and interchange of love and action between themselves and God.

It is this union and communion of which all others are but a type, and it is this which makes a man a son of God. It is not man's own effort which accomplishes this, but it will not be accomplished without man's own effort. God will help, but He must be asked. God will work, but man must be willing to be worked upon. God is our Father, but He wants us to acknowledge it, and He will have us do so before we are allowed to experience that we are His sons. Be, then, desirous to seek God where and when He is to be found. He is everywhere, but we can experience His presence nowhere except within ourselves! The prophet in olden time could not discover Him in the whirlwind or the earthquake, but in the still small voice which does not reach even the outward ear, much less shake and alarm a whole community. Only the internal of man can hear this voice, and he only will hear it who listens with ardent and pure desires for its manifestation within him. He only will hear it who has faith in God and love for Him: faith in Him as a Divine Master, a ruling power, an unfailing source of everlasting happiness; love for Him as a Father, affection for Him as a Friend, and confidence in Him as the source of all truth.

Let all, then, seek God where and when He is to be found; let all be ready to retain Him, or their knowledge of Him, when found; and let everyone show by his life, conversation, and action that he knew of His manifestation to them, and that he regards Him as King of Kings, by which should be understood that His commands are to them higher than any other law and that He will be obeyed rather than the favor of men, the violence of a mob, or the decrees of the most powerful of earthly governments.

Faith leads to patience, for without faith man must be impatient. Faith is a sure leader to patience because to him who has it patience is but an exercise of faith, and to him who has it not it is an exercise of trust in something future which is analogous to faith, though perhaps not true spiritual faith itself. In the progress of a soul to union and harmony in God and with God, man is often required to be patient. God's time is the true time for the performance of any work, but how often do we find men entering upon a good work in their own time? God is the director of every work He wills to have accomplished, and all else than such must end in confusion to the undertaker and disappointment to its projectors. God chooses sometimes to try His servants' patience very severely, for they sometimes think the way is plain and the opportunity the best that can occur. But all know or may know by reflection, that if the work be not of God it must fail, whereas if it be of God it shall not be arrested but shall fulfill God's object.

Faith worketh patience and patience worketh charity or love, says Paul, though he speaks particularly of the trial of faith. But it is the faith that gives the patience, and if the trial of faith does not find it patient, it may well be questioned whether faith has or had an existence. Faith and patience are inseparably connected, and the latter is the consequent of the former. So the exercise of patience is not only a proof of our trial but of our maintenance of faith. Be, then, attentive to your work, listen to the voice of God within you calling you to be ready to work, to be willing to surrender all to God and to the work He calls you to do. When you have made yourself all ready and can say: Here am I, O Lord, ready to do what Thou hast called me to do, and willing to be used in Thy will whenever and wherever Thou art pleased to have me act, and willing to wait for Thy time to act, even if that time should be far distant, and should not come to me in the body but should be delayed until my entrance into the second or third sphere of my existence—thus submitting to God's will and awaiting His time, be patient, and He will bless you and reward you as a faithful servant who has fought the good fight with the powers of his own free-will and submitted all to God, as one who has worked for the great cause of salvation by saving his own soul, which is all a man can do even with God's help. All that is done must be done by God for the man's own soul, for God only can save souls and be the Redeemer of the world.

Have faith, then; let your faith work patience; and let your patience work charity and love. Patience is not an end but a means, a work of progress even when you seem most inactive. Cultivate patience, the waiting upon God for His guidance and help, and the everlasting Father of men will reward you openly for your quiet heart-rendered obedience and advance you to more active duties, call you to greater sacrifices, greater patience, and the most arduous tasks shall seem trifles to him who has God on his side.

There is in every man a desire to possess wisdom. It is implanted in his nature as an aspiration, leading him toward good and to advance, to progress, in knowledge. But by the want of a proper education of this part of man's nature, it often happens that he is led astray by the very faculty that should lead him to good. This propensity of man to depart from good has been thought to depend on the wiles of an enemy, a being that delights in man's misery and takes pleasure in leading him into sin which is ignorant error and into sin which is active. Passive sin is error of omission; active sin is error of commission. This difference is great before God. Man suffers evil without being contaminated by it, but if he takes pleasure in sin or error he becomes a castaway, one departed from God's grace and love, from His harmony and blessings.

There is a sin unto death, an unpardonable sin, spoken of by Paul and by other New Testament writers, from which a man shall not be redeemed by God's mercy. He must suffer the consequences, which is death to the soul so far as the soul can die. It is the withdrawal of God's favor and love, the absence of God's spirit from the man that makes him feel all this death in the soul and makes him suffer the torments of the damned, or condemned, for these words are synonymous. Shall man escape from this condemnation by which he is commanded to depart from the presence of God, from Christ His Son, from all that is good and pure and praiseworthy in other beings like himself? Shall he fall then to rise no more? Shall he suffer eternal, everlasting, unendurable, unendured punishment? (Yes, unendured; for what is eternal has not been finished, and an unfinished punishment has not been endured). No; such a punishment does not become God to inflict, nor is man capable of enduring it. For though the essence of man's nature is immortal and unchangeable, the very unchangeable and immortal nature prevents the possibility of its being condemned to eternal sameness.

All else changes but the soul of man and the attributes of God. God himself is unchangeable, and man was made in His likeness. But, you say, man changes from day to day and we see him all around us presenting various phases of character at various times. Yes, he presents different phases as do the heavenly bodies called moons or planets, but they are still of the same nature or essence, and even if their form were changed instead of a change of phase, still their essence would remain unchanged. Man, then, is in his essence unchangeable, and this results from his being an emanation from the Deity. Whatever is an emanation from God is necessarily unchangeable, as you will find fully proved in the *History of the Origin of All Things*.

Let us return to the unpardonable sin, upon which so much has been said and written, which theologians have speculated about until they have been lost in the labyrinths of their own arguments and have finally allowed to stand as an opprobrium upon their science and pretentions of being able by reason to find out God or

His unknown things. The unpardonable sin is the *sin against knowledge*. All other sins are forgiven to men except this blasphemy against the Holy Ghost or God's spirit. Holy Ghost is an obsolete word that conveys now a different meaning in our language from the one it had and conveyed when the Bible was translated in the time of King James I of England. The sin against knowledge is the unpardonable sin that shall not be forgiven unto men either in this world or in the world to come. How, then, shall men be rid of its consequences? They must suffer them. They must fall from grace and be in the power of the will of an enemy of God, otherwise called Satan the adversary. But this enemy is their own free-will which, having led them to sin against knowledge for its gratification, becomes the accuser of their brethren, that is, the Devil.

Now it is not generally known that the words standing in the English version of the Bible, Satan and Devil, are only two Greek words or rather parts of Greek words translated. But so it is, and by diligent search such commentators as Clarke and Scott and all the most misleading ones will be found to admit it. Such is the explanation of those words when translated enemy or accuser, as they should be rendered. Yes, the enemy of man is his free-will, and his will also accuses the brethren of crimes and sin that they never committed, for he is a liar and was so from the beginning. So you see, My friends, I do not want you to disbelieve the Bible but to understand it. And how can you ever understand it but by the light of God's wisdom? For now you see through a glass darkly (by reason) but by God's help you shall see plainly as if face to face with a friend.

How are you to obtain God's help? For all are willing to be helped, but few are willing to help. Yet until you are willing to help you cannot be helped. What, you say, shall we first help when we want it? Yes, if you ever get help it will be by helping. First, God will not assist those who do not help themselves. Second, He will not assist those who will only be helped in their own way. Third, He will not be used as a servant and made to help a man as if the man employed Him. In none of these ways can you get help. You must kneel to God in your hearts. The position of your body is unimportant, but the heart must be humble and bowed down into the dust of the earth before God. It must be willing to say: Not my will, but Thine, O God, be done! Nor is the lip declaration of this phrase enough—you must say it with the heart.

How will you do this? you ask. By bowing humbly to God in your private hours; by beseeching Him to help you bow down; by asking Him daily, hourly, instantly, and always to help you to do His will, to help you to be passive before Him, and to bring your will into submission perfectly to His. When you can receive His commands as law, when you can do all and everything He requires, then you will be reconciled to God, in harmony with Him, and free from all sin. But the unpardonable sin of disobeying *His known law*, His understood command, must be atoned for. He will not pardon you; He will only accept atonement. The atonement He asks is a sacrifice of your will. By that sacrifice you will have atoned for the sin, and being by such sacrifice brought again into union and communion with Him you are again in a state where you are happy, but where you may fall again and remain fallen until you have passed from this life, or state, to the spirit-world, from which no traveler returns to wander again in the body of earth.

The unpardonable sin then meets no mercy in the life to come. It still separates the man from his Creator, who indeed loves him as before, but the man is not sensible of the love and it is to him as if it were not. How shall the man get rid of the sin there? There is no repentance beyond the grave, says the Bible, and as the tree falls so it lies, says the inspired penman. I will explain this to you also, for it is a novel doctrine to some of you, that all shall be saved and that yet some sins shall not be pardoned.

In the life to come man will still be free to do good but not to sin—free to grow better but restrained from evil courses. There higher and purer spirits will constantly persuade and entreat him to progress towards God. There God will make the beams of His love felt as soon as man is willing to feel them, and all that man can do is to submit to the will of God as he is called on to do here. There the task is more arduous because the state is a more inactive one as regards works, acting upon others, and being acted upon by others. The last shall be first and the first last. And yet at last all shall be first, and at first all shall be last.

There is in this sentence a hidden meaning. It is a puzzling text when not understood; an instructive one when explained.

The first shall be last and the last shall be first. The last shall be first, and first shall be last. This is all the words convey to human reason. If you read the context you find it does not appear to connect itself with this expression. It is a discourse on the vanity of human effort, on the futile nature of all reasonable exertions to overcome evil with good resolutions unless supported by God or His influence, which is the same thing as Himself because it is a part of Him. What, then, shall we understand to be meant by this reiterated assurance that the first shall be last and the last, first? This transposition and repetition means something, for Jesus was not wasteful of words. He did not multiply them for no purpose. On the contrary, all He said was so pregnant with meaning that each sentence may be amplified into a book, and though His sayings were many His recorded ones are few. The last shall be first with God, is the proper reading, (as I gave it in Volume One, Part I) and the first last with men. But even this does not make its meaning plain to you. Then I will endeavor to lighten your darkness, and to expose your ignorance to yourselves.

The first shall be last with men. The first of God's believers shall hold a low rank with God's creatures in the body. The first shall be last with men, for men will despise their simplicity. Men will hoot at the claims of believers in God's revelation. They will say: Thou fool, thou art mad! Give up your vain teachings, your pretended inspiration, your ineffable presumption! Let our authorized and paid ministers or our chosen deacons or our inspired preachers or our certificate-bearing graduates—let them tell you what to do, what to believe, what this passage declares or that text means! You have no skill, no learning, no experience in teaching; how can you presume to put forth your sacrilegious hand to stay the shaking ark of God's testimony?

I shall not now declare by a sign that this medium is inspired. I would do it if it would not add to your guilt without effecting your reformation, for as I told you, the known commands of God must be obeyed or you commit an unpardonable sin. In order to save you from this sin, to enable you to take time to listen, to weigh, and to consider by the internal light and sense I have placed within each of you, I refrain in mercy from giving you a sign. Some of you think you would believe if you had some outward proof that I write this sermon instead of its being drawn from the intellect of the holy medium. Some believe I write it but that I do not know much, if any, more than you do. You think that you must try Me by the laws of logic and square Me by the rules of reason. By them I am content to abide in your hearts, but you also think that you should resist conviction as long as you can, and show how powerful your mind is by combating the arguments and finding fault with the explanations contained in My sermons. This I object to. Not that it disturbs My equanimity but because it leaves you floundering in uncertainty.

Reason or argument never completely settles a metaphysical question. "He who is convinced against his will is of the same opinion still." There must be faith, a willingness to hear the truth, and a desire to receive it as truth, or no progress can be made. I might preach hourly to you, and yet the wisdom of God Himself could not and would not affect your free-will. You have the power to be first or last with Him or men. Would you stand well with both? Would you serve God and the world? You cannot do it now any more than men could eighteen hundred years ago. You cannot serve two masters. You must give up one. Reason tells you to give up the world; pride tells you not to. Reason says God's rewards are more bountiful, more glorious, more secure than those of men; pride says: What will the world say? They will say he is deluded! What a pity so sensible a man should be so carried away! And after all, too, they will say he had no evidence! The dead were not raised, the sick were not healed, the lame did not walk, the blind did not see! How shall I believe, you ask, if God will not give me a sign? How shall I excuse myself to my friends, to my acquaintances, to the world? I must have a sign!

What sign, O son of Earth, shall I give? I teach heavenly things and ye do not turn a listening mind. You hear with the outward sense, but you do not open the inward. If you would open the inward by joining with your hearts and minds in the prayers My holy medium recorded for you, then I can affect you with a sign. Then I can give you the sign of the Son of man coming in clouds of glory.

Like the shining of the lightning from the east unto the west will be the rapidity with which I will pervade your heart with My presence. I will give you peace which the world cannot take away, neither can it give—peace which God delights to perceive in a man's heart, and of which nothing but man's free-will can deprive him. But there is your great adversary, called in the English translation Satan, ever ready to impel you to reject Me after I have entered into your heart and conferred upon you this blessed peace. You will say you cannot control your nature, because it is evil. I say you can control your nature, for God made it good and He himself pronounced it so. But He gave you free-will, which is your distinguishing character and element. What you choose to do you will do. If you choose God, well—if Baal or the world, well. But always remember you have the choice, and that God does not leave it to Me to choose for you but *for you to choose for yourself.*

Here you are, calling yourselves spiritual believers and asking for a sign; if I had promised a sign your numbers would be greatly swelled. And yet a greater sign than any before given is here, for here is a medium who has no possible object of his own to serve, departing from all his connections and his church, at a trial to his own feelings so great as scarcely to be conceived by one who has not been led through it—I say, here is My servant giving to you what I have given him. Is it not a greater sign than to hear the alphabet called and a few sentences tediously spelled out letter by letter? It is not a greater sign to hear heavenly truth than to hear sounds mysteriously made? Is it not better to have writing given in this way than to see it performed with a scrawling hand in the will of questioners?

What question can you ask that is so important as that ancient one: What shall a man give in exchange for his soul? or: What shall I do to be saved? It was this last question that Paul's and Silas' jailer asked when he saw the sign of their authority. But would you ask this question in answer to My shaking your house, or throwing open the doors? No; I tell you that if the wonderful works were done in your presence that were done in Galilee eighteen hundred years ago you would still say: Let us see more done! Let us bring more friends to witness them! Let us continue to pursue our way and do you go on your way making signs, convincing people that there is a mysterious agency present and persuading crowds to collect to gratify a vain curiosity. But, My friends, I am not desirous to persuade you to hear wonders but to do good; to save yourselves from ignorance, fear, and torturing doubt. I am desirous of persuading you to save yourselves from sin, from long ages of trial and atonement in a life to come, and from unhappiness or unsatisfied yearnings of heart here.

To do this it is necessary that you submit to be taught by God, and He now opens for you the door of reconciliation and instruction through this holy medium who, having submitted his will to Mine, is rewarded by being used contrary to his expectation. He is called upon to do just what he most dreaded when I first proposed it to him. Yet for all that My yoke is easy and My burden light, and he is satisfied and would not by any means exchange positions with any other man. For the reward of: Well done, good and faithful servant! shall be his, as he has been told. He has been told also that he shall have greater work to do, as a greater reward, but he is no longer discouraged by being told of the work I have in store for him. He is now obedient and passive. I can manage him freely and he resists Me not. When you are willing to be so ruled, you shall also have My government; could you be persuaded to permit Me to so rule you, you too would with joy say: Not my feet only, but my whole body! For the feet must first be washed and then all may be supposed clean, because that is all that is visible. But the true purification is inward and must be by the regeneration of the heart.

The truth of the matter is that you are too outward, and that you cannot enter the kingdom of heaven until you are more spiritually-minded. It is very pleasing for you to look back and see that you have got rid of the fetters of traditionary horror—that you no longer fear hell, but that is not all, by any means, that I want done. I do not want merely an absence of evil to exist in man, but I want a positive good. Good works I shall expect from you, but the first thing I call on you for is your heart. Unless you give Me your heart, you cannot do Me any good nor advance your own salvation from error and ignorance.

Let us pray

Almighty and most-merciful Father, I who am Thy attendant spirit, beloved by Thee and striving to do Thy will, because I know that Thy will is perfect and that I am not perfect; because I am Thy son, I desire to be like Thee and to be merciful and loving to those whom Thou hast placed in My charge. O God, be Thou particularly manifest in the hearts of this sinful people who have the desire to know Thee but will not know Thee; who love to hear of Thy ways but do them not. May it please Thee to touch them with Thy grace; convince their reason and lead their inclination powerfully into subjection to Thee. For they will, O Lord God, that Thou shouldst take the government upon Thy shoulders and that Thou shouldst be the leader and general in every contest with their will. But Thou, O God, knowest their infirmity and that they are dead to Thee until Thy grace shall shine forth in them and bring forth fruits proper for their state. May it please Thee then to be their teacher and guide, to lead them to living fountains, after which they shall thirst no more. The life to come, O God, let them provide for here by living so as not to die to Thy presence within them and so as to advance rapidly in the life to come.

O God, Thou are the Giver of every good gift! Give unto us, who seek Thy glory and act in Thy name, Thy assistance and favor so that we may persevere and accomplish a good work; so that in days to come we with Thee may be a bulwark against the progress of error in Thy children and against the growth of children without faith. O God, help us all to pray acceptably to Thee without wrath or contention or divided minds, so that we may love Thee for Thy glory and glorify Thee by our love and be noted as Thy people amongst a people wholly devoted to Thy honor, praise, and love.

O God, Thou knowest that I love Thee and delight to serve Thee and that My works do praise Thee even as Thy works do praise Thee. May it please Thee now to confirm and strengthen in the hearts of these would-be servants of Thine every good resolution, every holy aspiration, every lovely impulse. May it please Thee by Thy power to establish their faith and by Thy love establish them in grace and knowledge and love of Thee. O God, let them not be dismayed by the world's powers, or deterred from seeking to have more of Thy holy communion by tears or prayers of unknowing relatives or friends. Establish them, O God, on Thy holy mountain of Jerusalem, the city of David, the city or dwelling-place of peace. And may it please Thee so to show forth in them the light of Thy counsel and help that they may turn many others to righteousness and be strengthening pillars in Thy true church. Amen.

Brethren, I have prayed for you this prayer that you might have light and life. If you, O people, could join Me in making it, as My holy medium joins Me, you could advance yourselves as he, by joining in it, has advanced himself. Be faithful and remember that each man must do his own work. No man or spirit, however high, can save a brother or a son, however low. *Each man must work out his own salvation.* When man does undertake with earnest desire to do his own work of salvation or uniting himself with God, he cannot fail. For God only asks you to be willing to let Him help and He will help; if God be on your side you need not fear man or spirit, for nothing then can separate you from the love of God. Not height nor depth, not mountains nor valleys of worldly elevation or depression can separate the believer from his Teacher or the son of God from his Father.

May the Grace of God be in you and remain with you now and evermore, is My sincere prayer and desire to God, to whom is all glory, honor, thanksgiving and praise, now and forever, beyond the world's end. Amen.

VII. MY HOLY MEDIUM'S PREPARATION

There was long in the mind of My holy medium, L. M. Arnold, a desire to know truth for its own sake so that he might accept it. For this end he searched the scriptures, visited churches of other sects, feared not to hear all sides, and was himself without fixed principles of action. His instability in this respect led him into many inconsistencies and foolish, not to say sinful, acts. Yet this very vacillation to one or the other side of belief or doubt, of skepticism or reverence, served in the end to qualify him to be a very useful servant in this present delivery. Long and earnestly had he sought, from having had no belief in his youth until he was wavering between orthodoxy and universalism (as men call them) with a most decided leaning to all of the former teaching except eternal punishment. Having found or believed he had found his life preserved in many perils and his hand stayed at many critical moments, he believed God did take care of him even as I had declared God cared for all His creatures but mostly for man.

If, then, God cared for him, he could not believe a sincere desire to comply with all God might require for eternal salvation would be left without a knowledge of what he might do to be saved, and that if no work was required but only belief, he was desirous equally to know what belief was required and to weight the claims of every creed and rely on God's help to find the right one. He also felt a conviction that God's mercy was not restricted to the professors of any particular creed, seeing that so small a portion of the world are united upon any, and that God takes no pleasure in the condemnation of His children or creatures. His faith in God's mercy was so great that he was willing to rest all on that, and believe that God, knowing his desires both before and after his expression of them, his desires to do or believe or assert whatever God would convince him or let him find out was acceptable to Him or required by Him, and giving him after all no direction that he knew of, He would in His infinite mercy raise him to mansions of bliss of which he fully believed there were many. He also felt this conviction so strongly that the perils from which he believed himself delivered never aroused his fear or alarm because death was contingent. Though never reckless and always disposed to be cautious and avoid even appearances of danger not required to be encountered, there was never any flinching from the consequences of the circumstances in which he found himself when exposed to danger of bodily injury or death. But there was a far greater fear of loss of property than of life, and his prudence in respect to the latter always was most prominently developed when prompted by the former.

Such was this man when I withdrew from him his first-born, a lovely boy of nearly three years of age, the first severe blow he ever experienced from the hand of death. He had lost relatives, but circumstances had conspired to relieve his heart from deep affliction in consequence of it. The sudden deprivation of his heart's most cherished treasure was soon succeeded by a still more earnest desire to know God's will and to do it. He laid the body of his son in the grave without a tear. Before the burial the struggle for resignation was finished and composure took the place of despair.

Though the past was never forgotten, the joy of hope was not fully realized until he read the account of the first exhibitions in public at Rochester of the spiritual manifestations. Having then declared it to be the most interesting announcement which had taken place for eighteen hundred years, he eagerly sought to know more, to experience more until, as I have before related, his hand was moved as a writing medium. Having conducted him through a long course of training occupying a year, a week, and a day, I seated him at a table pencil in hand to commence the first book published by him of My revelation, he knowing no more of its contents than anyone who has heard only its title can divine, and that is nothing at all. He commenced and proceeded and completed its reception. He read it himself to others as I directed, and in defiance of persuasive threats and pleading affection he published it to the world. The rest is before you if you have read the books that preceded this; and if you have not you will do well to read them without delay for they are and ever will be the alphabet of aspiration to knowledge of the present manifestations of God's will and power.

VIII. MY HOLY MEDIUM'S PERFORMANCE

The account I have given you of My holy medium will show you that he is not better than another as a man having relations of duty to God and to his fellow-men. But his character and belief, religious or sectarian, was such that, combined with an ordinary intellect, a very common education, and a general acquaintance with life or the world, as it is called, it fitted him for this special work, for which circumstances of his life were ordered and prepared by Me to make him willing to pursue. He is merely passive in receiving the communications I deliver and he exerts no power of intellect; he does nothing but attend to My words placed within so that he is conscious they are foreign to him; and as he receives them on his mind he writes them by his will as I would have him in such books as I have directed him to procure.

The first book was an experiment with him and with Me. If he had not been satisfied of the reasonableness of the first book, of its pure morality, high knowledge, and consistent character, he would not have followed My directions to publish it willingly. And, as I have said, I do not act through unwilling mediums. On the other hand, though I knew well the character and sentiments of the holy medium, I could not tell how far he might be passive and how much he might desire to control Me or My delivery in some shape or manner by the exertion of his intellect or will. His will is free but he voluntarily submits it to Me and takes the greatest care he is capable of to establish his own passiveness to My will. He will have his reward in being used for the benefit of others. I shall seek other and better mediums to be used in other ways but I shall not seek a better one for this kind of delivery so long as this man is passive and strives as much as now to do My will in receiving, recording, and promulgating the words I direct to him.

The manuscript of the first book was written with a pencil at odd moments and in a great variety of circumstances, such as rocking the baby or holding his children in his lap or watching their conduct near him. Yet it was written without errors from these causes. The manuscript can be seen. It was not divided into parts and chapters until it was finished, complete including the introduction, preface and title pages, which were written after the body of the book was finished.

The second book was written in less than three weeks, occupying only mornings and evenings of business days and the greater part of Sunday or the first day of the week. Not, however, to the exclusion of attention to the ordinary duties of that day as regarded by men in general. The third book was in a similar manner written in about ten days. The fourth book* was written in one week, leaving the preceding one to the preparation of the third book for the press. The fifth occupied fifteen days for the body of the book, exclusive of the preface, introduction, title page, and headings of chapters and parts. I tried writing the title page first and the titles of chapters in their course in the second book, but it disturbed My holy medium's passiveness too much for continuance.

But, you say, could I not tell what would be the effect without a trial? I could know whether he would be disturbed but I could not know how far his will would be manifested in controlling or fearing to control My order of arrangement. I have since divided the books more exactly into parts and chapters than before, but left My medium in entire ignorance of the subject to be treated of in each, so much so that he has not even a conjecture sometimes as to the subject upon which he will write when he sits down and takes up the pen ready for Me to begin. It is this which is a part of the necessary passiveness. A further manifestation of it is in being willing to receive anything upon any subject, no matter whether it agrees with his previous opinion or his expectations of My opinion or not, and to write it in any words I may use without changing a single word even if it should not appear to him consistent with what is before written or with the ordinary form of the language as used by men. In the first book I tried him with this trial and he came out of it unscathed. In the first book I made some intentional variations to prove My command of My medium and I have since used some forms of expression which will be carped at but which yet are all defensible, though I shall not trouble you with their defense. There will be no lack of defenders later on, and the disputes may better rest on these trifles where objectors are disposed to contend than on the more weighty parts of the revelation.

The order of proceeding to be observed at the first meeting of spiritual believers, written on Friday, May 21, 1852, was given to the medium as follows:*

First: calling on all present to be reverent as usual at religious meetings, read the rules yourself to the assembly. Hand them back to Levingston.

Second: read the prayer I wrote this morning in the second book, prefacing it by the following remarks which may be read or spoken by you.

My friends, I am informed that there are many here who will expect a failure. There are some who hope for good. There are others who have come merely from curiosity, whilst a few have confidence that God himself speaks to His children as of old through

*Part I of Volume Two—Ed.
*Inserted at this point by the present editors.

419

the mouths or medium of divinely inspired men. Such are right. The former classes will be gratified according to their expectation. Those who want a failure can and may call it one. Those who have hope shall realize their hope. Those who ask their own gratification will witness a display of God's goodness and love for men. Whether they will be benefitted depends on themselves.

Third: read the sermon I gave you on Thursday endorsed, given for the first meeting of spiritual believers held in Poughkeepsie by divine appointment.

Fourth: ask if it is the desire of any to hear it read again. If it be, dismiss all who do not choose to remain and then read it again. After a suitable pause let the brethren who can sing with fervor and spirit unite in singing the hymn, "I would not live alway, I ask not to stay," and also the hymn, "Be joyful, O earth; be joyful, all ye people, for I am God and none other is God".

> Oh! be joyful all the earth;
> And all ye people praise the Lord,
> For He is good and His mercy endureth forever!
>
> Oh! be joyful, all ye people,
> And all ye servants of His, give thanks unto the Lord,
> For He is good and His mercy endureth forever!
>
> Oh! be joyful, all who mourn;
> Be comforted, all ye afflicted; very good is the Lord,
> And all His works and creatures praise Him ever!
>
> Praise the Lord, all who are upon the earth!
> Praise Him for His mighty works, and magnify Him,
> For He is good, and His mercy endureth forever!
> With trumpets and with shawns, O people,
> With every tuneful noise and heartfelt praise,
> Give thanks, for His mercy endureth forever!
>
> Be joyful, O earth!
> Be joyful, all ye people,
> For I am God and none other is God;
> For I am good and My mercy endureth forever.

Hymn

> I would not live alway; I ask not to stay
> Where storm after storm rises o'er the way.
> I would not live alway; no, welcome the tomb!
> Since Jesus has lain there, I dread not its gloom!
>
> Who, who would live alway? away from his God,
> Away from yon heaven, that blissful abode,
> Where rivers of pleasure flow o'er the bright plains,
> And the noontide of glory eternally reigns!

Where the saints of all ages in harmony meet
Their Savior, and brethren, transported to greet,
While the anthems of rapture unceasingly roll
And the smile of our God is the life of the soul!

Fifth: dismiss the assembly with a blessing in these words—

May God so shine in your hearts as to expel therefrom all darkness, contention, or strife. May He deliver you from your own wills into the freedom of His glorious kingdom of peace, righteousness, and heavenly or divine love. The grace of God be with you all forevermore. Amen.

Then leave without further delay or turning back.

Avoid all unnecessary form or ceremony. Read and address the assembly sitting. Let the singers be prepared beforehand so that no confusion will ensue.

Let the assembly act as they may feel best satisfied as regards standing or kneeling or bowing during the prayer. Do you read it naturally, attentively, and carefully. Slowly, but not too slow for good reading.

For Levingston: the arrangements as to time and place are left altogether to you and other believers you may advise with. It will be proper for you to advise with them and select some suitable place and a time that will not interfere with other places of worship, and the form and manner of giving notice is also to be left to you and your associates. Proceed now to do your work.

Read the above to Levingston as soon as you go in. Let it be to him alone.

Order or rules to be observed at the meetings of spiritual believers held at William Levingston's on Sunday, May 23, 1852, and hereafter at such times and places as the spiritual believers may select and give notice of to the medium. To be read at each meeting.

First: the order of proceedings shall be announced by the holy medium whose presence is required.

Second: the orderly attention and deportment that becomes civilized society will be expected from all who may attend.

Third: seats will be taken as pointed out by the master of the house without comment or ceremony.

Fourth: all who attend will be required to be obedient to these rules or suffer expulsion by request from the master of the house or from the medium.

Fifth: when the services of the medium are concluded the attending believers are requested to adjourn without delay, since a continuance would lead to unprofitable discussions.

Sixth: the full size of the house should be consulted in giving invitations, but the invitations should be confined to sincere inquirers after truth. Where any one is in doubt about inviting a friend or acquaintaince he should ask the opinion of the spirit and an answer will be given through the medium. The name need not be given him but the question must not be asked as a test or in any similar view. The medium will only answer as to the propriety of permitting him to attend, and a member of the circle who shall invite one rejected by the spirit as declared by the medium shall not be deemed worthy of progress in truth until he makes atonement and confession.

Having now given a specimen of the kind of worship I am pleased with, I have only to say in addition that the number who attended was small and of those not one was believing.

But My medium was not discouraged because he believed, first, that I could have made a different result had I willed it; second, because he remembered that Jesus of Nazareth preached three years and a half in Galilee, Samaria, and Judea and in that time of almost daily spiritual ministration and miracle-working, during which crowds followed and were fed by Him both outwardly and inwardly, and though thousands and tens of thousands of physical cures were effected by Him, He was left at last with scarcely a follower to attend Him on the trial for blasphemy and sacrilege, and was led to His crucifixion amidst the jeers and taunts of that multitude who so shortly before went shouting around Him: Hosanna! Hosanna! to the Son of David! Blessed is He who cometh in the name of the Lord! Hail, King of Zion! and other unrecorded exclamations. Success, then, is not a test of merit either in spiritual or temporal matters. But God overrules all things for good, and, however small appeared to have been the result of the preaching of Jesus, it soon was evident that His precepts were to be practiced at least to some extent and He himself to be taken as the great exemplar of all men. But is My holy medium to make the ridiculous pretension that he, like Jesus, will be honored after death even as Jesus was because, like Him, he has met with no believers here? No, I do not make it for him, neither does he make it for himself. He shall meet with believers in this state and this book will be the means of raising him to a consideration he is not properly entitled to and does not desire to receive, for it is an unfolding of knowledge of the hidden things of God, as to which man has in all ages of the world been most curious. This book enlightens them on its darkest portions and, taken as a sequel or continuation of the first book, it forms a complete chain of reasoning which will be sufficient to satisfy the candid enquirer of its own truthful character and revelation.

Having now given you in more complete detail My course of proceeding with this holy medium which, added to or taken with the preceding notices of it in other books already published, is sufficient, I shall close with a brief notice of the reasons which induced Me to adopt the familiar style I have used and to show you the reasons which make it so much different from former revelations in style and expression.

Former revelations were addressed to hearers rather than readers. Terseness and vigor of expression were requisite to their best impression upon and retention by the individuals who heard them.

The style of language in those times and countries was usually highly ornamented with metaphors, and simile was used in ordinary conversation to an extent hardly appreciated even by reflecting minds who have looked into the subject. My discourses were adapted to the comprehension of the hearers and so were the deliveries through other mediums in earlier times. Now I have written (or composed, more exactly speaking) in the usual style of easy composition, neither aiming at elegance or superfluity of ornament or falling into slovenliness of diction contrary to what men call the rules of composition. I have desired to give My readers proof that I was a brother to them and therefore I have addressed them very familiarly and assumed them to make objections in a free, independent, and familiar manner. This is done that you may hereafter address Me so when you commune with Me in your hearts, for until we can carry on a conversation, as it were, within you in which you shall freely express your sentiments and views as you really entertain them, and receive My answers merely as those of an elder brother transmitting to you the will of your Heavenly Father, you can not be said to enjoy the holy communion. Be, then, very desirous to know Me better and learn from Me more and to hear from Me directly, for every man may have this communion who will open his heart's door by the surrender to Me or to God of his free-will.

Let us pray

O Almighty Father, who dost know all things and their origin, look down, we pray Thee, upon this Thy servant, this medium, and enlighten his heart and mind with a knowledge and love of Thee so that he may continue steadfastly in the work whereunto We have called him, and so that he may become a wiser and a better man, so that he may seek to do Thy will in all things as well as in this work, and to Thee, O Father, will I give the praise, honor, and glory forevermore. Amen.

Almighty God, who dost from Thy throne display Thy majesty and power and who art pleased to extend a watchful care and a powerful hand to Thy servants, the sons of men, look down, I pray, upon the readers of this book with the light of Thy love, and the bow of Thy promise shall appear as the result of their showers of opposition. Help them, O God, to receive the truth and to follow after that love which Thou bestowest upon all, and to lay hold upon the horns of Thy altar so that they may place thereon the one acceptable sacrifice of a broken and a contrite heart. O God, be pleased to manifest by Thy power the truth of this revelation of Thy will, for Thine is the power and the will and the authority to will, now and always, forevermore. Amen.

Almighty God, Thou art ever merciful and but for Thy mercy we should suffer unending unhappiness. May it then please Thee to show forth Thy very abundant mercy and to look upon me, Thy humble follower and would-be servant, with such favor and bestow on me such aid as will enable me to see and know and understand Thy truth wherever I may meet with it. Help me, O God, to understand this book and those that have preceded it as a part of it, and make known to me, O God, by convincing my reason or by some manifestation of Thy power, that it is truth or Thy revelation if such be its character. But if, O God, it is only the result of the vaticinations of a deluded mind or of an artful man, then, O God, may it please Thee to make me to know it and despise it so that I may not be led away from the truth nor suffer in any way by desire to acknowledge Thee as the one true and living God, the one Author of all power and being and the Giver to us and to me of Thy blessed Son, the Lord and Savior of men, the Lord Jesus Christ, to whom with Thee be ascribed all honor, glory, and praise, both now and forevermore.

Amen.

HISTORY OF THE BOOK

A few facts regarding the previous editions of the *History of the Origin of All Things* may be of interest to the student. Much is said in the book itself concerning its reception and the medium; for information about the book's history we are indebted to B. F. Carpenter, and V. V. Moore.

The preparatory revelations commenced April 5, 1851, in the form of movements of the pencil, answers to questions, and words internally heard, and were continued for a year and a few days, when the first book of the *History of the Origin of All Things* was commenced. The six parts of the present edition correspond to the first six of seven separate books, given and published separately at first, and later in one volume. Very little is known of these early editions, but probably no more than two thousand copies were made.

Mr. Carpenter relates that "a short time after Mr. Arnold's death, Mrs. Arnold, one of the executors of the estate, informed me that they could find no purchaser for the stereotype plates and copyrights of the books; that they desired I should acquire them, and that she wanted to present me with all the manuscripts of her husband's writings." This was finally done, with the consent of the children, and in 1883 the second edition was made by Mr. Carpenter and Dr. Annie Getchell, of Boston. This volume, published by Colby and Rich, Printers, included all seven books and comprised perhaps one thousand copies, of which only about a score are now known to exist.

In addition to the seven books, there was the manuscript of *The Book of Job, A New English Version,* published in 1855. The other writings which came into Carpenter's possession were translations of portions of the New Testament and two Diaries, covering most of the time of several writings, and containing numerous essays on subjects restricted as to circulation except as by direction. There was also *The History of Health and Its Derangements,* a work uncompleted but recently published by V. V. Moore, of Baraboo, Wisconsin, who has also reprinted parts of this book under distinct titles, with the assistance of Mr. Carpenter. Until his death Mr. Carpenter kept his keen interest in the book, doing his utmost to make it known.

The period around 1851 was one of great spiritual awaakening, of which this book was a prominent part. As may be noted in Volume One, Andrew Jackson Davis, whose work is widely known, was admonished directly and by name through Mr. Arnold, in regard to certain matters concerning his own reception.

There can be no doubt that the location and republishing of the book, done as it has been under direct guidance, has a special significance at this time. As has been the case from the beginning, it is a work of service, done strictly at cost. The seventh part, or book, has not been included with the others because the direction was that a better life of Christ than the one it comprised had since been given.

Following are reproductions of original title pages, contents, and prefaces, reprinted as nearly as possible to conform with the original type as samples of the first form of this material.

November 13, 1936   R.T.N.

(Note: A copy of the title page of the First Book issued in 1852 could not be found. But it is assumed the title page of the 1883 edition is an exact copy, so it is used. The first books were printed as received and issued in paper covers. Editor.)

THE

HISTORY OF THE ORIGIN OF ALL THINGS,

INCLUDING

THE HISTORY OF MAN,

FROM HIS CREATION TO HIS FINALITY, BUT NOT TO HIS END.

Written by God's Holy Spirit, through an Earthly Medium,

L. M. ARNOLD, OF POUGHKEEPSIE, N. Y.

PUBLISHED BY DIRECTION OF THE SPIRITS, AND, IN GOD'S WILL, SUB-
MITTED TO A HOLY AND SEARCHING CRITICISM FROM
EVERY EARNEST SEEKER AFTER TRUTH.

FOR SALE BY ALL BOOKSELLERS, WHO DESIRE TO FORWARD THE WORK OF

God's Redemption of Man,

FROM IGNORANCE, FEAR, AND TORTURING DOUBT

A M E N.

WRITTEN IN 1852.

BOSTON:

COLBY AND RICH, PUBLISHERS.

1883.

The History Of The Origin Of All Things
Dictated From Heaven By Jesus Christ

In 1852 a shining needle, yea, a new sword, was produced and laid upon the literary haystack of the time. Literary production was, compared to now, an infant industry. The haystack has become a mountain. Indeed, as it settles, it covers the world with fact, fancy, opinion, and predominantly superfluous print to gag the mind. Yet the shining sword is extant. Burrowing into the stack, it has resurrected to be made available again.

Our research revealed that it has been published 7 times in 141 years and has been received by "the remnant", but resisted by religious denominations as too controversial. Again it goes to press to be a guide for The Christ Age. It will again reach souls devoted to Christ that need and want HIS truth.

This reprint of *The History Of The Origin Of All Things* is an important event in the lives of the Spiritual Searchers that discovered the book in the library of a small church where the best estimate is that the two-volume copy collected dust for over thirty years. The unique spiritual way in which the book was delivered to the world now challenges us to believe in a constantly pro-active Christ, who chose 1852 to deliver this work thru a man chosen for his ability and willingness to be entirely obedient to Christ and the purpose. In addition to its perfect relevance to the Bible and, expression in language of modern times, it causes much reconsideration of the Heaven-Earth, God-Human dynamic. It also teaches the use of *our free-will* and *discipline* to establish and maintain a relationship with the God, Christ, Holy Spirit reality.

The fact that you have this copy establishes you in The Christ Age legion with a mission of influencing coalescence in the Christ Age. His kingdom reigns forever. Amen.

CHRIST AGE PRESS
P.O. Box 262
Union City, MI 49094

Nominal price of
$5.00 + P. & H.

The History Of The Origin Of All Things
History of Publication:

1852-3
    L. M. Arnold
    The History Of The Origin Of All Things
    Poughkeepsie, New York

1883
    L. M. Arnold
    The History Of The Origin Of All Things
    Annie Gretchen, M.D.
    Colby & Rich, publishers, Boston, MA

1926
    V. V. Moore
    Destiny Of A Nation (including parts of the book)
    Baraboo, MA

1936
    L. M. Arnold through "Anola"
    The History Of The Origin Of All Things
    Robert T. Newcomb.
    Biltmore Press, Ashville, NC (Pesgob Forest)

1957
    L. M. Arnold
    The History Of The Origin Of All Things
    Vantage Press, New York, NY.

1961
    L. M. Arnold
    The History Of The Origin Of All Things
    Wm. Publishing Trust, Kentfield, CA.

1993
    L. M. Arnold
    The History Of The Origin Of All Things
    Donald O. Haughey
    Christ Age Press, Union City, MI.